FRANCO-IRISH MILITARY CONNECTIONS, 1590–1945

Franco-Irish Military Connections, 1590–1945

Nathalie Genet-Rouffiac
& David Murphy

EDITORS

FOUR COURTS PRESS

Typeset in 10.5pt on 12.5pt EhrhardtMT by
Carrigboy Typesetting Services for
FOUR COURTS PRESS LTD
7 Malpas Street, Dublin 8, Ireland
e-mail: info@fourcourtspress.ie
and in North America for
FOUR COURTS PRESS
c/o ISBS, 920 NE 58th Avenue, Suite 300, Portland, OR 97213.

A catalogue record for this title is available
from the British Library.

ISBN 978–1–84682–198–1

SPECIAL ACKNOWLEDGMENT

This publication of this volume was made possible by subventions
from the Service Historique de la Défense and the Ireland Fund for France.
The editors and contributors gratefully acknowledge the financial
assistance provided by these two organisations.

Printed in England
by MPG Books, Bodmin, Cornwall.

Foreword

On 10 June 1917, the first American soldier to set foot on French soil was Colonel Daniel E. McCarthy, an Irish-American appointed Chief Quartermaster by General Pershing under the terms of General Order no. 1, American Expeditionary Force.

While inspecting the coast of France in order to select the ports of disembarkation for the American troops, Colonel McCarthy came across many Frenchmen who claimed to be of Irish extraction. At Bordeaux, an officer of the French Engineer Corps embraced him effusively, stating that his grandmother was a McCarthy and pressed him to visit her at the family château near Bordeaux. At a luncheon in Paris, he was introduced to a Capitaine d'Arcy who boasted proudly of his Irish ancestors of the same name. And General Peltier, whom he complimented for his mastery of the English language, a rare achievement in those days, answered with a smile 'Don't be surprised, my mother is Irish'.

Shortly after his arrival in France, Colonel Daniel McCarthy received a letter from a French aristocrat which ran as follows:

> My dear cousin,
>
> As the descendent of one of the Irish patriots who came here several centuries ago to fight for France, I welcome you as a worthy representative of the Irish race, and as a McCarthy, who has come from America to fight for France and the liberty of the world. Hoping to have the pleasure of meeting you soon, I remain,
>
> Your cousin,
> Pol, Comte de Blarney Carty

These reminiscences of ancient Irish military links with France would have been unfamiliar to French ears. In the republican political culture of the nineteenth and early twentieth century it was custom and practice to denigrate the royalist cause by vilifying the *ancien regime* for its blatant sectarianism and persecution of the Protestant Huguenots. The assertion would have been largely devalued had there been equal exposure of the brutally sectarian treatment of Irish and English Roman Catholics.

Thus, the economic, social and military history of the Jacobite exiles in France was largely ignored and so was the contribution in the field of the ten of thousands Irishmen who served the kings of France for a century in the Irish Brigade. Later occurrences of the Franco-Irish military relationship were also victims of this blind spot.

Fortunately, times have changed. Royalism in this country is now an obsolete dream and the fear, real or perceived, of the Royalist cause providing ammunition for the enemies of the republican ethos has vanished altogether. New fields of research have opened up and transnational historical cooperation has been extremely fruitful in this regard. In Ireland, the pioneering works of Richard Hayes and a handful of like-minded historians have been re-assessed and expanded by the Military History Society of Ireland and the learned contributors of its scholarly publication the *Irish Sword*. In France, I am proud to recall that the scholarly journal *Études Irlandaises* founded by Patrick Rafroidi has been, for the last thirty-five years, at the forefront of the 'rediscovery' of those numerous Irish soldiers who crossed the water to fight in our wars; and of the handful of French soldiers who sailed in the opposite direction to fight on Irish soil in 1690 and 1798. The impact of the Jacobite settlers on French society is now fully recognized and in-depth studies of their contribution to the well-being of their country of adoption are readily available.

It was felt that isolated initiatives, thorough and brilliant as they maybe, should be drawn together to liberate our memory and foster a better understanding of this shared experience which once linked our countries in war, just as they are today linked in peace within the European Union.

The launching of the Société d'Etudes Militaires Franco-Irlandaises (SEMFI) on 6 September 2007, was particularly timely and appropriate in this respect. The contributions which are to be found in this volume are the proceedings of the two days SEMFI symposium on 'Franco-Irish connections: military aspects, seventeenth – twentieth centuries' which was hosted at the Château de Vincennes, near Paris, by the French Service Historique de la Défense. They offer a major academic overview of the extraordinary interlinking of French and Irish military history. Hopefully it will be followed by many other lectures, conferences and publications on this fascinating and rich subject.

Shane Leslie once wrote that the Irish soldiers who died on far away battlefields in the service of France were known as the 'Wild Geese' because the people back home believed that their souls would return to Ireland with the shape and sounds of these migratory birds. Let us hope that, after pining in dire need of historical recognition, the souls of the Wild Geese will at long last be able 'to take flight to the wind-scarred paradise of the Western Isle'.

PIERRE JOANNON
Former editor of *Études Irlandaises*
Président d'honneur de l'Ireland Fund de France

Contents

Acknowledgments

This volume of proceedings evolved from a conference that was held at the Service Historique de la Défense in Vincennes in 2007. The editors would first of all like to acknowledge the support of Admiral Louis de Contenson and Mrs Catherine Oudin, *conservateur général du patrimoine*, who were then at the head of the SHD. From the moment that the conference was first suggested to them, they gave it their full backing and the SHD provided not only a venue but also practical assistance and funds. This level of support ensured that the conference was the success that it was.

The editors would also like to acknowledge the assistance that was given by the staff of the SHD. André Radkoto, chief of the SHD's international relations section, served as the third organizer of this conference. Many others at Vincennes helped along the way and these included Jean-Marie Linsolas, Commandant Antoine Boulant, Thibaut Mazire, Lieutenant Aicholzer, Cyril Canet, Marie-Camille Mainy, Bertrand Fonck and Adjudant-Chef Philippe Pons. We would also like to thank Charles Bertin, of the University of Bourgogne, who organized the translation facilities, and the Centre international d'études pédagogiques for their help to find the appropriate equipment.

The staff of the Irish embassy in Paris also gave considerable assistance and the editors wish to thank Cyril Brennan, Nicolle Porte, Laetitia Dufour and Orlaith Fitzmaurice. We are especially grateful to the Irish ambassador to France, Her Excellency Anne Anderson, who generously hosted a reception for the conference delegates at the embassy. We also wish to thank Lara Marlowe of the *Irish Times* for bringing the results of this two-day conference to a wider audience. We are extremely grateful to Pierre Joannon, president of the Ireland Fund for France and a long-time scholar of the Franco-Irish connection. Not only did he provide a fitting Foreword for this volume but we also benefited from his practical advice.

In bringing this volume to publication, we have benefited from the help and enthusiasm of various people. The editors would like to thank everyone at Four Courts Press for their patience and practical help, particularly the late Micheal Adams, Martin Fanning and Aoife Walsh. As usual, they have carried out a form of publishing alchemy. Finally, we would like to especially thank Seán Ó Brógáin for allowing us to use one of his paintings as a fitting cover image for this volume.

Introduction

In 2006 we decided to hold a conference on the theme of the Franco-Irish military connection since the sixteenth century. We were aware that the were various people carrying out research on different aspects of this historic connection and we hoped to gather at least some of them in one location to share their research while also engaging in wider discussion. The experience of Nathalie Genet-Rouffiac, who works at Vincennes, had shown that the SHD would be the best and most obvious possible location as so many historians researching the Franco-Irish connection habitually used its archives in their research. Such a conference had been suggested before but different factors combined to make it happen in September 2006. The response to the call for papers was encouraging and we eventually managed to assemble enough speakers for a two-day conference.

The proposed speakers included not only 'professional historians' but also people from a wider sphere, some of whom were engaged in purely private research. As a result we felt that the conference programme presented a truer impression of the population of researchers that use the archives at Vincennes. The ultimate idea was that these two days would result in a new understanding of the research that was then in progress in this field. With few exceptions, these same speakers have devoted more time over the last year to work their lectures up into a more publishable form and this book of proceedings is the result.

One result that emerged from the conference was an increased awareness among the participants of the importance of the material in the French archives. It is clear that these sources still have much to offer historians. They can be used to examine aspects of the Franco-Irish connection and to provide a French viewpoint on aspects of Ireland's history and also an insight into the Irish themselves. The French archives include the vast collections of the Archives Nationales and the Service Historique. Less well-known collections, such as those held in departmental and municipal archives, also have much to offer as has been shown in this collection.

The essays in this collection have been organized on a chronological basis. Quite simply, they have been arranged by order of the date of the events that they are focusing on and the earlier subjects appear in the first chapters. Apart from being used as a means of organizing the essays in this volume, we also hoped that it would make long-term patterns, developments and turning points more apparent. Readers who are interested in Irish military history in general, and in the Franco-Irish link in particular, will be familiar with some of the subjects examined in this volume. But we would hope that, while the subjects themselves might be familiar, these essays represent the latest research in this field while also highlighting new sources. In some cases, new research on aspects of the Franco-Irish military connection is being introduced to a wider readership for the first time.

The earlier essays in this collection concentrate on the history of the Wild Geese – those Irish officers and men who fled with their families to France in the late seventeenth century. A second group of essays within this volume that deal with the Revolutionary and Napoleonic period. Finally, the later essays in this volume deal with the Franco-Irish military connection after 1815 and these cover the period up to 1945. Ultimately, it could be argued that there are three main themes in the essays of this collection. Firstly, the history of the involvement of Irish soldiers in the French army since 1590; secondly, the interest of French governments in Irish affairs through the centuries and how this interest was reflected in French strategic plans; thirdly, the theme of the creation of Irish identity through the context of service abroad. Certain crucial events in French history – such as the French victory at Fontenoy in 1745 – later came to be seen as formative moments in Irish history. It could be argued that the French military reputation was often founded on the basis of the service of foreigners in the French service while, for the Irish, their military identity was based on the concept of military service abroad.

It is hoped that this volume will also encourage further research. The last number of years has seen renewed interest in the Franco-Irish military connection and this has been facilitated by the activities of larger research programmes such as the 'The Irish in Europe' project overseen by Dr Thomas O'Connor and Dr Mary Ann Lyons. The research currently being organized by Professor Ciaran Brady of Trinity College, Dublin, on the involvement of Irish men in the French navy of the *ancient regime* has opened up new areas of possible research. It seems certain that the history of the ancient Franco-Irish connection will become richer as further avenues of research are explored.

This connection is indeed an ancient one and the two countries have maintained strong links to this day. It is a connection that has been examined by generations of historians and it is connection that is still developing; indeed, in the aftermath of the recent rejection of the Lisbon Treaty in Ireland, Ireland is once again re-examining its links with Europe. From an historic point of view it could be argued that the Irish embraced European ideals centuries ago. Many thousands of Irish people have lived in Europe through the centuries. At the present time there are thousands of Irish men and women living in France and the Irish have always proved themselves to be adept at integrating into French society. The Irish have assimilated themselves into every aspect of French life, despite the fact that many Irish families arrived as refugees in the late seventeenth century. Further political and economic upheaval drove later generations to French soil. By the twentieth century men and women who were Irish-born or of Irish descent could be found across France and were engaged in all manner of activities. They could be found embedded deep in the political, economic and social life of France. In the context of this volume, it must be remembered that, for many Irish people, it was military service that first brought them to France. From the late seventeenth century, it was through military service that Irish soldiers and their families established themselves in France and this military service facilitated their later activities and integration into French society.

It is also obvious that the history of the Franco-Irish connection was always the story of a *ménage à trois*, an expression that even when expressed in English is given a French spelling. England was the cause of many hesitations in French policy through the centuries. When France eventually succumbed to the Entente Cordiale with Great Britain in 1908, Ireland was left without its historical ally and faced new temptations from Germany.

Apart from this volume of proceedings, the 2006 conference also resulted in the establishment of a new society – the Société d'études militaires Franco-Irlandaises (SEMFI). At present it is a loose association of scholars who are interested in the shared aspects of the France and Ireland's military history. It is hoped that further lectures, conferences and publications will follow.

NATHALIE GENET-ROUFFIAC & DAVID MURPHY
Paris, July 2008

Irish soldiers and regiments in the French service before 1690

ÉAMON Ó CIOSÁIN

The crossing of the Wild Geese to France in 1690–2, called in French the '*passage*' of the Jacobites in many contemporary French sources, has traditionally marked the starting point for studies of Irish military service in France, ranging from O'Callaghan's history of the Irish brigades to more recent standard works on the French army. While several such authors acknowledge the presence of Irish soldiers on French soil or in French service prior to that landmark date, the activities and destinies of earlier soldiers form a mere backdrop to the substantial theme of the exploits of the 'Wild Geese' in the eighteenth century. However, the life of Justin McCarthy, commander of the Mountcashel brigade, which was sent to France in exchange for the French contribution to James II's army in Ireland in 1688, is in itself proof of long-term Irish engagement with the French service. Justin had commanded a family company in the Hamilton regiment which Charles II seconded to French service from 1672 to 1678, and which had fought on the Rhine. Furthermore, his father, Donough McCarthy, had sent a regiment from Munster to France in 1647, and Justin's elder brother Cormac had held the command of the regiment into the late 1650s, having been educated in Paris.[1] While Irish historians have in the main concentrated on the post-1690 exiles, some studies by French historians and recent articles by Irish researchers have opened up perspectives on the earlier century and pointed to the Irish presence in French armies as a 'long-term phenomenon', as French historians are wont to say.[2] Indeed, it could be argued that this phenomenon begins to develop at much the same time as Irish military migration to Spain – Brendan Jennings' major review of sources *Wild Geese in Spanish Flanders* takes 1582 as its starting point.[3] Irish military migration to France also follows, at a remove,

1 R. O'Ferrall and R. O'Connell, *Commentarius Rinuccinianus, de sedis apostolicae legatione ad foederatos Hiberniae catholicos per annos 1645–9*, ed. S. Kavanagh, 6 vols, Dublin, 1932–49, ii, 638; v, 100. 2 P. Gouhier, 'Mercenaires irlandais au service de la France 1635–1664', *Irish Sword*, 7 (1965–6), 58–75, reprinted with few changes in *Revue d'histoire moderne et contemporaine* (1968), 672–90; R. Chaboche, 'Les soldats français de la Guerre de Trente Ans, une tentative d'approche', *Revue d'histoire moderne et contemporaine* (1973), 10–24; E. Ó hAnnracháin, 'Some early Wild Geese at the Invalides', *Irish Sword*, 22 (2001), 249–64; É. Ó Ciosáin, 'A hundred years of Irish migration to France, 1590–1688', in T. O'Connor (ed.), *The Irish in Europe 1580–1815* (Dublin, 2001), pp 93–106. 3 B. Jennings,

the formation of the first Irish ecclesiastical community in Paris in 1578, with initial numbers being small.

In late medieval times, Irish military service in France did not necessarily mean the service of the French king. Two early appearances of Irish troops on French soil involved forces raised by the Butlers in Ireland for the English king: the first occurrence being during the Hundred Years War, the second during the reign of Henry VIII. In 1418, up to 1,600 Irish troops were among the English army besieging Rouen, and their appearance and behaviour drew comment from writers on both the French and the English sides.[4] The French source accuses the Irish troops of marauding in the countryside near Rouen, and claims they were poorly armed, with long knives and strange horses. The sixteenth-century episode saw some 600 Irish soldiers at the siege which Henry VIII's forces laid to Boulogne (1544–5).[5] (The various Butlers and their allies were to be important in the seventeenth century: troops stationed in Anjou in the mid-1640s mention Kilcash as their native place, and James Butler (Ormond) was involved in directing the Irish troops in France in the 1650s.) A further example of Irish participation in an English/British campaign in France occurred when Piers Crosbie's companies – a few hundred men – participated in the siege of La Rochelle (1627–8) under Buckingham's command, on the side of the Huguenots of the town who were opposed by the French royal army commanded by the cardinal-minister Richelieu. These expeditions were short-term and hostile to the French interest, but they nevertheless form part of a longer-term pattern insofar as English royal involvement was to be a factor in the arrival of various Irish regiments in France in the seventeenth century, whether it be Charles I's assent to the departure of Irish regiments for Richelieu's army in the 1630s or Charles II's detaching a regiment to assist his cousin and supporter Louis XIV in the 1670s. The marriage of Charles I and Henrietta Maria, daughter of Henri IV, provided a French connection to this royal nexus, and the recruitment of Irish troops to France in the 1650s parallels Henrietta's efforts to raise troops on the continent for her husband's interest in the 1640s.

The study of Irish military service in France raises another question, relating to the documentation which is available. Given that the Irish were seen in France as subjects of the English king (notwithstanding the fact that his dominion was constituted of three 'kingdoms'), one might expect that there could be difficulty in identifying specifically Irish individuals or bodies

Wild Geese in Spanish Flanders, 1582–1700 (Dublin, 1964). 4 Enguerrand de Monstrelet, *Chronique (1400–44)*, ed. L. Douët d'Arcq (Paris, 1859), iii, 284–85; English sources quoted by R. Hayes, 'Irish soldiers at the siege of Rouen (1418–19)', *Irish Sword*, 2 (1954–6), 62–3, who also mentions Enguerrand de Monstrelet's account. Further discussion of the Rouen appearance in J.-M. Boivin, *L'Irlande au Moyen Age: Giraud de Barri et la Topographia Hibernica* (Paris, 1993), pp 142–3. 5 For an account, see D.G. White, 'Henry VIII's Irish kerne in France and Scotland, 1544–45', *Irish Sword*, 3 (1957–8), 213–25.

of men. French official documents and parish or institutional registers which contain mention of soldiers establish a clear distinction among the subjects of the three kingdoms, however, and it is quite rare to come across an Irish soldier being termed '*anglois*' (English) in the records. Scottish and English forces and individual soldiers are also clearly distinguished according to their origins; if anything, evidence of family names may indicate that there were possibly occasional cases of Englishmen (maybe recent arrivals in Ireland) being classified as Irish in the records of the Hôtel Royal des Invalides, but this does not happen to any significant extent.

The chronology of Irish service in the Spanish dominions begins in the 1580s, and given the realities of military life at the time, it would not be surprising to find strays and stragglers crossing from the Spanish forces to the French side. A number of Irish companies served with the Spaniards in Brittany during the wars of the French Catholic League in the 1590s. Dominic Collins, who was later executed by the English, commanded one such company; other Irishmen served under Dom Juan del Aguila (of Kinsale fame) also in Brittany,[6] and with the Catholic dukes of Mayenne and Mercoeur during the war.[7] The rare cases of Irish soldiers on the French side may be instances of desertion from the Spanish lines, or Irish soldiers who formed part of the regiments sent by Elizabeth I to assist Henri IV of France.[8] Arrivals of Irish combatants in France were thus a result of either English or Spanish interventions in France rather than French recruitment in Ireland or direct entry by Irishmen into the French armies. When Irish migration to France surged in 1602–3, one could expect to find more substantial figures of Irish soldiers or recruits on French soil. Some Irish chiefs offered their services to the French king, one example being Charles O'Daly, who in 1606 asked the archduke of Flanders to intervene on his behalf with the French king so that he might have a commission to command the 200 or so soldiers he had at his disposal in the Paris area.[9] The surge in so-called Irish 'beggars' and 'vagabonds' in France in the years 1602–9 may be partly explained by the orders given to companies of Irishmen or Irish soldiers in Spain to remove themselves to Flanders, notably in 1605 and 1606. Spanish sources indicate that several such companies travelled overland through France and not by sea, nor by the 'Spanish road' from Genoa across the Alps and northwards through Lorraine to Flanders. This contradicts R.A. Stradling's description of France as 'a kind of traffic island' in the midst of Habsburg territories, and is further evidence of Henri IV's policy of allowing *transitus innoxus*, neutral

6 Deposition of Collins before Sir G. Carew, 9 July 1602, *Calendar of State Papers Ireland 1601–03*, pp 439–40, relating the story of his life in France. 7 *Calendar of State Papers Ireland 1592–96*, pp 63–7, mentions 17 men serving under Mercœur. 8 For further analysis of English support for Henri IV and its context, see M. Lyons, *Franco-Irish relations 1500–1610: politics, migration and trade* (Woodbridge & Rochester, 2003), p. 165. 9 Jennings, *Wild Geese*, p. 134; date established by G. Henry, *The Irish military community*

transit, as applied to the ordinary Irish and the Ulster earls at this time.[10] In June 1606, for example, the O'Driscoll and Stanihurst companies left Galicia for Flanders en route by land through France.[11] A number of groups of Irish soldiers are reported as arriving in Flanders from France in the years 1605–7.

Movement could go in the opposite direction: the first Irish company in French service which research has hitherto revealed is that of 'Rodrigo' O'Donoghue, who had served under the Spanish in Flanders. His body of some 200 men formed part of the army raised by France's cardinal-minister Concini in 1614–16 to fight rebellious nobles' forces in Picardy. O'Donoghue's men appear to have fought only one campaign, in 1616, and he subsequently sought to be reintegrated into O'Neill's regiment in Flanders.[12] Occasional mentions of Irishmen in the French army during the following decade are to be found in various sources: a Thadeus Carti arrived in Flanders in 1622 having served the king of France for 'several years', and Robert Barry and John Bayes, Irishmen, had served in the army of the king of France for 6 years before being arrested in 1626 while attempting to return to Ireland.[13] The sudden arrival of a large Irish 'band' in Brittany in 1622 may correspond to an emergency landing of soldiers being transported to Spain or Flanders; a further mention the following year in Rennes links 'Irish' and soldiers.[14] The poet Padraigín Haicéad would appear to suggest that France was a destination for Irish soldiers, although his poem to Captain Cary's band (1628) does not clarify their allegiance (to the English or the French king?).[15]

This hitherto small-scale migration underwent a sudden acceleration in 1634, when the king's lieutenant, Wentworth, authorized the departure of

in Spanish Flanders 1586–1621 (Dublin, 1992), p. 71. **10** For a detailed analysis of this phenomenon, see É. O Ciosáin, 'Les Irlandais en France 1590–1685 : les réalités et leur image', doctoral thesis, Université de Rennes 2 (2007), pp 139–41; see also R.A. Stradling, 'Military recruitment and movement as a form of migration', in S. Cavaciocchi (ed.), *Li migrationi in Europa Secc. XIII–XVIII* (Florence, 1994), p. 478. Here Stradling follows G. Parker's views on the workings of the Spanish Road. **11** C. O'Scea, 'The significance and legacy of Spanish intervention in West Munster', in T. O'Connor and M.A. Lyons (eds), *Irish migrants in Europe after Kinsale, 1602–1820* (Dublin, 2003), p. 63; see also details on land crossings in O'Scea, 'Irish emigration to Castile in the opening years of the seventeenth century', in P. Duffy (ed.), *To and from Ireland: planned migration schemes c.1600–2000* (Dublin, 2004), pp 17–37. **12** Jennings, *Wild Geese*, pp 149–50, documents of July and August 1616; E. Fieffé, *Histoire des troupes étrangères au service de la France* (Paris, 1854), i, 134–5; according to Susane's *Histoire de l'infanterie française*, v, 230, the 'regiment' was called up in August 1615: see ibid., regiment no. 410. **13** For Carti, see Jennings, *Wild Geese*, p. 180; for Barry and Bayes (Hayes?), *Historical Manuscripts Commission Report XV, Appendix 2*, Hodgkins Papers, Miscellaneous, pp 289–90. **14** Documentation for the 1622 'invasion' (*sic*) in P. Parfouru, 'Les Irlandais en Bretagne XVIIe et XVIIIe siècles' *Annales de Bretagne* (1894), 525–7; the 1623 mention is in a declaration of the Parlement de Bretagne concerning the city of Rennes, in Archives d'Ille-et-Vilaine, 1 B 303, *arrêt sur remonstrance* dated 13 December 1623. **15** 'Chum cuideachtan Chaptaoin Cairi ó Londain go Canterbury 1628', in M. Ní Cheallacháin (ed.), *Filíocht Phádraigín Haicéad* (Dublin,

regiments from Ireland for the armies of France's chief minister Richelieu. The interests of both men coincided in this matter. Richelieu's papers and letters contain sufficient evidence of his interest in Ireland and Scotland as recruiting grounds; he also envisaged attracting Irish colonels from the Spanish service as part of his drive to constitute a large army capable of facing the might of the Habsburg 'house of Austria'. Recruiting Irish troops from the enemy brought a double benefit, which his successor Mazarin and others would articulate in turn: each body of Irish troops b(r)ought over from the Spanish would simultaneously strengthen the French army and weaken the Spanish.[16] In terms of 'push factors', various reasons can be advanced for the willingness of Charles I to allow open recruitment of Irishmen for France and their transit across England (a route which continued to be used until the late 1630s, the practice was discontinued thereafter in favour of direct maritime transport to France). Among these are the government's usual wish to clear Ireland of so-called 'idle' or rebellious elements. Furthermore, in a period where the phenomenon of local militias was on the rise, arming a militia which was bound to be Catholic (if only for demographic reasons) and keeping it in Ireland was bound to be problematic. Finally, there was on the part of the crown a desire to maintain good relations with France, the native country of Charles I's wife Queen Henrietta. In return, French envoys and ministers stressed the necessity of maintaining Ireland under English rule, and preventing it falling into Spanish hands at all costs.[17] These considerations came into play notably in 1640 when it was decided that the Irish army raised to support Charles I would not cross to England: the levies were directed towards France, Christopher Bellings being among the colonels who successfully raised men for France from this army. Bellings, formerly an officer in the Spanish service, had been approached in London by the French ambassador P. de Bellièvre.[18] Bellièvre played an important part in raising soldiery in Ireland and Scotland for Richelieu and later for Mazarin: his stays in England correspond to the peaks of French recruitment in the Three Kingdoms in the 1630s and 1640s. Contact was made from 1634 with Irish colonels in Spanish service with a view to bringing them over: overtures may have been eased by the openly-expressed dissatisfaction of Munster and Old English captains in Spanish Flanders, who smarted under the command of Ulstermen, if they are to be believed.[19]

1962), p. 5; English translation in M. Hartnett, *Haicéad* (Loughcrew, Oldcastle, 1993), p. 20. **16** Gouhier, 'Mercenaires irlandais' (1968 version), p. 687. **17** For ex. Bellièvre to Brienne, 6/16 December 1646, in J.G. Fotheringham (ed.), *The diplomatic correspondence of Jean de Montereul and the brothers Bellièvre French Ambassadors to England and Scotland 1645–48*, i, 350. **18** See R. Bellings, *History of the Irish Confederation and the war in Ireland*, ed. J.T. Gilbert (Dublin, 1882), i, 58–63. Gouhier, 'Mercenaires irlandais', confused the soldier Christopher with his brother, the author Richard, later secretary of the Confederation of Kilkenny. **19** See statement of Col. MacSuini, 1635, complaining of the

Six Irish regiments were raised between 1635 and 1640, totalling perhaps 7–8,000 men when reinforcements are included. (Gouhier bases a higher total of 10,000 on official figures, whereas comparison with local records in Brittany indicates that the theoretical figure of 1,000 men per regiment should be reinterpreted as 1,000 persons in all, including family and other persons. The military historian J.A. Lynn has also questioned French official figures for the period in general, and gives probable real figures of combatants which are lower.)[20] The six regiments in question were raised by mainly Catholic colonels of Old English/Anglo-Norman extraction, and were predominantly composed of soldiers with Gaelic family names. On the evidence of French records, and the local origins of the colonels, most of these regiments appear to have been raised in the southern half of Ireland. As organizers and commanders of military levies, the Old English were perceived by the English administration as loyal and reliable, and were allowed to raise troops for both France and Spain at this time. Several among the colonels appear to have fitted the pattern of younger sons who sought employment in the army, one case being that of the Wall family of Coolnamuck near Carrick-on-Suir, whose regiment was highly thought of in the French army between the time of its arrival in 1639 and its becoming the duke of York's regiment (that is, James Stuart's) in 1654, and later the Royal Irish regiment under the command of Louis XIV himself. Five of the Wall sons had military careers in France. The Sinnott regiment was probably raised in the Wexford area, given the strong presence of the family in the county. Other regiments were those of Fitzwilliam, Bellings, Cullen and O'Reilly/Tyrell. The variation in the name of the latter regiment may derive from the fact that its colonel was Thomas Tyrell whereas the men may have been raised in the present Co. Cavan, where Richard Tyrell, a veteran of the Nine Years War, had lands among the O'Reillys; further research is required on both commander and men to establish this with certainty. Sir Piers Crosbie, who had participated in the siege of La Rochelle on the English side, later raised men for France in the late 1630s. His recruits served under Fitzwilliam when on French soil.[21]

scorn shown towards others by Ulstermen, which prompted MacSuini to accept a commission in France: Jennings, *Wild Geese*, p. 316. **20** J.A. Lynn, 'Recalculating French army growth during the Grand Siècle, 1610–1715', *French Historical Studies*, 18 (1994), 881–906. For analysis of two cases where local documentation gives useful indications on the real strength of regiments, see Ó Ciosáin, 'Les Irlandais en France 1590–1685', pp 268–72. **21** John Butler and Owen O'Sullivan are also mentioned in English and French papers as being among the superiors of the Crosbie men. The prowess of Irish soldiers in the 'Fischwilain' (Fitzwilliam) regiment is praised by Abbé Robert Arnauld d'Andilly, *Mémoires d'Arnauld d'Andilly*, tome II in *Collection complète des mémoires relatifs à l'histoire de France* (Paris, 1824), pp 212–13, year 1643; English translation in C. Duffy, 'Irish in the service of Louis XIII', *Irish Sword*, 9 (1969–70), 238–39. Other sources quoted by Gouhier and especially D. Parrott, *Richelieu's army* (Cambridge, 2001), are less positive, in particular as regards discipline, but it should be borne in mind that complaints of

Crosbie was a member of Queen Henrietta's circle and a Protestant. These regiments followed the model of the colonel-proprietor current in the French army at the time.[22] They were part of the 'armée d'Allemagne', that is to say, defending the northern and north-eastern frontiers of France as they were then. The Irish were mainly sent to fight on the northern borders, where they faced fellow-countrymen, among others, on the Spanish/Flemish side.

When war broke out in Ireland in 1641, some officers returned to Ireland, albeit on a temporary basis in some cases. Christopher Bellings sailed to Ireland, but returned to France in 1642; Colonel Richard Cullen was recalled to France in 1643. This reflects the policy adopted by the French administration, under Richelieu in particular, towards the Irish colonels. They were regarded as French officers, which explains the return of some of them to France while war continued in Ireland. However, Richelieu allowed officers to leave at the beginning of the war, but not in such a manner as to create diplomatic difficulties for his government. In most cases, Franco-Irish officers returning to provide manpower and expertise for the Confederate war effort appear to have been accompanied by few rank and file, and not to have carried any substantial quantities of munitions from France. Their return, albeit temporary in some cases, explains the sudden halt in recruitment for France in Ireland after 1640. The civil wars in Scotland and Ireland, and Charles I's weakening position also hampered French recruitment. The end result in France was a fragmentation of the Irish corps, and declining figures caused several regiments to be disbanded, accompanied by the routine complaints by administrators about the fickleness of soldiers. The Irish were criticized for their supposed tendency to continually move and were compared to 'birds of passage' in one instance.[23] The Tyrell, Sinnott and Bellings regiments were disbanded or reformed between 1640 and 1643. However, there is evidence in the Invalides entry registers that individuals continued to serve in France even after regiments were disbanded or reformed. Mazarin envisaged forming new Irish regiments from among the Irish in France on more than one occasion in 1643–4, thus indicating that many of the rank and file stayed on after the departure of the officers.[24] Several soldiers whose careers are stated by them to have begun in the 1630s and early 1640s applied to the invalid hospital in the course of the 1670s.

indiscipline among French soldiers are commonplace in the administrative papers of the time. **22** Parrott, *Richelieu's army*, contains several mentions of the Irish regiments, and also describes the structure and organization of the French army at the time. **23** 'des hyrlandois estant des oyseaux de passage tantost forts et tantost foibles', Du Hallier to de Noyers, 31 March 1640, quoted in Parrott, *Richelieu's Army*, pp 307–8, which gives Archives du Ministère des Affaires Etrangères, Correspondance Politique Lorraine, vol. 31, fol. 304 as its source. **24** For example see A. Chéruel (ed.), *Correspondance de Mazarin*, ii, 77, 22 September 1644, concerning Colonel Wall who was to incorporate his countrymen.

However, foreign recruitment did occur in Ireland during the time of the Confederation of Kilkenny (1642–9), and contrary to what has been claimed, the French administration was more successful than the Spanish in this regard, spending less and recruiting substantial numbers at a time when the Irish Catholic Confederation was at war. A regiment paid for by the French agent La Monnerie sailed from Galway at the end of 1644, and was in Anjou the following April. This regiment was commanded by Arthur Magennis (Iveagh), and various captains of Old English extraction (Barnewall, Blake, Fleming, Plunkett).[25] Its fate is not clear, as French archives do not mention it at any later stage. It arrived in Brittany, and the probable breaking-up of the regiment may provide an explanation as to why strenuous orders were issued by Mazarin later in 1645 to round up several hundred Irishmen in Nantes and elsewhere in Brittany and put them under military command.[26] Some of these men may have been drafted into the Wall regiment. One interesting fact is that of 21 Irish applicants to the Invalides who stated that they entered the French service in the 1640s, some 11 or 12 claim to have joined in 1644–5, precisely the year of the arrival of the Magennis troops.[27] France succeeded in recruiting two further corps in 1647: in May 1647, some 500 men under James Preston, who were originally destined for Spain, changed tack in Waterford harbour and sailed for France.[28] The Prestons, father (Thomas) and son (James), Old English officers who had long served the Spanish, took the French side thereafter. Preston's men were incorporated into the Wall regiment. In the autumn of that year, the Muskerry regiment was sent to France by Donough McCarthy. When French regiments were being stood down after the treaty of Westphalia (1648), MacCarthy Muskerry was guaranteed his regiment would be maintained, and this was indeed the case.[29]

As the Confederation disintegrated and the Cromwellian campaign took hold of Ireland, recruitment for France ceased until 1652. However, several royalist commanders left the country for France, notably the earl of Ormond and Murrough O'Brien, who sailed together to France in December 1650. Ormond remained a powerful figure in exile, and was credited by the French

25 Some material concerning this regiment can be found in Jennings, *Wild Geese*, pp 595–6. TCD MS 3276 contains a muster roll of a company of foot of this regiment drawn up in Ponts de Cé, Anjou, dated 1 April 1645. Parish records in this location in the same month mention other individual soldiers and their families. 26 *Correspondance de Mazarin*, ii, 202 and 212, letters of 11 July and 18 August 1645. 27 These men include Pierre Aubrin, the first Irishman to enter the Hôtel des Soldats Estropiés (as it was called prior to 1674) on 1 October 1670 with 25 years' service; Mathieu Grigelle, nicknamed St Louis, 23 April 1673, 28 years' service, Hugues Lacy nicknamed Lacy, 3 August 1675, 30 years' service, Simon Bernan (Brenan?) nicknamed La Montagne, 1 December 1685, 40 years' service. 28 This is based on sources listed by Jennings, *Wild Geese*, and R.A Stradling, *The Spanish monarchy and Irish mercenaries: the Wild Geese in Spain, 1618–1668* (Dublin, 1994). 29 Brevet à Donough Maccarthi vicomte Muskrey, 7 July 1648, Bibliothèque Nationale de France, Collection Chatre de Cangé, vol. 6, fol. 191; copy in

with drawing troops to the French royal party, and subsequently blamed for drawing them away to Flanders when Charles II was expelled from France under the terms of the 1655 agreement between Mazarin and Cromwell. O'Brien remained for some ten years in France, and commanded regiments in various theatres of war. His conversion to Catholicism in June 1657 did not facilitate his career, and he seems to have been regarded with suspicion by the French authorities, not to mention the detestation of his fellow-countrymen, who were mindful of his tactics in the war in Munster.

Initially, the Cromwellian clearance of Ireland led to Irish royalist and Catholic forces being shipped to Spain and to Flanders in 1652 and 1653, as Spain had recognized the English Republic sooner than France. However, France was the ultimate beneficiary of the transportation of thousands of men from the armies of the various provinces. The way in which this came about is a remarkable historical episode, an episode which merits a study in its own right. Part of the tale has previously been told by R.A. Stradling, using Spanish documentation, and some sources for events in France are noted by Jennings.[30] Further evidence for France can be found in French sources used in Gouhier's important article on the period and in the final section of the *Commentarius Rinuccinianus*. The complexity of the Irish military crossings into and out of France requires more space than can be devoted to the matter in this article, but the principal events of the period 1653–7 are at least reasonably clear. Several contexts came into play in the course of these events, the most important being the final stages of war between the Royalist forces and the rebels of the 'Fronde' in France in 1652, the ill-treatment of Irish soldiers arriving by sea in northern Spain in 1652–3 (detailed by Stradling), and the collapse of Spanish state finances, with the consequence that troops were unpaid. The international dimension of the phenomenon of the Irish movements is underlined by the movements of Charles II between various kingdoms, which was an important factor in what happened in France.

In 1653, Charles arrived in France. Irish troops began to leave the Spanish armies on the Catalonian front and cross to the French, notably the regiments of Grace and O'Ferrall. The McCarthy Reagh regiment serving with the combined Spanish-Condé-Conti forces in Aquitaine did likewise, and a garrison under Sir James Dillon which held the place of Lormont (near Bordeaux) for the Spanish-Frondist forces handed the fort over to the French Royalist forces.[31] Irish commanders led men across in some cases (for example, Richard Grace), while in other instances the commanders remained in Spain but their men changed allegiance; this was the case with the

Archives Nationales, Paris, O[1] 12, fol. 40. **30** Stradling, *The Spanish monarchy and Irish mercenaries*; Jennings, *Wild Geese*, pp 609–15. **31** These statements are based on Gouhier, 'Mercenaires irlandais' for French sources, *Commentarius Rinuccinianus*, which follows the fate of the Irish abroad after 1650 in some detail, and English papers cited in Jennings, *Wild*

O'Ferrall regiment. Indications are that Irish soldiers crossed to the French forces, whether in companies, regiments, or in a spontaneous movement of small groups of men. On arrival in France, these combatants were grouped in regiments under the command of Royalists such as Murrough O'Brien, Sir George Digby, Richard Butler, Cormac McCarthy, Daniel O'Neill etc.[32] A further indication of the shift in Irish allegiance came in the summer of 1653 when Dermot O'Sullivan Beare brought his regiment to Brittany rather than to Spain.[33] The O'Sullivan Beares had been staunchly allied to the Spanish and leading members of the family had achieved comfortable positions in Spain since the aftermath of the battle of Kinsale; the fact that members of the family chose to go to France at that point is indicative of the situation created by the mixture of Irish disenchantment with the Spanish military authorities and the powerful attraction created by the presence of the Stuart king on French soil. The following year, 1654, saw Irish troops crossing to the French on the northern front, from the armies of Condé and Lorraine: these were men which Munster general Murtough O'Brien, George Cusack, former governor of Inishbofin, and others had brought to Flanders over the two previous years.

Full and detailed calculations remain to be carried out to determine the number of Irish soldiers and cavalry in the French armies at the peak which was reached at the end of 1654, but if one takes it that there were nine regiments (according to both Gouhier and the contemporary authors of the *Commentarius Rinuccinianus*) and if one calculates on the basis of sample percentages of Irish soldiers in documentation concerning invalids in subsequent years, an estimated total of some 6,000 men is reasonable, perhaps even conservative.[34] This figure includes those who were already in France before the mass movement from the Spanish side, however, and the number of fresh soldiers may have been 3,000 or 4,000. William Petty's often-quoted estimates whereby 18,000 persons migrated from Ireland to Spain and 16,000 to France in these years require checking against the evidence from France, as there may be a partial overlap between the two figures as far as the military are concerned. A further presence in France in the mid-1650s was that of Irish Stuartist privateers in Brittany. This group comprised some Irish captains

Geese. 32 A list is given in *Commentarius Rinucinnianus*, v, 207 which corresponds to those named in French sources. 33 *Commentarius Rinucinnianus* provides useful and accurate information on this arrival and other events concerning Irish military corps in France, Spain and Flanders. For the arrival of the O'Sullivan regiment in summer 1653, see *Commentarius*, v, 107. 34 For percentage estimates based on lists of invalid soldiers, see A. Corvisier, 'Anciens soldats, oblats, mortes-payes et mendiants dans la première moitié du XVIIe siècle', *Actes du 97e Congrès des Sociétés Savantes, Nantes 1972, Section histoire moderne et contemporaine* (Paris, 1977), i, 7–29, and a revised analysis of the documentation mentioned by Corvisier in relation to Irish soldiers, in Ó Ciosáin, 'Les Irlandais en France 1590–1685', p. 314, giving rates of 5% and 6% of the total of soldiers listed in two orders,

who had operated out of Wexford in the previous decade in cooperation with Dunkerkers and Randal McDonnell's ships.[35] They had constituted the 'naval arm' of the Irish Royalists and fulfilled the same function with regard to the young exiled king. Basing themselves in Brest from 1653 on and sailing under *lettres de marque* of Charles II, they captured many prizes which were sold in Brest, Morlaix and Saint-Malo; some of the proceeds went to Ormond and to the impecunious Stuart court.[36] Their numbers dwindled after the departure of James Stuart from France in 1656. As with the soldiery, the issue of allegiance among these Royalist exiles became important when the Stuart court left France: the evidence from the years in France is that some viewed their primary allegiance as being to their king (Charles II), others to the French into whose service they had entered, others still to Spain – those who remained in Spain and Flanders. This problem of allegiance was to manifest itself again from 1655 onwards.[37]

In terms of the flow of Irish soldiers, France in the years 1652–6 could be compared to a canal lock: the men flowed in first from the southern fronts, and then came directly from Ireland and later from Flanders. The sluices opened towards the north in 1655 when Charles II was obliged to leave French territory under the agreement between Cromwell and Mazarin. His brother James was ordered to join him, which he did the following summer. They were followed by many Irish colonels, and indeed men – some of the long-serving Muskerry regiment for example, although their colonel remained behind in Paris for some time. The personal contacts some soldiers had with James duke of York and their respect for his ability as a commander appear to have motivated some persons to leave, and it may be that the exceptional attachment of the Irish to the person of James (the future James II) dates from this shared exile in France. The Ormond circle worked to rally men from the French ranks to those of the young king in Flanders. The transfer of allegiance left a sour atmosphere among French leaders: Mazarin's letters of the time contain aspersions on the Irish national character which he paints as fickle and changeable. While Carte's *Life of Ormonde* claims that some five or six regiments were rapidly constituted in Flanders from the Irish who crossed over,[38] the number of men who changed camp remains open to question, as the well-known penury of the exiled Stuart court would militate against its maintaining a large army.

In fact, a number of regiments remained in France for financial or political reasons. Some commanders had little expectations of favour in Spain – James

dated 1657 and 1660. 35 J. Ohlmeyer, 'The Dunkirk of Ireland: Wexford privateers during the 1640s', *Journal of the Wexford Historical Society*, 12 (1988–9), 23–49, and 'Irish privateers during the Civil War', *The Mariner's Mirror*, 76 (1990), 119–34. 36 *Commentarius Rinuccinianus*, v, pp 242–3. 37 The *Commentarius* highlights this issue of allegiance and the problems caused by the Irish switching in this decade in v, 103–5. 38 Carte, *Life of Ormonde*, v, 654, quoted in Jennings, *Wild Geese*, p. 614.

Preston had defaulted in 1647, and Murrough O'Brien remained in France in the hope that his conversion to Catholicism might procure him some advancement. (This was not to be, as Mazarin avoided giving him commands in order not to antagonize the then English régime.) Gouhier counts no less than six regiments remaining in the French army in 1658: Dillon, Royal Irlandois, O'Brien Inchiquin, Preston, Muskerry and Digby. The peace concluded between France and Spain the following year in the Treaty of the Pyrenees led to the disbanding of further regiments over time: Inchiquin left for Portugal, was taken prisoner by the Moors and eventually returned to Ireland. The Royal Irlandois is said by Gouhier to have been dissolved in 1661, and Muskerry followed the restored king to London, as did others. However, Gouhier's statement that no Irish troops served in France between the dissolution of Dillon's regiment in 1664 and 1688 is inaccurate in several respects.

Irish soldiers continued to serve in a variety of regiments throughout the 1660s, among the native French (Carignan, Maine, Ferté regiments ...), in English or Scottish regiments such as Monmouth and Douglas, according to the evidence presented by soldiers entering the Hôtel Royal des Invalides. It is likely that some of these men were part of Irish companies within non-Irish regiments. Furthermore, it has recently been noted that local archives in Angers mention 'two regiments of Irish infantry' in 1667, indicating at the very least an Irish presence in such regiments, if not Irish command and regimental name.[39] Another Irish regiment is mentioned under the name 'Roscommon': this regiment was constituted in 1669,[40] and was merged into the Hamilton regiment in 1672. The Dillon connection was again prominent, as the colonel of the Roscommon corps was Wentworth Dillon, a gentleman said by sources not to have been an effective military leader. Sir George Hamilton's Irish regiment was raised from 1671 onwards, partly in Ireland, partly among Irish soldiers previously in France,[41] and with regiments such as Douglas and Monmouth was part of the army Charles II sent to the aid of Louis XIV in 1672. The Hamilton (later called Dongan) regiment was constituted of companies whose surnames are indicative of long-term Irish connections with the armies of France: Butler, Dillon, Walsh, Henessy, Macnamara, etc. There was also a MacCarthy company; several McCarthys served in the regiment (Justin McCarthy took over command in 1676), as did

39 I owe this information to P.-L. Coudray, who is researching the Irish presence in Anjou in the early modern period. 40 Hayes, *Manuscript sources for the history of Irish civilisation*, under 'Irish abroad: France', gives references to papers relating to the Roscommon regiment, 1671–72. The raising of the regiment was earlier, if the evidence of former soldiers who apply to enter the Invalides is to be believed: all mention 1669 as the year their service in it began. Nine men call the regiment Hamilton, and a tenth soldier names Roscommon. 41 For Wentworth Dillon and Hamilton's regiments, see C.T. Atkinson, 'Charles II's regiments in France, 1672–78', *Journal of Army Historical Research*, 24 (1946), 53–64, 128–36 and 161–72.

Hamilton's sons, one of whom, Antoine, later became a writer of note. The Invalides registers and those of the Hôtel-Dieu in Nantes provide between them the names of 100 of the Hamilton-Dongan soldiers, making a regimental analysis possible. Having served on the Rhine with Turenne, the regiment was recalled by Charles II who came under parliamentary pressure to do so in 1678. However, Irish soldiers again remained in France after the official departure, as they had done in the 1650s. The Invalides registers indicate that Macnamara, Callaghan and Hamilton companies were part of German regiments: from 1678 to 1681 they were in the (German) Furstemberg regiment. This regiment became Konigsmarck (also German). Some of the Irish candidates for the Invalides served in the Swiss regiment of Erlach in the 1680s, others served in French regiments such as the Gardes Françoises in the decades after 1660. When the Jacobite regiments arrived in France from 1688 onwards, some of these men were transferred to the regiments of their fellowcountrymen, thus ensuring an unbroken chain of Irish units serving under the kings of France which began in 1635 (leaving aside the individual cases and small groups which can be found prior to that date).

This history is rich in its human variety, in the Irish contributions to many French campaigns and in the dramatic migratory surges which happened in the late 1630s and again in the mid-1650s. Documentation is reasonably plentiful, continuous, and often surprising in its detail – one can find Irishmen in German or Swiss regiments in the later century, or Irish cavalrymen serving under a German colonel in 1644, for example. In purely numerical terms, the total numbers of Irish soldiers arriving in France during the period studied here range from 15,000 to 18,000, if one retains conservative estimates of 6,000 for each of the two main surges, 1,000 or so of Magennis' men in 1644, another 1,000 of Hamilton's, which were replenished by further recruitment in Ireland (evidenced in Nantes local records), and sundry others such as the O'Donoghue company of 1616. These figures relate to those who served in Irish units or regiments; there were however additional numbers of Irishmen serving elsewhere in the French army (see below). Some of these thousands moved on, such as O'Donoghue's company and a portion of the Stuart soldiers of the 1650s, but many remained permanently in France. The social history of those who remained can be constructed from military, administrative and local records, especially those concerning invalid or crippled soldiers (Invalides registers and decrees from 1657 and 1660) or parish registers, which provide information on soldiers' wives and families. It is surely indicative of successful integration of the Irish that they make for the very French institution of the Hôtel Royal des Invalides, which was watched over by minister Louvois himself, in order to end their days there.

Several other phenomena in the area of the social history of these soldiers are worth noting. While the Invalides registers do not indicate the appearance of soldiers, unlike the post-1716 individual records, they do provide a considerable amount of valuable information in the statements of the applicants. They indicate their regional origins, religion, marital status, place of residence in France (twenty-one cases), occasional references to their profession and of course their record of service and certificates which indicate the regiments and companies they served in, and in some cases major battles or sieges in which they participated. The Invalides registers are also very important in that one can trace careers back from the 1670s or 1680s to the early 1630s, using the method proposed by Chaboche in his study of the veterans of the Thirty Years War.[42] Thus, we can identify six Invalides Irishmen who joined the French forces in the course of the 1630s, and increasing numbers for each following decade. This implies that one takes the statements of soldiers as being reliable as far as the years of service are concerned. There are several reasons for doing so, in spite of the healthy measure of critical acumen an historian always has to apply to documentation (for example, one Raymond Reilly who applied in 1671 said he was about 100 years old, and the son of a Colonel Reilly who had died at 152 years!). The soldiers carried certificates in many cases, which could be checked, and applicants were examined and questioned by the council which granted entry. Dates of events and names of colonels are verifiable now as they were then. One striking case of an Irishman whose long service record proves to be very accurate is the statement of Eugene Maccauulÿ, who came before the council in 1686: he was

> Irish by nation, 61 years old, sergeant under Sr Windisch, Konigsmarck regiments where he says he has served for 4 years, and before that 4 in Furstemberg, 10 in Hamilton, 5 in the Royal, 8 in Douglas, 5 in O'Brien and 3 in Mouskry. It is marked in his certificate that he has served the King[43] for 30 years, and is a Catholic.[44]

If one calculates the years in reverse, his narrative corresponds to various important dates mentioned in earlier parts of this article: the change from Furstemberg to Konigsmarck in 1682, then four years earlier, the incorporation of former Hamilton men in the German regiment when Hamilton was recalled in 1678; a ten-year service in Hamilton (give or take a year, 1669 being the year of the formation of the Dillon unit which became Hamilton); service in the Royal (*Royal Anglois* or *Irlandois*?) from 1663 to 1668, a change in 1655 from O'Brien to the Scottish regiment of Douglas, at the time of the passing of many Irish troops to Flanders and the breaking up of some Irish units, and interestingly, service with Muskerry from 1647 to 1650, with

42 R. Chaboche, 'Les soldats français de la Guerre de Trente Ans'. 43 The king of France. 44 Service Historique de l'Armée de Terre, série XY, vol. 10, council of 26 April 1686.

accurate recall of the year the Muskerry regiment first arrived in France. Maccauulÿ's statement also suggests that some of Muskerry's men were given to Murrough O'Brien on the latter's arrival in France in 1650. This soldier singlehandedly outlines a historical record which concerns many of his fellow-countrymen over four decades, and his testimony proves the value of the records at our disposal.

Further facts emerge from the pre-1688 Invalides registers, such as the fact that the Irish were the largest foreign group seeking entry for eight of the 18 years between the opening of the Hôtel des Soldats Estropiés in 1670 and the arrival of the Jacobites. In those eight years, the Irish were ahead of the Swiss, France's great foreign troops, and rivalled them statistically at other times. They were also the most numerous of the subjects of the Three Kingdoms in the Invalides records. Another phenomenon which clearly emerges from French military records concerning invalids is the exceptional proportion of Irishmen who bear military or regimental nicknames: roughly 30% of those who joined before 1688 and who appear either in the Invalides registers or the two decrees from the 1650s have nicknames in the French language. It is extremely rare to find invalid soldiers from the Swiss cantons or the empire carrying such names; the rate is one nicknamed man out of 31 Scots for the years 1670–1688, and only two Englishmen from the pre-1690 period. All the Irishmen who bore nicknames had served in regiments other than those of their nation, and one in six of the overall number of Irish applicants to the Hôtel Royal mention no Irish regiment whatsoever in their statement, suggesting that they may have spent their entire career in non-Irish units or regiments. Even if one takes into account the fact that long-serving soldiers who had been integrated into French regiments were likely to be more aware of the institutions of the army such as the hospitals, and therefore gravitated towards the Hôtel Royal des Invalides, the possible over-representation of Irish soldiers with regimental names does not affect the fact that the proportion among them bearing such names is much higher than other foreign nations.

Most of those who have nicknames had long service records,[45] and this obviously facilitated familiar relations between them and their French comrades in arms, and possibly competence in the language. The attribution of nicknames, whether by officers or fellow soldiers, indicates participation in French military life and the world of the French armies, where these men may have been valued as battle-hardened veterans among the recruits of French native regiments.[46] There may have been other factors behind the exceptional rate of surnames among the Irishmen: they were almost all Catholics and

45 The average length of service for the Irishmen who entered the Gardes Françoises between 1650 and 1669 was 15.75 years. 46 Some discussion of regimental nicknames in J.A. Lynn, *Giant of the Grand Siècle: the French army 1610–1715* (Cambridge, 1997), p. 200 and J.-P. Bois, *Les anciens soldats dans la société française au XVIIIe siècle* (Paris, 1990), p. 131.

Royalists, unlike the Scots or English, and this may have distinguished them from the other subjects of the three kingdoms and constituted common ground with the French. The historical tradition of Scottish service in France does not appear to manifest itself in familiarity in this particular case, so it may be that the factors at work in attribution of nicknames were contemporary. The language they spoke may have also come into play, although one would expect English and Scottish soldiers to have had a higher rate of nicknames if this was the case. One notes the almost total absence of nicknames among the Swiss soldiers, who were mainly German-speaking and Protestant. All of these considerations may point to religion and individual length of service in French regiments as the main factors of this phenomenon. There may also have been some cultural factors at work among the Irish, such as the practice of epithets in Gaelic names (Red Hugh, Conn Bacach O'Neill – the lame ...). The nicknames we find in the registers are interesting by their very ordinariness: *La Garenne* (the Warren), *La Rose*, *La Forest*, or *Sans Soucy*, *Sans Peur*, *Marcheatirre* which reflect military qualities (if not used ironically), or *Saint Patrice*, *Saint Martin*. Some names are of course connected to Ireland: *Espérance d'Irlande*, *l'Ibernois*, or simply *Irlande*. In their lack of exoticism, the choice of names follows normal French practice and may be an indicator of familiarity and integration. This phenomenon of nicknames parallels the frequent occurrences of local godparents of Irish children in parish registers in parts of France, and points to acceptance of the Irish in France on a more general level during the seventeenth century.

The history sketched in this article and the sources on which it is based have implications for the historiography of Irish service in France. The length of service and the recurrence of certain family names point to the necessity of taking a long-term view of the subject: Butlers served abroad – in France in particular – in the fifteenth, sixteenth and subsequent centuries, to take but one example. The chronology of Wild Geese history-writing must be modified to take account of the sizeable numbers involved and the considerable activity which took place in France throughout the seventeenth century. Both the time-span and the scale of the phenomenon deserve to be placed on the same level as the more familiar Spanish service during this period. French military historians have hitherto displayed curiosity and an open mind in relation to Irish material in the period covered here, and it is to be hoped that they will continue to contribute to the topic. Much work remains to be done on constructing a history of individual units, especially the various regiments. While this work is both valuable and necessary in order to understand the evolution of Irish involvement in the French armies, individual prosopographical analysis is also necessary in order to take heed of the fact that one in six Irish soldiers presenting in the Invalides in this period did not mention service in an Irish unit. This implies that there must have been a sizeable number of Irishmen scattered in various regiments in the French armies, and these men deserve the attention of historians no less than their comrades in arms who followed the many Irish captains and colonels in France before the arrival of the Jacobites. In terms of research, following a comparative approach in the context of the history of other foreign troops in the French service will also enable a fuller assessment of the nature and importance of the Irish contribution. The history of Irish military men and Irish units must therefore be placed in the broader context of French military history in order to understand it fully and over a long timespan: as we have seen, this story is composed of many 'passages' of soldiers long before the celebrated '*passage*' of the Wild Geese who followed James II into exile, and made of many deaths on far foreign fields before the great campaigns of the Irish brigades.

The Wild Geese in France: a French perspective

NATHALIE GENET-ROUFFIAC

On 10 June 1688, Mary Beatrice d'Este, the second wife of James II, king of England, Scotland and Ireland, gave birth to a son named James Francis Edward. Although the birth was a public event and courtiers were present as was the custom of the time, few births have given rise to such bitter controversies or had such momentous consequences. Within six months, a successful Dutch invasion had secured the throne for James' son-in-Law, the prince of Orange, and his daughter Mary and ensured that Britain would be a Protestant camp against Louis XIV. James II, those near him and his supporters, the 'Jacobites', made their journey into exile in France, which constituted the 'other refuge' at that time, to use the expression of the French historian Bernard Cottret. Louis XIV placed at their disposal the château of Saint-Germain-en-Laye, near Versailles, and it was here that the British court settled. Until 1715, Saint-Germain became one of the centres of intrigue in British political life.[1] The first exiles, comprising mainly English and Scottish courtiers, were joined by the Irish who had fought for James II from 1689 to 1692 in Ireland.

After the treaty of Limerick and the 'flight of the Wild Geese' to France in 1692, the history of Jacobitism was still full of English plots and Scottish military attempts in Scotland, but Ireland was seemingly quiescent. The fidelity of the Irish to the Stuarts took a different and specific form. Even before William of Orange promulgated severe anti-Catholic legislation in Ireland, they chose to follow their master into exile and continued to serve him by fighting in the army of his staunchest ally, the king of France. Although they were paid by Louis, the Irish regiments always rejected any association with mercenaries by stressing the political and religious nature of their allegiance: 'Our dutie to our King will make us serve the French King with all zeale and faithfulness that he can expect', so wrote Captain Rutherford to Henry Browne, secretary of Mary of Modena, from Lille, on 21 August 1691.[2] This quotation reveals clearly the nature of the presence of the Jacobite Irish regiments in the French army after 1690. This presence was a clear proof of their commitment to the Jacobite cause, but it also led them to

1 Our knowledge of the life at the court is manly due to the exhibition held at Saint-Germain in 1992. See E. Corp (ed.), *La cour des Stuarts à Saint-Germain-en-Laye* (Paris, 1992); *L'autre exil: Les Jacobites en France au début du XVIIIe siècle* (Montpellier, 1993).
2 British Library, London, MS Add. 37662, fol. 244.

their integration in the French society. Yet in this process they never questioned their 'Irishness' which was always highly proclaimed.

As a French observer, I won't dwell on the heroic actions of the Wild Geese in France and on the Continent, but seek to sketch an outline of their presence in the French army in an attempt to understand the stake and the unique position it gave them, not only in the French army, but also in the French society. My aim is to also underline the subtle balance and game of influence that it involved for Versailles, Saint-Germain and the Irish themselves.

THE WILD GEESE IN FRANCE

It is very significant that the Irish Jacobite exile in France was part of the broader movement of Jacobite exile in France, and that it also gave a new dimension to an old phenomenon, the presence in France of Irish soldiers. For the 'flight of the Wild Geese' was a new reality, as well as being an aspect of the Jacobite exile in France and an episode of the long-term relationships between France and Ireland.

Soon after 1603, some Irish soldiers had entered the service of Spain and France and this service had continued throughout the seventeenth century. The French armies always attracted the poor from economically backward countries like Ireland. Louis Cullen has already depicted a constant flood of an average of 900 Irishmen each year between 1601 and 1701, that rose to perhaps 1000 annually between 1701 and 1750.[3] Eamon Ó Ciosáin's article in this book also examines importance of Irish individuals and units in French army before 1688.

But the 'flight of the Wild Geese' occurred at a much bigger scale and as a collective move. Their number is difficult to establish from the seventeenth-century archives and has been the centre of many scholar debates. The Irish scholar John Cornelius O'Callagahan and the French historian Guy Chaussinand-Nogaret gave a number of 19,000 men, but Guy Rowlands, the English specialist of Louis XIV's military administration, who has spent several years in the French archives, doesn't think than more that 14,000 Wild Geese, with 4,000 women and children, came over to Brittany.[4] Éamonn Ó Ciardha has made an excellent summary of this whole debate.[5]

3 L. Cullen, 'The Irish diaspora of the seventeenth and eigteenth centuries', in Nicholas Canny (ed.), *Europeans on the move: studies on European migration, 1500–1800* (Oxford, 1994), pp 139–40. 4 J.C. O'Callagahan, *History of the Irish Brigades in the service of France* (Dublin, 1854), p. 29. See also G. Chaussinand-Nogaret, 'Les Jacobites au XVIIe siècle: une élite insulaire au service de l'Europe' in *Annales économiques, sociales et culturelles*, 28, p. 1098; Guy Rowlands, *An army in Exile; Louis XIV and the Irish forces of James II in France, 1691–1698*, Royal Stuart Society, LX (2001), p. 5. 5 Éamonn Ó Ciardha, *Ireland and the Jacobite cause, 1685–1766: a fatal attachment* (Dublin, 2002), p. 32.

We know more about their geographical, social and professional origins by examining the sample of 1020 soldiers admitted at the Hôtel des Invalides between 1689 and 1715, which we can consider as 5% to 8% of all the Wild Geese. English and Scottish soldiers are only a very small minority (only 3.5% and 3%) whereas the Irish represent 93.5%.[6] Among them men from Munster (47.1%) are predominant, then Leinster (25.4%) and Ulster (20%). Only 7.5% came from Connacht, which fits both the demographic weakness of this area and its isolation from the political context. Except for those who already were in the military service, 10% of the newcomers had had a previous professional occupation, most of the time as craftsmen.[7]

When it comes to their integration to Louis XIV's army, it's important to keep in mind that they were spread in two different kinds of Irish regiments: the Irish regiments *of* the French army (the Mountcashel brigade) and later the Jacobite Irish regiments *in* the French army (the Wild Geese regiments).

MOUNTCASHEL

The Mountcashel brigade had been organized early during the war in Ireland. In May 1690, James's situation had been weakened by William's of Orange landing. Louis XIV agreed to send new French troops but only if Irish infantry regiments were integrated in his army. The French king had been warned by his agent d'Avaux, to be extremely demanding as to the quality of these troops, because most regiments in Ireland had been recruited by 'gentlemen who know nothing of war [...] the companies are made of tailors, butchers or shoe-makers, who keep them at their own expenses and are the captains'. The statement was not purely rhetorical for the registers of the Hôtel des Invalides, show that one third of the Irish that had had a previous profession had been clothes makers or tailors![8]

But Louis XIV made his conditions very clear: he wanted men of experience and noble birth. The negotiations between the two cousins were rather tense, and the position of the Comte d'Avaux often turned out to be rather uncomfortable.[9] Eventually, James II had to give in to his cousin's demands.[10]

6 Service Historique de la Défense, Vincennes, archives de la Guerre, 2Xy 10–19. Graphs and maps are available at the end of this chapter. 7 Clothes trades and craftsmen: 36%; leather trades and leather craftsmen: 16.8%, wood trades and wood craftsmen: 11.2%, iron and stone trades and craftsmen: 10.4%. There are also to be found one painter, four surgeons, two gardeners and a shepherd. There are no merchants but there are three bakers and an ale brewer. 8 Service Historique de la Défense, Vincennes, archives de la Guerre, 2Xy 10–19. 9 'Better regiments could have been given, Milord. The truth is that we find here some bad will, and the fear of losing troops, but even more the wish to make Milord D'avaux's negotiations fail, for he is not well loved in this court': letter from the *commissaire* Fumeron to the marquis of Louvois, Cork, 28 March 1690 (Service Historique de la Défense, Vincennes, archives de la Guerre, A^1 894, fol. 174). 10 'One has to

On arrival in France, these troops were sent to Bourges and organized in three regiments, Mountcashel, O'Brien and Dillon, as the 'Mountcashel brigade'. They were given the red uniform with white socks and the hat known as the 'chapeau lampion'. Each regiment consisted of two battalions, including fifteen companies of 100 men each, and the colonel's company, with a various number of cadets.[11] According to Arthur Dillon, the brigade numbered 5,371 officers at its creation and 6,039 in 1697. All the officers were commissioned by Louis XIV, and the regiments were paid like French foreign regiments, one sol more a day than the French troops. Besides their own wages, the colonels were given a *sol* for each *livre* paid to the soldiers of their regiment, and Mountcashel one more *sol* for the men of the two other regiments. These financial conditions were granted against the backdrop of a wider European context where war had become something of a business and each recruiting officer was a businessman, more or less successful depending on how he administered his unit.[12] In every way the regiments were French regiments composed of Irish soldiers. But the feature was not unusual in a French army where 13.8% of the *maréchaux de France* were foreigners.[13]

The arrival in France of the Mountcashel brigade did not put an end to the transactions between the two courts. The war on the Continent made France more and more demanding. On 30 July 1690, the marquis of Louvois informed the duke of Lauzun of the new requirements laid down by the king of France.[14] James II needed persuading, and the French minister had to become more insistent with the duke of Tyrconnel in May 1691: 'I'm very surprised to see that, in spite of your promises, you are very cold in the sending of new recruits.'[15] James persisting in ignoring his cousin's demand, the French became far more explicit. 'I give you the order', wrote the marquis

consider that the king of France had demanded four regiments of 1600 men each, but the ambassador has said that five regiments of 1000 men would be enough, that colonels from the Service were not necessary if they were of good condition. On this, King James has designated the regiments of Mountcashel, O'Brien, Butler, Dillon and Fielding, that should make 5,000 men of the best kind the country can give' (Mémoire sur les troupes irlandaises, July 1690, Service Historique de la Défense, Vincennes, archives de la Guerre, A[1] 1082, fol. 74). 11 Orders from the Marquis of Louvois to the *commissaire* Bouridal, 4 May 1690 (Service Historique de la Défense, Vincennes, archives de la Guerre, A[1] 960, fol. 254). The Mountcashel regiment counted twenty cadets, the Dillon regiment sixteen. 12 Kiernan, 'Foreign mercenaries and absolute monarchy' in *Past and Present* (1957), 131. Quoted by P. Melvin, 'Balldearg O'Donnell abroad and the French design in Catalonia', *Irish Sword*, 12 (1977–9), 114, footnote 13. 13 A. Corvisier, *L'armée française de la fin du XVIIe siècle au ministère de Choiseul*, p. 951. The percentage grows with the rank: 5% of the *maréchaux de camp*, 12.6% of the *lieutenants généraux* and 13.8% of the *maréchaux de France.* 14 'Her Majesty would be very pleased and expect from Milord Tyrconnel's affection that you may embark three, four or even five thousands Irishmen in four or five regiments and that you embark at least eight hundred, and even one thousand or twelve hundred men as new recruits for the Irish regiments that are already here' (Service Historique de la Défense, archives de la Guerre, A[1] 960, fol. 325). 15 Letter from

of Louvois to the commissaire Fumeron, 'to tell the duke of Tyrconnel in a private discussion, that if he fails me on the recruits matter, he will find me failing in all matters concerning Ireland.'[16] The message was clear enough. As soon as 21 May, Louvois could thank Tyrconnel for having sent 1,200 more men to the Mountcashel brigade, 'which is a small present in comparison with all the expenses the king makes for the sake of Ireland'.[17]

THE WILD GEESE

The situation was rather different for the later Wild Geese regiments that Patrick Sarsfield brought with him. After the fall of Limerick, and even before the treaty was signed in February 1692, three-quarters of the Jacobite army in Ireland were convinced to follow their officers to the Continent at the expense of William of Orange. It was a simple decision for the treaty would have allowed them to stay and, while the Williamite anti-Catholic legislation had not yet been passed, a retreat to the Continent seemed to be the best way to defend James II's interest.

The maintenance of such a large amount of troops represented a huge expense for the French, and it was essentially a foreign army that was maintained in French territory. But Louis XIV seems to have accepted for four main reasons: a feeling of moral obligation toward his cousin, the demands of James II, the need for the French army to have troops in reserve and the need to make sure that the Irish entered the French service before they could be recruited by an other catholic country, such as Austria or Spain.[18]

All the Irish troops were gathered in Brest, under the command of Dominick Sheldon. James II came with Berwick and declared himself to be very satisfied with their fidelity. In fact, they were on a very poor condition. The voyage had finished the Irish army: at least 1,500 men were sick and the French had to supply them with hats, shoes, shirts and socks. Until the battle of the Hougue, on 24 May 1692, the Irish troops were expecting to re-embark to Ireland. The Hougue defeat sealed the fate of the Irish regiments. In the second week of June, Louis XIV unilaterally decided to distribute them

Louvois to Tyrconnel, 12 March 1691 (Service Historique de la Défense, archives de la Guerre, A[1] 1065, fol. 65). 16 Letter of Louvois to the *commissaire* Fumeron, 27 March 1691 (Service Historique de la Défense, archives de la Guerre, A[1] 1065, fol. 71). 17 Service Historique de la Défense, archives de la Guerre, A[1] 1065, fol. 85. These troops left Limerick on 18 July (Service Historique de la Défense, archives de la Guerre , A[1] 1065, fol. 116). Only in December 1691 did Barbezieux let M. d'Usson know that 'there is little evidence that now than more troops can be expected from Ireland' (Service Historique de la Défense, archives de la Guerre, A[1] 1065, fol. 153). In January 1692, when the Irish Jacobite regiments arrived on the Continent, 600 men were detached to reinforce the Irish regiments of the French army (Service Historique de la Défense, archives de la Guerre, A[1] 1065, fol. 178 et 184). 18 Rowlands, *An army in exile*, p. 5.

between his armies. The officers were given their travel warrants to rejoin the French armies.

The reorganization of the regiments was extremely contentious. James II had to reduce his army to two regiments of horse guards, two cavalry regiments, two foot dragoon regiments, eight infantry regiments and three independent companies. Many regiments that had fought in Ireland were disbanded and many officers lost their ranks, but they all were commissioned by James II. As foreign soldiers in the French army, they should have received higher wages than the French troops, but it was agreed that they would be paid the same, 50,000 French *livres* each month for the whole contingent. James II promised them he would pay them back the difference as soon as he recovered his throne.[19] No doubt this reduction in wages was very unpopular. An Irish officer complained to his son: 'No sooner had we arrived in France than King James made an arrangement with Louis XIV by which he had put us on a French footing, reserving for himself the difference in our pay for his own upkeep and that of his house.'[20]

THE DISBANDMENTS OF 1697

The 1692 organization lasted until the peace of Ryswick in 1697. In addition to the political failure, the king of France had to proceed in a major reform of the Irish troops and this had extremely heavy repercussions for the economic situation of the Wild Geese and their families. As early as September 1697, the 25 battalions had to reduce the number of their companies from 16 to 14, and each of them had to disband half their men.[21] A broader revision occurred in February 1698. Only the three regiments of the Mountcashel brigade, which were already French, and the marine regiment, now the Albemarle regiment, were totally spared. The two regiments of Limerick and Dublin were disbanded, and the others merged into five new regiments (Sheldon, Dorrington, Galmoy, Lutrell then Bourke and Berwick). The two cavalry regiments were amalgamated into a new Dorrington regiment. Three fifths of the Irish soldiers were disbanded, somewhere around 5,000 men. Being forbidden to go back to Ireland by the British government, many turned to Saint-Germain for help and charity. A Scottish exile, Walter Inness, wrote a striking description to his fellow Charles Leslie on 24 March 1698:

19 Historical Manuscript Commission, *Stuart Papers*, i, 66. **20** Bibliothèque Nationale, Paris, MSS Français 12161. **21** Bibliothèque Municipale, Nancy, MS 305 (423): *Notes concernant les régiments irlandois depuis leur arrivée en France en may 1690 recueillies par milord Clare, maréchal de France.* Quoted by Yves Poutet in 'Jacques II, MacMahon et Kennedy', *Revue d'histoire de l'Amérique française*, 21:3 (1968), 432, footnote 8.

> It would pitie a heart of stone to see the Court of Saint-Germain at present, especially now that all the [...] Irish troops are disbanded, so that there are now at Saint-Germain not only many hundred of soldiery starving, but also many gentlemen and officers that have not cloathes to putt on their backs or shoes to putt on their feet or meat, etc.[22]

But in spite of his good will, James II could not sustain all of them. After their arrival in France, many Irishmen were already dependant on their wages as soldiers, or on the charity of the court. The king of France gave James II a yearly pension of 600,000 French *livres* for his own maintainence. In his *Memoires*, James II assessed that he had managed to 'relieve an infinite number of distressed people, ancient and wounded officers, widows and children of such as had lost their lives in his services.'[23]

But the condition of the Wild Geese, uncomfortable from the beginning, became extremely difficult after 1697. Many French archives record the poverty of the Irish soldiers' families. A 'list of the poor who lives in two houses in Saint-Germain and are in very poor condition' mentions 71 poor Jacobites in the first house, 33 being soldiers or relatives of soldiers, and 43 widows ('whose husbands have been killed in the service of the king of France') and their daughters.[24] On a list recording 46 Irish living in the parish of Saint-Sulpice in Paris, the first name was a man of quality: 'Henry Canty, Irish captain, broken. He is cousin to Mr Plankett [*sic*], archbishop in Ireland.[25] He has a wife and three daughters. Very poor, has to sleep on straw.'

Many were admitted at the 'hospital general', usually arrested in 'waves'. The largest number was in June 1704, followed by a second in May 1705 and a third one from February to June 1709, '*l'année terrible*'.[26] The institution of La Charité, rue Jacob, admitted 77 diseased between 1702 and 1715[27]. It seems, however, that the beggars were usually sent to Bicêtre. These latter included men such as Charles Carny, an Irishman, who was described as 'rotten'.[28] In spite of the prejudice of the time against poverty, it seems that Jacobites were better treated than the other patients, their condition being 'legitimized' by their exile. Some of the disbanded soldiers also used the facility given by the French military administration and entered the Hôtel des Invalides. Actually the number of Wild Geese to be admitted in the Invalides

22 Scottish Catholic Archives, Edinburg, Blairs Letters BL 2/38. 23 Clarke, *Life of James II*, ii, 472. 24 Bibliothèque nationale de France, Paris, MS Français 20866, fol°106, 107 et 113. Also quoted by Y. Poutet, op. cit., p. 433. 25 Patrick Plunkett, bishop of Ardagh from 1647 to 1669, then archbishop of Meath, from 1669 to his death in 1679. 26 Archives Nationales, Paris, O^{I} 48, fol° 99, 105; O^{I} 49, fol° 77 and 101 ; O^{I} 53, fol. 16, 24 v°, 26 v°, 40 v°, 71 and 226. 27 Archives de l'Assistance publique, Paris, La Charité: 1 Q 2/1, 2 et 3: registres of admissions 1702–7/1707–14/1714–19. 28 Archives de l'assistance publique, Paris, Bicêtre, 4 Q1: registres of admissions of beggars: 1716–26.

increased by 150% between 1697 and 1698, with 107 men being admitted that same year.[29]

In Saint-Germain, the extreme poverty of the Irish seems to have caused particularly acute problems, and contributed to a climate of endemic violence. Some disbanded soldiers had stayed together in small groups and the description of Irish misdeeds in Saint-Germain soon became part of Williamite propaganda. Dr Doran, by example, wrote that the town had become virtually un-inhabitable because of the sanguinary violence of Jacobite bandits.[30] It can also be connected to the tradition of the 'rapparrees' described by Eamonn O'Ciardha among the Irish troops during the war in Ireland.[31] And it is indeed obvious in French judicial archives of this period. It's not exactly surprising that the Irish were a source of anxiety to the French authorities. It was for this reason that La Reynie, the lieutenant of police in Paris, exercised constant surveillance on the exiles in France, in co-operation with the ministers at Saint-Germain. In 1707, Middleton and Torcy organized a kind of 'census': 'all subjects of HBM [His Britannic Majesty], who are not so well known that there is nothing to fear from them, should proceed or send to Saint-Germain to identify themselves as faithful subjects and to provide themselves with certificate [... after having provided] good sureties for their conduct and in point of loyalty'.[32] For such was the situation of the Wild Geese in France, torn between the court of the Stuart king they were faithful to, and the administration of the king of France they were serving.

VERSAILLES, SAINT-GERMAIN AND THE IRISH

To understand better the situation of the Wild Geese, it's important to start with their position in the first generation of the Jacobite diaspora in France. The first wave of arrivals from 1688 to 1691 consisted almost exclusively of English people and was the immediate consequence of political events in England. In 1692 there was a spectacular increase in the number of the exiles (more than 230%), easily linked to the arrival of the 'Wild Geese'. The numbers swelled until 1698, particularly in the years 1696–7 and then slowed down and continued at a steady rate until 1701. This phenomenon was

29 Thirty-three men were admitted in 1696, forty-eight in 1697 and no less than one hundred and seven in 1698 ; the worst year was 1700 with one hundred and forty-seven soldiers admitted. (Service Historique de la Défense, Vincennes, 2Xy12). The 1696 disbandment was the end of the pre-Wild Geese generation of soldiers who were now merged in the new 'Jacobite regiments'. The oldest men were not kept in the regiments but sent to Les Invalides. The percentage of soldiers having already been in the service of France before 1690 is at the highest between 1698 and 1700. 30 Quoted by O'Callaghan, *History of the Irish brigades*, p. 189. 31 O'Ciardha, *Ireland and the Jacobite cause*, p. 67. 32 Letter from Middelton to Torcy, 14 June 1707 (Bodleian Library, Oxford, MS Carte

especially noticeable among the Irish and may be explained by the regrouping of army families close to Saint-Germain after the disbanding of the troops. After 1701 the population decreased. The process was identical for the English and the Irish and was accompanied by a decrease in the number of the marriages and baptisms, a sign that the population which came between 1688 and 1692 was then getting older and that it 'escaped', mainly because the disbanded soldiers left for other catholic countries with their families.

The composition of the exiled community stabilized after 1692 and remained the same until 1715, a date which saw the second wave of arrivals and coincided with the second generation of exiles. The Irish formed the majority, around 60% of the exiles, as against 35% who were English, with only 5% of Scots, even though the political influence of this last group was in inverse proportion to its numbers.

First of all, the links between Saint-Germain and the regiments were maintained by the frequent presence of the officers at the Stuart court, especially in winter times. Some of them, and even more their relatives, were members of the king's household. Dominick Sheldon, colonel of the regiment of the same name, was one of the main figures of the court of Saint-Germain where he was vice-chamberlain and under-preceptor to the prince after 1697. His brothers Ralph and Edward were esquire to the king and to the queen. Sheldon had gained a reputation of great bravery at the battle of the Boyne. The French courtier Saint-Simon, usually more sparing on compliments, declared him to be 'one of the rarest, finest and broadest minds in Britain'.[33] In his youth, he had belonged to the troops Charles II had lent Louis XIV before he came back in England in 1675. He became brigadier in 1693, *maréchal de camp* in 1694 and served as aide de camp of the duke of Vendôme in 1708.

The link between the regiments and Paris was also entrusted more specifically to a small numbers of representatives, also called 'the agent of the Irish troops in the service of France'. One of these agents was Morgan Scully, captain in Regiment of Lee, who is described as such an agent in a document of 1695.[34] A letter of the duke of Berwick is more explicit: 'the habit has always been, for the 18 years the Irish troops have been in France, to have someone at court to take care and manage the interests of the regiments of this nation, for the expense of a single agent to maintain [there] is infinitely lower than [that of] many officers we should have to send'.[35] This agent was in charge of their interests both in Versailles and in Saint-Germain. At the same time, the king maintained written relationships with the regiments, usually

238, fol° 186). 33 *Mémoires de Saint-Simon*, quoted by Jacques Dulon in *Jacques II Stuart et sa famille* (Paris, 1897), p. 142. 34 Leave from Colonel Talbot, 30 January 1695 (Archives Nationales, Paris, Minutier central, étude LIII/ 112). 35 Letter of the duke of Berwick, 9 February 1710 (Service Historique de la Défense, Archives de la guerre,

with the colonel. In 1691, in Lille, his correspondents were Captains Knighteley and Rutherford. The responsibility was so heavy that on 31 December 1691, Knighteley asked the court to send him help for he was overwhelmed by his task "to that degree that I am harassed to death".[36]

THE LINKS WITH SAINT-GERMAIN – A DOUBLE FAILURE

And yet, the relationships between the Irish Wild Geese and the Jacobite court of Saint-Germain ended up with a double failure. James II and his family were the focuses for the exiles, and the importance of any exile could be determined according to his social distance from the monarch. The whole community in exile can be likened to four concentric circles around the Stuart sovereign, and all those living in Paris and Saint-Germain can be assigned to one of the four.

The first circle consisted of the aristocrats with political power and social leverage. Around 65 nobles can be placed in this first circle and they composed a narrow political elite. To the second circle belonged the gentry and all the nobles who had an office in the royal household; these totalled about 200. Beyond this, in the third group, were the servants and all those who lived on grants and pensions – perhaps 300 Jacobites. Finally, in the fourth circle, are found those who did not live at the court but were occasionally linked to it; 77% of the Jacobite community in exile around Paris constituted this outer circle and 85% of the Irish families belonged to it.

The closer to the centre, the more Scots and English are to be found. The more distant, the more Irish. The outermost circle was in fact composed mainly of Irish Jacobites. Since political influence was determined by social status. The Irish, despite their numbers and fidelity, exerted little influence. The main problem was the weakness of the Irish social patronage. In Saint-Germain, like in all courts, patronage was very important. Receiving grants from the king could determine ranking and sometimes simply ensure survival. It was essential in the exiled Jacobite world to rely on patrons, naturally the leading aristocrats of each national group. If the Scots were extremely well protected by the duke of Perth, the Irish patronage soon appeared too weak for their needs. The deaths first of Tyrconnell and then, at the battle of Nerwinden, of Patrick Sarsfield, deprived them of any prominent patron at the court. Protection was given by Tyrconnell's and Sarsfield's widows, and the duke of Plowden who was never more than a second-rank courtier to James II. Another patron was the duke of Berwick, James II's illegitimate son, who had married Sarsfield's widow.[37] The duke is known to have been very

Vincennes, A^{I} 2262, fol° 109). **36** Letter to Henry Browne from Lille (Westminster Diocesan Archives, London, Browne Papers: B7 n°106). **37** Lady Honora Bourke,

fond of his Irish soldiers whom he commanded, and his private clerk, Patrick Farely, was Irish. But the duke was more successful in the French army than at the court of Saint-Germain, and after his second marriage, to Lady Bulkeley, he seems to have been closer to the English group.

The question was also one of the social distinction between the Irish members of the courts, including the officers, and the rank and file of the Irish regiments. The soldiers, most of the time, belonged to the 'Old Irish' families while the officers were more often 'New Irish'. As a very simple, but striking, evidence, their names were different. The Irish on the parochial registers of Saint-Germain and the Irish soldiers in the registers of the Invalides, at the very same period, simply don't have the same first names. You find names like 'Oine' and 'Oile', 'Darby', 'Mortagh', 'Dawson' or even 'Eoghan' among the soldiers. These names are not to be found in Saint-Germain.

The second failure was due to the fact that Saint-Germain missed the point with the Irish regiments. When it came to the disbandment of the regiments, for example, the decision had been taken by the court of Versailles and the consequences had fallen on the court of Saint-Germain. For all, the fate of the Irish regiments from their landing in Brittany had been a matter of tight bargaining between the French and the Jacobite court. The stakes were high. For the French king, the Irish regiments were a weapon against England that he never found to be wanting. Until 1697, the main danger for William of Orange still would have been a decisive military success for Louis XIV. For the Irish, service in the French regiments was the only way left in which they could fight for their king and their country. But for James II this was not only proof of a loyalty to his person, but it also offered proof that he still was a sovereign.

It is important to stress that Louis XIV made a major political gesture towards his cousin when he left him the right to commission the officers of the Irish regiments. For keeping an army was the prerogative of a sovereign, and it is a sign of the depth of Louis's commitment to his cousin that he let him do so on the soil of his own kingdom. And the fact that the Irish regiments were *de facto* parts of the French army doesn't lessen his gesture for no one was more aware of the political significance of all these measures than Louis himself. The Irish regiments, as the army of James II, were a proof that he still was a king. They gave James and his son a different status from all the other exiled princes protected by the court of Versailles. 'As long as there is a body of Irish Roman Catholiks troops abroad,' Charles Forman wrote to the

daughter of William, earl of Clanricarde, married Berwick on 26 March 1695. The duke acted as the guardian of Sarsfield's son. The French courtier, Saint-Simon, admired the duchess as 'une très belle femme, touchante et faite à peindre, et qui réussit très bien à la cour de Saint-Germain', quoted by A. Rohan-Chabot, *Le Maréchal de Berwick* (Paris,

Walpole government in 1720, 'the chevalier will always make some figure in Europe by the credit they give him.'[38]

At the court of Saint-Germain, the two kings had agreed that James could keep a military administration proportional to the size of the Irish regiments, under the direction of Sir Richard Nagle, James's secretary of state for Ireland and his general attorney, who became secretary of war until his death in April 1699. James II appointed the officers, could suspend them and decided on their replacement. He supervised the court martial sessions and reviewed them several times. Most of the administrative questions were managed at the court of Saint-Germain or in the regiments. Nagle often had to settle quarrels between Irish colonels, and he created the loan system which made them able to pay their troops. But Saint-Germain's authority and control on the regiments was weakened by the early death of Tyrconnel and Sarsfield, in spite of the many disagreements between the two men.

Guy Rowlands in his study of the Bureau de la Guerre, the military administration created by Louvois, has shown that the management of the Irish regiments was a huge administrative issue between Saint-Germain and Versailles, the importance of which has often been underestimated. Until the death of Nagle and the peace of Utrecht, a regular correspondence and means of cooperation was settled with the French minister, and James II gave Barbezieux several audiences.[39] From the very beginning, the French minister gave the officers their travel warrants and the regiments their orders to move; and his commissaries could inspect the Irish regiments, even if Dominick Sheldon had been made general controller. Above the rank of brigadier, the Irish officers ceased to be given their commissions by James II and could be promoted by Louis XIV only.[40] It was a good way to settle his patronage on the Irish officers whose career, eventually, depended on the Versailles patronage, not on their relationships to Saint-Germain. James II was very eager to display at Saint-Germain the image of a royal court, but he never quite understood that keeping up a royal military administration could also have been a proof of his royalty.[41] In fact, when it came to administration, the Jacobites were sheer amateurs compared to the French professionals of the Bureau de la Guerre. When the time for official reverence to James II was over, the professionals attempted to take over the Irish regiments from Saint-Germain.

1990), p. 57. 38 C. Forman, *A letter to the right honourable Sir Robert Sutton for disbanding the Irish regiments in the service of France and Spain* (Dublin 1728), quoted by Rowlands, op. cit., p. 1. 39 Richard Nagle stayed in charge with the War department for James II until his death in April 1699, but the Versailles Correspondence was given to the State Ministers Melfort and Middleton. 40 On the matter, Louis XIV did not always give way to the wishes of James (Rowlands, op. cit., p. 8). 41 Rowlands, op. cit., p. 13.

LINKS WITH FRANCE

But taking absolute control on the Irish regiments was another matter. In fact, the French never quite did. The first decision of the French administration was to try to teach them order and discipline. Previously in Brest, Louis XIV had sent officers of the gardes françaises to the Irish regiments in an effort to try to teach them French drill. M. de Famechon, French *brigadier* in Ireland, had warned the marquis of Louvois about the difficulty of the task: 'I have no doubts that in other places and under the command of other officers, they could become good troops, for the Irish have good dispositions for war, but they first have to be taught how to be soldiers.'[42]

The temper of the Irish and their lack of discipline had already surprised Louis XIV's agent in Ireland. In France, they incited the indignation of the French commissaires des guerres. Commissaire Bouridal complained to Minister Barbezieux: 'I would never have believed old troops could lack discipline so badly, and the officers are even worse than the soldiers.'[43] Barbezieux had strong convictions about the best way to deal with the Irish temper 'the only way to avoid the disturbances commited by Irish soldiers, horsemen and dragoons is to have the most guilty of them hanged, as I've already told you. That should put an end to the troubles and the troops will get used to the French discipline.'[44]

Even after Barbezieux was put in charge of the inspection of the Irish regiments in 1696, his agents found it difficult to understand the way the Irish officers were managing their regiments, especially in matters of wages. The Irish kept 'their own way', often opening the way to financial problems or extortions. In 1695, 100,000 french *livres*, seven months' wages of a battalion, were stolen, and four regiments were almost driven to bankruptcy.[45] The Irish regiments therefore remained exceptions in the French military administration and were maintained in a very ambiguous position.

ASSESSMENT OF IRISHNESS

The Irish regiments were a kind of little Ireland on the Continent. For these men who had initially thought that they would back in Ireland soon and knew nothing of life in France, the regiments offered a reassuring home. The colonels of the regiments were the landowners they had known at home and

42 Letter of Ignace de Belvalet de Famechon to the marquis de Louvois, Limerick, 9 August 1690 (Service Historique de la Défense, Vincennes, archives de la Guerre, A[1] 1082, fol. 80). 43 Letter of 28 December 1691 (Service Historique de la Défense, Vincennes, archives de la Guerre, A[1] 1082, fol. 145). 44 Letter of the marquis de Barbezieux to the maréchal d'Estrées, Versailles, 14 February 1692 (Service Historique de la Défense, Vincennes, archives de la Guerre, A[1] 1065, fol. 191). 45 Guy Rowlands, op.

usually the men were gathered according to the area they came from. The regiments were also a world in itself because, very often, the families of the poorest soldiers could not afford to stay away from their relatives and followed them during the military campaigns.

French archives help us follow them. On 6 October 1712, John Gernon, one of Mary of Modena's (James II's wife) chaplains, gave to a lawyer in Paris a 'leave', written in his favour by 'Catherine Sloakes', widow of a captain in Berwick regiment. She was said to be with her children at the regiment's winter quarters at Manosque in Provence, and the letter had been written by a local lawyer. Two years later, Gernon gave the Parisian lawyer a second letter from the same woman, written by 'Diego Garcia Calderon, lawyer in Malaga' in Spain, where the regiment was then stationed.[46]

The singularity of the Irish were certainly reinforced by the strong animosity that existed between the three national groups of the Jacobite diaspora. It seems that suspicions between the three groups were very intense. Indeed, they hated each other. D'Avaux, Louis XIV's agent in Ireland, had already discovered before 1692 that 'the Irish love the French and are the enemies of the English so that if they were let loose, it would not take them long to slaughter them'.[47] Animosity still was particularly strong in the army where the Irish at first refused to serve under officers but their own countrymen.

Even in Paris and Saint-Germain, the tensions were in evidence. The Princess Palatine summarized the French point of view by complaining that the exiles were all fighting like cats and dogs.[48] The French parochial records reveal the extreme reluctance with which the different communities accepted marriages between members of the three national groups. Only a very few can be traced in the Saint-Germain registers during the first generation. The phenomenon may seem very obvious, but actually it was something new for the Irish gentry. Before their exile, it was among the daughters of the English Catholic gentry that they had looked for suitable alliances rather than among other Irish families. As their period of exile developed, it could be argued that the restrictions imposed on their lives by the confined sphere of the court of Saint-Germain made them more Irish than they had been before their exile as rates of marriage within the Irish gentry families increased.

The French military administration maintained the procedures and regulations of the French army and made it very clear that the Irish had to be organized in the same way and that Irish soldiers should be gathered in Irish regiments. The *ordonnances* of 29 November 1694 and even more of 12 February 1702, made it a strict obligation 'for all the English, Scottish and

cit., p. 15. 46 Archives Nationales, Paris, minutier central, étude LIII/152 and 163. 47 D'Avaux, *Négociations en Irlande*, pp 50–1. 48 A. Joly, *Un converti de Bossuet: James Drummond, duc de Perth* (Lille, 1933), p. 317.

Irish that are in France to enter the Irish regiments on the King's service' on pain of being prosecuted for vagrancy, which could have resulted in being sentenced to the galleys.[49] We know the measure was effective for after 1694, all the Irish soldiers at the Invalides belonged to Irish regiments.[50] In March 1702, the *intendant* of Franche-Comté wrote to Chamillart that the magistrates and sub-delegates had been looking for British in his area, and had found none.[51] The actual meaning of such a measure was less to find new soldiers, when it had been necessary to disband troops in 1697, than trying to keep control of these troublesome foreigners in this century of the *grand renfermement*. This could explain the threat of using vagrancy laws and sentencing men to the galleys. In the same way, in 1702 the minister of war ordered that British soldiers were to be sought out in French regiments so that they could be sent to the Irish regiments.[52]

The other main source of recruitment of the Irish regiments still was Ireland. The Catholic clergy and the Catholic gentry played a major part in the process of enlistment during the eighteenth century, creating rituals giving a strong religious signification to the departure of troops for the Continent. This became a main theme in Jacobite literature in Ireland, as well as the main form of displaying commitment to Jacobitism in the eighteenth century. The owners of the Irish regiments kept agents associated to their former estates, maintaining a local patronage. But the French also kept recruiting officers in Ireland, before and after 1692 and a 'mémoire touchant des moyens pour avoir des recrues d'Irlande. 1693' stipulated that 'an agent was to be established at Dublin, who was to have agents to act, according to his directions, in the several Counties. They were to enlist recruits and to facilitate their escape to France.'[53] The danger was real and Sir James Dempster, Mary of Modena's secretary, could complain to Cardinal Gualterio about 'the severity with which are persecuted recruited and recruiting staff of our Irish regiments. It is most unlikely that the English government tolerates this enlistment for it's a deadly crime to enlist people in a Prince's estates without his leave.'[54] After the failure of the 1745 rebellion in Scotland, a register of Jacobite prisoners shows that most of the rank and file were Irish natives and a *register de contrôle de troupe* for FitzJames cavalry in 1737 records 80% of men actually born in Ireland.[55]

49 Isambert, *Recueil général des anciennes lois françaises*, xx, 405. 50 Until 1697. After the disbandment, some officers tried to stay in the French service by hiding in French regiments. 51 Letter of 26 March 1702 (Service Historique de la Défense, Vincennes, archives de la Guerre, A[1] 1581, fol. 48). 52 None could be found in the cavalry regiments 'where they have already been taken, so that it prevented the colonels to take others': letter from de Courtebourne to Chamillart, 9 March 1702 (Service Historique de la Défense, Vincennes, archives de la Guerre, A[1] 1561, fol. 92). But the Irish officers in charge of this inspection could find 87 men in Flanders, 120 in Artois, 96 in Alsace and 6 in hospitals (Service Historique de la Défense, Vincennes, archives de la Guerre, A[1] 1561, fol. 260; A[1] 1581, fol. 52 and fol. 46*). 53 O'Callaghan, op. cit., p. 160. 54 Letter written on 12 August 1714 (British Library, London, MS 46495, fol° 140). 55 Eoghan Ó hAnnracháin,

It seems also that some of the Irish soldiers had previously deserted from the British army to join the Jacobite army, before or after the flight of the Wild Geese. The desertions, a rather common phenomenon in all European armies of the time, took on a different meaning when it came to Jacobitism. In fact, it seems that desertions from the British army was one of the most distinct proofs of commitment to James II.[56] William of Orange was advised in April 1691 not to visit his troops in Flanders, mainly Irish soldiers, for fear they would turn against him.[57] Very early in 1689 the desertions were taking place at such a scale as to facilitate French recruitment.[58] 'We have established meeting points on the Somme,' wrote Dangeau in February, 'to receive all the English, Scottish and Irish that want to enter the cavalry.'[59] The first reaction of the French was to send them to the regiments of Mountcashel's brigade at Bourges, that is to say to have them join the French army.[60] But the deserters were extremely reluctant to do so, for they intended to join James II's army in Ireland as soon as possible: 'Soldiers daily arrive from Flanders who want to go over to Ireland. I've offered them to join the regiments of their countrymen in France but they won't listen,' complained the commissaire Bouridal.[61] It seems 'they all apprehend it is not for our king's service'.[62] At this stage, the French could frame the movement, but they could not use it at their own profit. Saint-Germain was a kind of natural relay for the Irish on their way to Ireland, under the care of Henry Browne, secretary of the queen, to whom M. Talon, secrétaire of Louis XIV's cabinet, wrote on the 18 September 1690:

> The king of England has made me the honour to tell me he leaves at your care to examine the English, Scottish and Irish soldiers who leave the Provinces' service to enter His British Majesty's, so that when you have been made certain of their loyalty and good intentions to enter their duty, you can give them certificates upon which M. Michel, vicar

'An analysis of the FitzJames Cavalry Regiment, 1757', *Irish Sword,* 19:78 (1995), 254. 56 The opposite case could exist yet: in 1693, d'Auvergne reported than Catholic Irishmen had entered the service of William of Orange at the instigation of some Irish priests of Louvain. Two Dominican friars and an Irish priest were responsible for this. P. Melvin, 'Balldearg O'Donnell abroad and the French Design in Catalonia', *Irish Sword*, 12 (1977–9), p. 118. 57 Quoted by Melvin, op. cit., p. 119. 58 Instructions from the Marquis of Louvois to the *commissaire* Charlier, 7 November 1689: 'if some Irish officers among those who are in Lille want to stay at the service of the King, on the condition they come with enough men from the soldiers in Gand or in Bruges to create companies, you can assure them that His Majesty will be pleased to have them at his service' (Service Historique de la Défense, Vincennes, archives de la Guerre, A[1] 960, fol. 157). 59 *Mémoires*, iii, 311. 60 Instructions to the *commissaire* Sandrier, 18 June 1690 (Service Historique de l'armée de Terre, Vincennes, archives de la Guerre, A[1] 960, fol.294). 61 Letter from the *commissaire* Bouridal to the marquis of Louvois, Landernau, 5 June 1690 (Service Historique de l'armée de Terre, Vincennes, archives de la Guerre, A[1] 961, fol. 43). 62 Letter of M. Connelly to Henry Browne, Lille, 31 May 1691 (British Library,

> in Saint-Germain, will give them six *sols* a day for their living till they can come to Brest.[63]

Many deserters presented themselves to Saint-Germain for instructions and help, such as Colonel Frapp, who wrote to Henry Browne on 12 May 1691:

> I beg the favour that you will be pleased to acquaint the King of our being come out and to let us know his commands whether to stay here or go to Saint-Germain. If the latter you will be pleased to send our passes, if we stay here we all make it our request that you will represent our naked condition to His Majesty for there is hardly any here that have two shirts for their backs.[64]

In 1697, a new wave of desertions occurred, but in a very different context. Those who entered James II's regiments knew they were entering the service of Louis XIV, wishing that they could help the Stuart cause to prevail, and that a new military attempt of James could make them fight under his command. So was the case of James Flemming, 'Irish squire [...] bred in the Protestant religion by his mother; in the English service, left it before the peace of Utrecht to bring over to France, at the risk of his life, recruits he had paid at his own expenses, for the rumor had it that his legitimate king was about to make a descent on his estates; whose recruits were later on spread into the Irish regiments.'[65] In July, Middleton wrote to Barbezieux for James II 'being informed that most of the prisoners taken in Camaret and now kept at Nantes are asking to enter his service, he asks you to represent it to his Majesty [...] that he may order that the Irish should be given warrants to join the two regiments of Irish dragoons, and the English and Scottish may go to join the three companies that are in garrison in Schelestat [*sic*].'[66]

A new wave of desertions took place in 1702, involving mainly Irish soldiers. To face it, 'officers of that nation should be left on the boarder; they are coming in large number; twenty-two came over yesterday to Anvers from the camp the Dutch have at Rosendal'.[67] On 23 September 1702, Louis XIV gave Berwick a regiment made with the British deserters of 1702, and sent him to Bayonne on 4 December to gather the Irish deserters of the Cadix army.[68] Four hundred more joined him in 1704[69]. The very same year, Middleton planned to create a second regiment with the Flanders deserters, on the same conditions as the second Berwick regiment (27 *livres* and ten *sols*

London, Add MS 37662, fol. 141 v°–142). **63** British Library, London, Add. MS 37662. fol. 1 v°. **64** British Library, London, Add. MS 37662, fol. 118 v°. **65** Bodleian Library, Oxford, Egerton MS 1671. fol. 112. **66** Bodleian Library, Oxford, Carte MS 256, fol. 40. **67** Services Historique de la Défense, Vincennes, archives de la Guerre, A¹ 1561, fol. 236 v°. **68** Dangeau, op. cit, viii, 507 and ix, 56. **69** Melvin, op. cit., p. 119.

for each deserter). The *intendants* were supposed to provide these deserters with quarters. A captain and a lieutenant would have been in charge in the 'Grande armée', as well as a captain and a lieutenant in Brussels, Anvers, Gant and Namur, who were to bring the deserters and send them to Douai where they would have been gathered. On 31 March, Middleton was even able to give the names of officers who could be put in charge. But the plan was aborted and eventually *Berwick étranger* remained the only regiment created this way.[70]

THE WAY TOWARDS INTEGRATION IN FRENCH SOCIETY

The Irish were strangers and strange creatures indeed to French people and they were foreigners too. They were what the French called 'aubains', which means 'born elsewhere'. And indeed in *ancien régime* France only people born in France could be French. All the others had the rights that every human soul should be given, the *jus gentium*, but only people born in France could be allowed to follow the *jus civile*.

Foreigners were allowed to get married, to own property and to go to court, but they had no right to civil parenthood. This means that they could not inherit, nor bequeath their properties to their family members, except if their relations were French. All properties of a foreigner who died in France belonged to the king of France, as his 'droit d'aubaine'. Only the rents on the Hotel de Ville de Paris were not submitted to it, and that's why they soon became the Jacobites' favourite way of saving money, and the king of France favourite's way to tax them.[71] If the average tax was 200 *livres*, both the lowest and the highest taxes were paid by Irishmen: Patrice Montcolif had to pay 38 *livres* 3 *sous* et 9 *deniers* and the Irish banker Sir Daniel Arthur 18,000 French *livres*.[72] There were only two ways to escape the 'droit d'aubaine'. The king could give the inheritance to the children of the deceased by a letter of 'don d'aubaine'. Louis XIV seems never to have refused it to any Jacobite, but it never became an automatic gift – which makes, of course, the letters requesting that this custom be employed a very interesting source about the Wild Geese in France.

The second way to escape the 'droit d'aubaine' was to ask for a 'letter of naturalization' and to become French. At some point, Jacobites tried to obtain a general measure of naturalization for all of them. In November 1715, Louis XV gave French naturalization to all foreigners that had belonged to French army for more than ten years, rank and file as well as officers. In 1749 the

70 Bodleian Library, Oxford, Carte MS 238, fol. 81. 71 This is the basis of J.F. Dubost and P. Salhins work, *Et si on faisait payer les étrangers* (Paris, 1999). 72 See rolls of imposition on naturalized French (Archives Nationales, Paris, E 3706/11 et 12).

main aristocratic families of Irish origin (FitzJames, Tyrconnel, Lally, Dunkell) were advocating a project to make all the Irish Catholics living in France true French *naturels*. But it never was granted.[73] Unlike Spain, where Philip V in 1701 and 1718 gave all the Wild geese the same rights as the native Spanish, naturalization in France still was a personnel request and an individual grant.[74] And this makes the letters of naturalization also a very interesting source.

From 1689 to 1715, 508 Jacobites were naturalized as French, which makes an average of almost 19 per year, including 309 between 1697 and 1703, which makes 44 per year. Of these, 56% were Irish and 39% English, which is very close to the overall statistics for the Jacobite community. But these naturalizations did not occur due to the same reasons. The number of Irish naturalizations rose after 1697, the year of the treaty of Utrecht and the disbanding of the Irish regiments. The English requests for naturalization came mainly after 1701 and the death of James II. Some of them, starting by the duke of Berwick, found it necessary to ask for James II's leave to become French.[75] But the naturalization was in many ways more convenient than real. In French archives, the new subjects of the king of France are never called 'French' but 'naturalized French'. For the Wild Geese, service for France and loyalty to Ireland were never mistaken. In 1745, François Sarsfield was declaring 'letters of naturalization give us a right that we had not yet [...] but they can't break the old link to the homeland for this bound can't be destroyed.'[76]

One of the main concerns of the Jacobite exiles was to have their nobility acknowledged by the French. To do so, they had to ask to be granted a coat of arms of their nation at the court of Saint-Germain, which necessitated a letter from the king's secretary and then a letter of nobility from the French authorities. For 1700, no less than fifteen leaves were given in Saint-Germain. Among them was Nicolas Lukes, the son of Etienne Lukes of Waterford and Elizabeth Linch, both of noble birth.[77] Nicholas Lukes had come to France with his brother John and their cousins (the Comerfords of Waterford) and enlisted in the Irish regiments.[78] In Ireland, they were outlawed, their lands confiscated and their father sent to jail. John was killed in Catalonia, but Nicolas was part of an infantry regiment from 1688 to 1693. Sent to Guadeloupe, in the West Indies, he was saved by his Irish temper for he refused to get back on board a ship that he was convinced was unseaworthly.

73 Archives des Affaires étrangères, Paris, Correspondance politique, Angleterre, 425, pp 336–60. 74 Micheline Kearney-Walsh, 'Irish-Spanish links: a case study', a lecture delivered during the Trinity College Dublin conference on 'The Place of the Irish in British and European History', 14–17 September 1988. 75 Bodleian Library, Oxford, Carte Papers 209, fol. 6. 76 G. Chaussinand-Nogaret, op. cit., note 7, p. 1102. 77 Archives nationales, Paris, O[1] 221, fol. 102. 78 His cousin Joseph Commerford (Comerford) was given in January 1717 a nobility certificate (Archives Nationales, Paris, PP[1] 46bis, fol. 54). Joseph's son, Luc Commerford, had received his on 2 January 1700

This was the *Ville de Nantes*, which sank a few days later (220 men were drowned). He then became *cornette de cavalerie* under the order of the governor d'Amblemont and settled on the island. He married the daughter of a French nobleman and became French himself. (To make their names sound more noble in France, especially in Brittany, some Irish started to wear the French 'de' at the beginning of their names and the *Dictionnaires de la langue bretonne* records two Irish brothers in Brest called O'Brien who turned their names into D'obrien.)[79]

For the exiles, the Irish regiments also turned out to be the spearhead in the process of assimilation and service in them often marked the beginning of a later phase of prosperity for exiled Irish families in French society. One example was Arthur Dillon, who commanded the regiment that bore his name. He had come to France after several of his estates had been confiscated (2,800 acres in Mayo, 825 in Roscommon and 1,042 in Westmeath). Dillon was made a brigadier after the victory of Cremona in 1702, became lieutenant-general in 1706 and distinguished himself at the side of the duke of Berwick in the campaign of 1714, which was his last. When he retired from active service in 1730, he handed the regiment over to his eldest son, thus establishing the Dillons permanently as one of the great officer families of France. His youngest son, Arthur, became abbot of St Etienne de Caen, then archbishop of Evreux, archbishop of Toulouse and finally archbishop of Narbonne, which made him a leading cleric in the French Church. And yet, Dillon's career at the heart of the Mountcashel brigade never lessened his links with the court of Saint-Germain where his wife, Catherine Sheldon, was maid of honour to Mary of Modena. This link lasted after the departure of James-Edward Stuart (the Old Pretender) in 1713 and, even after the death of Mary of Modena in 1718, Dillon still was lodged at the castle of Saint-Germain, where he died on 5 February 1733.

In conclusion, it is useful to return to come back to these triangular relationships between Saint-Germain, Versailles and the Wild Geese. In the long term, the most natural connection, which linked the Irish Jacobites with their king, turned out to be the weaker link. Not that the link itself was not important – the Wild Geese were committed Jacobites and one has to remember what they had accepted to follow their king, not knowing yet what exile would bring them. But it simply happened that in the court of Saint-Germain, Ireland had ceased to be an option for military and political activity. For the Stuart court, the exiled Irish soldiers had no further purpose and maintaining them could sometimes look more like a burden than an investment for a future expedition to Ireland.

(Historical Manuscript Commission, *Stuart Papers*, i, 145). **79** Eamon Ó Ciosáin, 'La langue irlandaise et les Irlandais dans le *Dictionanaire de la langue bretone de Dom Pelletier*' in Université de Rennes II, *Mélanges en faveur de Léon Fleuriot* (1992), pp 57–8.

The French, however, got this point and, while Louis XIV protected his cousin, the professionals of the Bureau de la Guerre, without of course ever disobeying the official policy to the Jacobites, tried to make sure that *they* took over the administration of the Irish regiments, at least as much as was possible with the Irish.

For the Irish Jacobites, exile had been chosen out of loyalty to the Stuart cause but when the focus of Jacobite activity left France, they didn't follow. Eventually exile had prevailed on Jacobitism and the army gave them an useful way to become integrated in French society. Interestingly, this was done not by assimilation nor acculturation, but by maintaining the symbols of their identity, which says something very revealing of the French *ancien régime*.

The Wild Geese gave officers to the French army throughout the eighteenth century, and even provided some of Napoleon's generals. The Irish themselves, in spite of the privations in the early years, found larger opportunities for profitable careers, first in France then on the whole continent. With increasing confidence, they ranged across Europe and in the expanding overseas empires of European states.

APPENDIX 1: IRISH REGIMENTS IN FRANCE AFTER 1690

Mountcashel Brigade:

Régiment Mountcashel	*by comission*
Régiment O'Brien	*by Louis XIV*
Régiment Dillon	

Troops arrived in 1691–1692

Régiment de Lord Dover	Régiment de Berwick

Two cavalry regiments:

Régiments de cavalerie du Roi	Régiment de cavalerie de la Reine

Two regiments of Foot Dragoons:

Régiments de dragons du Roi	Régiments de dragons de la Reine

Eight Infantry regiments:

Royal Irlandais	
Régiment d'infanterie de la Reine	
Régiment d'infanterie de Marine	*by comissison*
Régiment de Limerick	*of James II*
Régiment de Charlemont	

Régiment de Dublin
Régiment d'Athlone
Régiment de Clancarthy

Three independant companies:
Company Sutherland
Company Hay
Company Browne

APPENDIX 2: BRITISH SOLDIERS AT THE HÔTEL DES INVALIDES, PARIS 1688–1715

Year	*English*	%	*Scots*	%	*Irish*	%	**Total**	Irish regt	French reg.	In service before 1689	In army after 1689
1689	1	*100*	0	*0*	0	*0*	**1**	0	1	1	0
1690	1	*20*	2	*40*	2	*40*	**5**	0	5	5	0
1691	1	*33.3*	0	*0*	2	*66.7*	**3**	1	2	3	0
1692	0	*0*	0	*0*	28	*100*	**28**	26	2	4	24
1693	2	*4.5*	2	*4.5*	40	*91*	**44**	38	6	7	37
1694	0	*0*	0	*0*	45	*100*	**45**	40	5	5	40
1695	2	*4*	2	*4*	41	*92*	**45**	45	0	13	32
1696	0	*0*	1	*3*	33	*97*	**34**	33	1	11	23
1697	4	*8.5*	3	*6.5*	41	*85*	**48**	48	0	11	37
1698	6	*1.9*	2	*5.6*	99	*92.5*	**107**	107	0	41	66
1699	0	*0*	2	*7.1*	26	*92.9*	**28**	28	0	11	17
1700	1	*0.7*	4	*2.6*	146	*96.7*	**151**	147	4	89	62
1701	0	*0*	0	*0*	14	*100*	**14**	13	1	1	13
1702	0	*0*	0	*0*	35	*100*	**35**	35	0	1	34
1703	1	*2.4*	0	*0*	41	*97.6*	**42**	41	1	5	37
1704	0	*0*	1	*2.9*	34	*97.1*	**35**	29	6	3	32
1705	1	*2.9*	1	*2.9*	33	*94.2*	**35**	32	3	5	30
1706	1	3.7	0	*0*	26	*96.3*	**27**	27	0	5	22
1707	1	*4.2*	2	*8.3*	21	*87.5*	**24**	24	0	2	22
1708	1	*5*	0	*0*	19	*95*	**20**	19	1	0	20
1709	1	*7.7*	1	*7.7*	11	*84.6*	**13**	10	3	1	12
1710	0	*0*	2	*7.4*	25	*92.6*	**27**	26	1	0	27
1711	0	*0*	0	*0*	17	*100*	**17**	17	0	1	16
1712	0	*0*	1	*12.5*	7	*87.5*	**8**	7	1	2	6
1713	1	*2.9*	0	*0*	34	*97.1*	**35**	32	3	0	35
1714	2	*8*	0	*0*	23	*92*	**25**	23	2	0	25
1715	9	*7.3*	4	*3.2*	111	*89.5*	**124**	121	3	4	120
Total	**36**	*3.5*	**30**	*3*	**954**	*93.5*	**1020**	**969**	**51**	**231**	**789**

APPENDIX 3: FRENCH NATURALISATIONS, 1689–1715

Year	English		Scottish		Irish		Total
	number	%	number	%	number	%	
1689	1	50	0	0	1	50	2
1690	1	25	1	25	2	50	4
1691	2	50	0	0	2	50	4
1692	7	76	1	12	1	12	9
1693	2	67	0	0	1	33	3
1694	1	20	0	0	4	80	5
1695	3	33	0	0	6	67	9
1696	1	20	1	20	3	60	5
1697	17	60	3	10	9	30	29
1698	18	19	2	2	74	79	94
1699	6	78	0	0	5	22	11
1700	28	45	3	5	30	50	61
1701	9	39	5	22	9	39	23
1702	40	77	0	0	12	23	52
1703	16	42	3	8	20	50	39
1704	6	36	0	0	10	64	16
1705	5	25	1	5	13	70	19
1706	2	33	0	0	4	66	6
1707	7	59	1	8	4	33	12
1708	1	11	0	0	8	89	9
1709	6	33	0	0	11	66	19
1710	3	15	1	5	16	80	20
1711	8	40	0	0	13	60	21
1712	0	0	0	0	5	100	5
1713	1	25	0	0	7	75	8
1714	2	40	0	0	3	60	5
1715	4	20	2	10	14	70	20
Total	**197**	***39***	**24**	**5**	**287**	**56**	**508**

Guests of France: a description of the *Invalides* with an account of the Irish in that institution

EOGHAN Ó HANNRACHÁIN

The registers of the Invalides in Paris, which are conserved in the château of Vincennes contain particulars of more than 130,000 veterans of the French army who applied for admission to that institution between 1674 and 1770. Of these, some 2% were Irishmen. For each applicant, the registers give the man's family name and first name, the names of his captain and regiment, his age, marital status, length of service, the nature of his wounds or disabilities, when he applied for admission and his fate. For some men, a trade is also shown. These registers give us very valuable information on a sizeable cross-section of the Irishmen who fought in the armies of France during the reigns of Louis XIV and Louis XV. The records also recall to memory nigh-forgotten European campaigns and struggles in which Irish soldiers took a prominent part – not only in the major wars but also in such events as the attempted recovery of Gibraltar and the campaign in the Highlands of Scotland with Prince Charles Edward Stuart.

Over the period in question, a considerable number of Irish travelled to France as students, clergy, businessmen and tradesmen. Several studies have been made of their considerable contribution to France.[1] Their impact overall was considerable and it has been stated that the arrival of Jacobite fugitives brought commercial and military advantages to France that went some way towards counterbalancing the flight from France which resulted from the revocation of the Edict of Nantes.[2]

Irish soldiers had been present in continental Europe from the beginning of the sixteenth century, for instance, in the service of Henry VIII. During the sixteenth and seventeenth centuries, they also served in the Spanish armies in the Lowlands. Pay in the armies of Spain was often in arrears, and Irish troops were among those who mutinied in 1594 and shut themselves up in Tirelemont. In view of their earlier good service and because of the serious nature of their grievances, a full pardon was sent to them by the Archduke Albert on 12 June 1596.[3] After the defeat at Kinsale, Irish troops were a

1 See for instance: Patrick Clarke de Dromantin – *Les Oies Sauvages* (Bordeaux, 1995), and *Les réfugiés jacobites dans la France du XVIIIe siècle* (2005); also, the archives of Baron Jean Le Clere conserved in Brives. 2 François Crouzet, *Britain, France and international commerce* (Aldershot, Hants., 1996), p. 224. 3 Joseph Lefèvre, *La Secrétaire d'État et de*

regular feature of the Spanish forces in the Lowlands.[4] When Arras was besieged by the French in 1640, the defending garrison was under the command of Owen Roe O'Neill. In the town hall of Aire-sur-la-Lys, there is a remarkable painting which shows the encampment of the O'Neill's regiment[5] on the very site that had been occupied by Irish troops in the pay of France some months earlier during the successful French siege; this is the earliest representation of an Irish camp on the Continent.

The decline of Spain in the seventeenth century resulted in Irish officers and soldiers transferring to the French service, particularly after 1630. However, information on the presence of Irishmen in the army of France is scant, prior to the establishment of the Hôtel Royal des Invalides. But when that institution came into being, more precise details on them became available. The present text examines the nature of the data on applicants for admission to this royal institution.

Estimates of the number of Irish who migrated to France during the seventeenth century are impressively high. William Petty gave a figure of 34,000 Irish who travelled to the Continent between 1651 and 1654. This high level is accepted by various commentators such as R.A. Stradling and Brendan Jennings. According to the reports of Oliver Cromwell's agents, 13,000 Irish left Ireland in 1652 and a further 12,000 left in six months of 1653.[6] These would have been accompanied – or followed – by unknown numbers of women, children and other dependants.

Irish soldiers were highly valued by Louis XIV and his entourage. These men were engaged in all the theatres of his wars – in Germany, Italy, the Iberian Peninsula, Flanders, even in Crete and north Africa. When Marshal Bellefonds was preparing his campaign to save Crete from the Turks, he regretted that he did not have enough time to replace his Italian troops with one or two Irish regiments, 'more solid and better trained' and he did not have time to bring them from Flanders.[7]

In the 'surprise of Cremona' on Saint Bridget's Day 1702, for instance, their action was declared by Quincy as being so memorable that it would be remembered for centuries and merited a special account.[8] Following that battle, Louis XIV had his minister for defence write to the French commander in Italy issuing the instruction that an exchange of prisoners should commence with Irish soldiers.[9] That same episode inspired Arthur

Guerre sous le Régime Espagnol 1594–1711 (Brussels, 1934), p. 234. 4 Matthew O'Conor, *Military history of the Irish nation* (Dublin, 1845), *passim*. 5 Ó hAnnracháin and Aubert, 'Irish tents in Spanish Flanders, winter 1641', *Irish Sword*, 24:95 (1641), p. 1. 6 Frank D'Arcy, *Wild Geese and travelling scholars* (Cork, 2001), p. 30. 7 Charles Terlinden, *Le Pape Clément et la Guerre de Candie* (Louvain, 1904), p. 162, note 2, quoting Vatican Archives, Nunziatura de Francia, vol. 138, f. 34. 8 Charles Devin, marquis de Quincy, *Histoire militaire du règne de Louis le Grand, roy de France* (7 volumes, Paris, 1726), iii, 612. 9 Service Historique de la Défense, château de Vincennes, letter from Chamillart to Revel,

Conan Doyle, the creator of Sherlock Holmes, to write a remarkable poem in praise of the tenacity and courage of the regiments of Dillon and Burke:

> Time and time they came with the deep-mouthed German roar,
> Time and time they broke like the wave upon the shore,
> For better men were there
> From Limerick and Clare
> And who will take the gateway of Cremona?[10]

Because of their sterling service in this specific encounter, Louis XIV granted an extra month's pay to the men of the two regiments as well as promotions and recompense to many of the officers.[11]

Royal appreciation was not based on a few isolated feats: Irish regiments fought with distinction for France at the battles and sieges of, among others: Almanza, Berg-op-Zoom, Calcinato, Cassano, Castiglione, Chiari, Fontenoy, Graebenstein, Hochstet, Lafelt, Landau, Landen, Luzzara, Malplaquet, Marburg, Marsaglia, Melazzo, Ramillies, Rosbach, Spire and Steinkirk. There are also many other major encounters referred to in O'Callaghan's work.[12] In all of these struggles, the Irish suffered heavy casualties and some of their disabled wounded sought admission to the Invalides. At frequent intervals, new studies on the Irish serving France in other war theatres appear – for instance, see O'Byrne's recent work.[13]

WHAT INSPIRED LOUIS XIV TO ESTABLISH THE *INVALIDES*?

The Invalides was, undoubtedly, the most useful and impressive construction in Paris during the reign of the great king. When the elaborate and grandiose structure had been completed, Montesquieu exclaimed that if he were a prince he would prefer to have created this building than to have won three battles.[14] Before we go into the details of the archives relating to the Irish who were admitted to the institution, we look at the origins of the place, how it functioned and the great care taken of the invalids in it. Prior to the reign (1643–1715) of Louis XIV, when a war ended, disabled veterans were in a terrible situation. There was no public social welfare system. Crippled soldiers wandered about, begging for alms and haunting the neighbourhoods of abbeys and churches counting on the generosity of the clergy and of the

14 February 1702, A1, vol. 1588, item 210. **10** *The poems of Arthur Conan Doyle* (London, 1922), p. 8. **11** Capitaine Malaguti, *Historique du 87e Régiment d'infanterie de Ligne, 1690–1891* (Saint-Quentin, 1892), p. 24. **12** John Cornelius O'Callaghan, *History of the Irish Brigades in the service of France* (Dublin, 1854). **13** Henry O'Byrne, *Des régiments Irlandais dans les Alpes, 1690–1713* (Val-des-Prés, 2006). **14** Montesquieu, *Lettres*

faithful.[15] A relatively recent study shows that, for the population in general, the giving of alms to paupers was regarded as a religious duty. In the Middle Ages, Christians saw in the dignity of the poor an image of the suffering Christ, and they also believed that giving charity to them could store up rewards in heaven.[16]

However, in times of failed harvests or general scarcity, the generosity of the faithful risked being insufficient. An additional consideration for Louis XIV was his awareness that the sufferings of indigent disabled veterans due to his wars was projecting a bad image of his reign, and that their sufferings resulted directly from his policies. The remarkably lucid summary of his motivations was set out by him in the enabling an edict which he signed in April 1674 and which was registered by the *parlement* in June of that year.[17] Not only did he write of his debt to disabled men, now often reduced to indigence, who had contributed to the defence of the monarchy but he specifically mentioned their entitlement to a certain level of subsistence and, further, reflected that their sad state might discourage other men from enlisting in his armies.

The founding of the Invalides represented a watershed in the attitudes of public authorities across Europe. Louis XIV was imitated by William of Orange who established the Royal Hospital at Chelsea. In Ireland, a similar hospital for veterans was founded at Kilmainham.

Work on the building commenced with the laying of the foundation stone on 30 November 1671 and the first veterans entered the hospital in October 1674 at a time when the great church had not yet been built. The construction of the soldiers' palace was closely supervised because of the intense interest that the monarch, then in his prime, took in the project. Louvois raised the construction funds and the architect Libéral Bruant and the entrepreneur Pipault were responsible for the building work. Louis XIV was not satisfied with the speed of the work and the young architect, Jules Hardouin-Mansart, was given charge of finalising the place. His plan for the church was approved, and the king inaugurated the great dome in 1706.

The objectives of the institution were set out very clearly: it was to be a worthy asylum for disabled, worn-out or indigent veterans; it was to help them prepare for eternity; it was to serve as a hospital for their maladies. As such, it was in advance of its time for there were cases of veterans being able to resume active service following their being treated there – though fewer than had been hoped for at the outset.

Louis XIV's sincerity in relation to the Invalides was underlined by the terms of his last will and testament in which he stated that, of all the works

persanes, ed. P. Vernière (Paris, 1992), p. 177, lettre LXXXIV. 15 For an account of the ancient origins of the links between old soldiers and monasteries, see Robert Burnand, *L'Hôtel Royal des Invalides, 1670–1789* (Paris, 1913), pp 5ff. 16 José Cubéro, *Histoire du Vagabondage*

done during his reign, none was more useful to the State than it. He went on to say that it was appropriate that soldiers who, because of wounds received in wars, their long service or their great age, were no longer able to work or earn their livelihood, should have subsistence assured for them during the rest of their days; moreover, many officers who had no property or fortune could, through this institution, find an honourable retirement; several motives should commit the dauphin and the succeeding kings of France to maintaining the place and he urged them that they should fully support this establishment and grant it their special protection.[18]

But this wise advice was not followed by his successors. Under Louis XV – many of whose actions helped to pave the way for the French Revolution – the Invalides was neglected, and he visited the place only once during his long reign. This negligence weakened the loyalty of the old soldiers – who were still military men capable of firing muskets. It was an aspect that became critical when the mob entered the place prior to the attack on the Bastille. Unwisely, during the reign of Louis XV also, the king's musketeers were stood down.[19] That was a major error for such a loyal formation of French troops could have been most useful to the monarchy in 1789 and 1790. On 14 July 1789, the old soldiers present in the Invalides allowed their arsenal of thousands of muskets to be pillaged without firing a shot or offering any resistance – a crucial passivity. Royal neglect of this possible fund of goodwill led to a loss of sympathy for the monarchy.

During the revolutionary period, the title 'Hôtel Royal des Invalides' became unacceptable to the anti-royalists and it was re-named 'Hôtel national des militaires invalides'. For his part though, Napoleon appreciated the value of the institution and visited the place after his campaigns in Prussia and Poland. During his visit on 11 February 1808, surrounded by veterans with whom he struck an excellent rapport, he noted the absence of the high altar which had been dismantled during the Revolution and he ordered that it be replaced immediately, adding that, in the eyes of old soldiers, religion, the recourse of one's last days, could not be too greatly honoured.[20]

THE IRISH IN THE *INVALIDES* – EARLY (PRE-1690) ARRIVALS

The registers[21] of the institution contain the succinct biographies of more than 130,000 soldiers. The great majority of these were French, but there

du Moyen Age à nos jours (Paris, 1998), p. 63. **17** The full text can be found in Anne Muratori-Philip, *Les Grandes Heures des Invalides* (Paris, 1989), pp 320–9. **18** Bibliothèque Nationale, Paris, Lb 37 4442 (*Testament du Roy*, 2 août 1714) cited in Jean-Pierre Bois, *Les Anciens Soldats dans la Société française* (Paris, 1990), p. 38 ; and, Robert Burnand, op. cit., p. 2. **19** Anne Muratori-Philip, op. cit., p. 106. **20** Ibid., p. 178. **21** These registers are conserved in the Service Historique de la Défense, château de Vincennes, series 2Xy.

were also Belgians, Bohemians, Danes, English, Germans, Italians, Poles, Portuguese, Spanish, Swiss and men from other regions. In that large number, the author identified 2,600 Irish veterans between 1670 and 1789. The vast majority of these arrived in France from 1689 onwards. However, 180 Irish veterans were registered in the registers prior to that date.

It has been estimated that, generally, about 1% of the manpower of a regiment had the characteristics that made them eligible for admission to the *Invalides*. A wounded man would have to survive the medical treatment of his time and the hospitals before he got to the application for admission stage. Moreover, he had to have a certificate of discharge from his colonel, and even badly injured men were kept on for as long as possible – particularly at times when recruiting was difficult. When pressure on places in the Invalides was great, men were considered for admission, only if they had served in their regiment for a certain minimum period which could run to twenty years. Men with a trade, even if they were no longer fit for active service, would prefer to avoid being institutionalized – even in so fine a residence. A feature of the registers of the Invalides is the frequent mention of elderly Irish veterans who quit the place soon after being admitted. Moreover, there are cases of Irishmen (and others) who seemed to be in a bad state and who held the necessary documentation being refused admission by the authorities responsible for the running of the institution.

Were it not for the registers of the *Invalides*, we would be less aware of the extensive presence of Irish soldiers in French regiments and in other formations such as Fabert's volunteers prior to the massive arrival of Irish troops in France loyal to James II at the end of 1691. These registers show that there were Irish soldiers in at least forty-seven French formations before 1690. And, of course, this does not cover Irishmen serving in other French regiments who never applied for admission.

The early registers show that, before 1690, the authorities tended to record Irish postulants for admission as '*Irlandois*' (Irish), and only rarely did they record the name of a town or county of origin. They also had difficulty with their Christian and family names. Following are translations of the entries for some Irishmen admitted in the earlier years. Until the *hôtel* was completed, veterans were kept in temporary accommodation. A sample of some of the entries for Irish soldiers contained in the registers is given below:

> **Peter** (*Pierre*) … [family name not entered]; Irish; soldier in the engineers' company, Royal Roussillon regiment; aged … [not entered]; crippled by a musket shot received at the siege of Douai; he held a certificate from his captain and from Major Ximenes; having been examined and accepted by the administrators, he was admitted to the *hôtel* on 1 October 1670.

John Malone (*Jean Malein*); aged 75 years; native of *Lumence* (Limerick?) in Ireland; he holds a certificate from *Sieur* Douglas dated 17 November (1670) which confirms that he had *bien servy* (served satisfactorily) in his regiment for 18 years as soldier; having been examined and accepted by the administrators, he was admitted on 29 November 1670.

Simon Dalton known as *St Simon*;[22] aged 40 years; native of Co. *Rosquement* (Roscommon) in Ireland; soldier in the colonel's company, Douglas regiment; he held a certificate dated 17 November 1670 from the Marquis Douglas; his right leg was crippled by a wound received at the recent siege of Gravelines; he had satisfactory service; on 5 December 1670, he was accepted and entered the *hôtel* on the following day.

Bernard Simon, known as *La Fontaine*; aged 65 years; native of Ireland; corporal in *Dorty* (Doherty's?) company, *régiment des Gardes* (the French Guards regiment) where he served for 25 years and had served previously for seven years in the Irish regiment commanded by *Sieur de Bely* (Bailey?); he had received several wounds; he held a certificate from his captain dated 19 September 1670; having been examined by the Council on 27 December in the presence of *Mr Lebret* and *Sieur* Sandoux, he was admitted to the *hôtel* on 28 December 1670, as a soldier; married.

Three Irish veterans, **Raymond Reylly** (aged 100 years), **Anthony Marcantil** (of county *de Corquit*)[23] and **William Peyre** were admitted during the year 1671. The third of these is of particular interest, because he survived in the institution for a relatively short period:

William Peyre; aged 48 years; soldier in d'Espagne's company, La Ferté regiment, where he had 18 years' uninterrupted service; his left arm was broken and he had suffered many other wounds, including a fluxion of the lungs, which made him unfit for service; he had a certificate signed by *Sieur* Mornas; he was accepted on 22 August 1671 and entered on the following day; he died on 6 November 1671.

Two Irishmen were admitted in 1672; one of these, **William Tracy** known as *La Chapelle*, had served for 40 years. The second man's story is

22 *Nom de guerre* (sword name); soldiers in the French regiments – as in the Foreign Legion today – were encouraged to adopt such names, so as to break with their previous life. In the Wild Geese regiments from 1689 onwards, this practice was virtually unknown. Any later Irish who declared a sword name came from French regiments. 23 Cork, pronounced after the Irish language manner: *Corcaigh* – as in the case of Roscommon,

interesting, for he was the first Irishman to abandon his right to a place in the *Hôtel*:

> **Jacob Ban**; Irish; aged 40 years; soldier in MacNamara's company, Hamilton regiment, where he had his right arm carried off (*emporté*) at the siege of Rhimberg; he had served previously in the regiments of Preston and *Encheniquen* (Inchiquin); he held a certificate signed by Lieutenant-colonel Dongan of the Hamilton regiment which was uttered at Zutphen on 2 September 1672 and he also had a note from *Monsr*. Saint Pouange; on 6 November 1672, he was admitted; *à renoncé* (he renounced his rights to a place in the institution). It is possible that the second name shown for this man was not a family name but a non-pejorative nickname: *bán*, or the fair-haired.

Six Irish veterans were admitted in 1673:

> **Matthew Grigelle**, known as *Saint Louis*; aged 61 years; 28 years continuous service.
>
> **Richard Tobin**, known as *Sans Peur*; aged 68 years; 42 years service.
>
> **Malachy Tierney**, known as *La Forest*; aged 42; soldier in Croyselle's company, the king's Guards' regiment, where he served 14 years; he was wounded by a musket shot to the right leg at the siege of Lille in 1667; he continued to serve until 1671 when the wound re-opened; having been in the Hôtel Dieu (a hospital) for over a year, it was found necessary to amputate the leg; as the wound did not heal,[24] he was presented to the Council of the Invalides on 13 May 1673 and it was agreed to admit him; however, he renounced his right to a place in the Invalides.
>
> **Denis Shaughnessy** known as *La Garenne* whose case is described below.
>
> **Baltazar Foster**; aged 36; virtually blind; sergeant in Count Hamilton's company, Jones' regiment, where he served six years; admitted to the Invalides on 3 September 1673 on the basis of a note signed by Louvois at Nancy on 23 August 1673.
>
> **Guelaprome Marquentin**, known as *La Fortune*; Irish; aged 60 years; soldier in Valois' company, Hamilton regiment; he had been in the king's service, without interruption, since the first siege of Gravelines in

above, and Inchiquin, below. **24** The medical services available in hospitals in the seventeenth century were so terrible that even badly-wounded soldiers tried to avoid being

1644, in several regiments; he presented a certificate signed by Turenne at the Vestelas headquarters on 31 July 1673 and also signed by Hasset; he had, as well, a certificate signed by Count Hamilton, lieutenant-captain of the English Gendarmes and by the colonel of the said regiment, signed at Thilhau on 12 July 1673; at the top of this certificate, there was an instruction given by the king at Nancy[25] on 20 August 1673 in conformity with which he was admitted to the Invalides on 4 September 1673; he was a Catholic and unmarried.

A REMARKABLE MILITARY CAREER

The following is the record of an Irish soldier who suffered much and had long service:

> **Denis Shaughnessy** (*Chosny*), known as *La Garenne*; Irish; aged 60 years; soldier in Valon's company (first captain of Hamilton's regiment); he had served the king without interruption since 1636, being in Marshal de la Masleray's guards, de la Magne's company, Normandy regiment, for 15 years; he was captured at Gigeri and remained a slave until the king had the kindness (*bonté*) to have him 'recovered'; he then served eight years in the Champagne regiment; during his service, he received many serious wounds, including a crippling injury to his left foot; he held many good certificates; presented to the Council the day after Pentecost 1673 and was admitted; he was unmarried.

This soldier is one of two among all those on the registers of the *Invalides* to have been ransomed by Louis XIV from the Turks or from north African Muslims. Shaughnessy's story relates to an effort made by the king of France to destroy the Algerian pirates by tempting them to attack a strong French military force. Louis XIV set up a base on the coast of Algeria and expected that the Algerians would be cut to pieces when they attacked it. Literally, it became an early Dien Bien Phu. Like the besieged base in south-east Asia in 1954, the strong camp at Gigeri was overrun and the French lost heavily. Denis Shaughnessy was one of the few survivors among the captives, rather like the solitary Englishman to emerge from Afghanistan in the nineteenth century ill-fated venture. The Gigeri disaster was hushed up, and only recently has it been the subject of a book.[26]

hospitalized. **25** It is remarkable that, in the case of this soldier, Louis XIV and the great marshal Turenne (1611–75) should have been involved in his paper work. **26** B. Bachelot, *Louis XIV en Algérie* (Le Rocher, 2003).

However, part of the expeditionary force was rescued, under fire. Another Irishman, Thomas Howells was wounded but was among the men who got away. His story was as follows:

> **Thomas Hoüels**, known as *La Tour*; Irish; aged 59 years; soldier in Villeclos' company, Picardy regiment, where he said he served eight years and previously served eleven years in the Navarre regiment; he had served the king for 34 years, in all; his left arm was crippled by a musket shot received at Gigeri; admitted on 16 March 1686.

Unlike Shaughnessy, who had evidently gone to considerable lengths in contacting former commanding officers so as to reconstitute his file, Howells was apparently unable to produce any papers but was accepted on his word – no doubt following close questioning by the administrators of the Invalides.

As mentioned above, one of the aims of the institution was to foster a religious dimension among what would have been a hard-bitten group of men. One Irishman gave the administrators full satisfaction on this front:

> **Simon Cotter**; Irishman by nation; aged 26 years; soldier in Macarthy's company, Furstenberg regiment where he said he served seven years; his right hand was crippled by an exploding mine; admitted to the Invalides on 24 May 1681; on 22 August 1687, he renounced his rights to a place in the institution in circumstances that resulted in the king giving him 200 *livres* to enable him to return to his own country as a priest having completed his preparatory studies in the institution.

Another objective which Louis XIV had set for the Invalides was the rehabilitation of wounded or ill veterans. Some of the Irish, like the following instance, did return to active service:

> **Anthony Rayel** (*Real*) known as *La Jeunesse*; Irish by nation; aged 48 years; sergeant in the Caraman company, French Guards, where he served for 17 years, according to the discharge given to him by his captain and according to the certificate issued by *Sieur* de la Rappée; he suffered from a number of wounds which made him unfit for active service; he was a Catholic and married in Paris. On 4 July 1689, he re-entered active service in *Sieur* de Rancher's company, the Royal infantry regiment.

Homesickness affected many of the Irish veterans. This sentiment was eloquently expressed later by the poetess, Emily Lawless:

In this hollow star-picked darkness, as in the sun's hot glare,
In sun-tide, moon-tide, star-tide, we thirst, we starve for Clare!

...

The whole night long we dream of you and waking think we're there,
Vain dream and foolish waking, we never shall see Clare.[27]

One early Wild Goose to whom the sentiment applied was:

> **Daniel Black**; Irishman; aged 45 years; soldier in Hamilton's regiment where he served since the creation of that unit; he received several wounds at the battle of Entzheim which resulted in his being crippled and rendered unfit for service; married in his own country, he had several children; he was admitted to the Invalides as a soldier on 15 November 1675. The governor wrote: 'This foreigner (*estranger*) was happy to leave active service and be admitted to this institution, but two months later, he wished to depart definitively; Mgr. Louvois, who happened to be in the hôtel, had him given 30 *livres* and he left on 24 January 1676.'

During the twenty years covered above, some 28% of the Irish who had been admitted abandoned their right to a place in the hôtel. Over this period, the numbers of Irish veterans applied for admission to the Invalides fluctuated considerably:

1674	6	1679	15
1675	9	1680	3
1676	44	1681	7
1677	15	1682	10
1678	17		

During the years to 1690, some 181 Irishmen applied for admission to the institution. But one of these, however, was stated to be *Anglois* or English by the scribe who compiled the register. Still, his name has an Irish ring to it, though of course he could well have been English-born:

> **Hugues Reilly**; English by nation; aged 63; gendarme and *sousbrigadier* (sub-sergeant) in the English Gendarmes' company where he had served for seven years; he had served previously for a very long time in the armies of the king of England[28] in the service of the king (of France); he

27 Emily Lawless (1845–1913), in her poem *Fontenoy 1745*, quoted in Kevin Haddick-Flynn, *Sarsfield and the Irish Jacobites* (Cork, 2003), pp 211–12. 28 For Louis XIV, English and Irish mercenaries in the pay of France were seen as being from the English

> was not disabled but his long service had caused him much inconvenience which made him unfit to continue in active service; he held a note from *Mons.* Hamilton; Catholic; unmarried; admitted on 13 April 1675.

Virtually all of the veterans admitted to the Invalides were Catholic. However, there were some Protestants among the Irish who were received, as the following shows:

> **John Kelly** (*Jean Kelli*); Irish; aged 33 years; soldier in captain Harbert's company, Hamilton regiment, where he served three years; he was crippled by a musket shot to the right side which he received during the recent campaign and which made him entirely unfit for active service; he held a certificate from his captain dated 16 May 1676; unmarried and 'of the religion which pretends to be reformed' (*de la religion prétendue réformée – R.P.R.*); admitted on 6 June 1676. On 21 December 1678, this soldier being perfectly healed (*parfaitement gueri*) was given 15 *livres* and was sent home (*chez lui*), presumably to Ireland.

Another Irish soldier suffered during the Rhineland campaign the infantryman's nightmare of being trampled by cavalry:

> **Robert Ashe**; Irish by nation; soldier in MacNamara's company, Hamilton regiment, where he served four years; he was quite unable to continue in service having had his back broken (*les reigns rompus*) by horses that trampled him underfoot at the crossing of the Rhine; he held a discharge from his captain which was given to him at Toulon 10 May 1676; since then, he had remained at that place; unmarried; Catholic; he was considered suitable for admission on 22 August 1676 but asked to be allowed to return to his own country; permission was given and he received 24 *livres*.

These Irishmen had served in the following French, Irish and English regiments and other formations, as follows: Alsace, Anjou, Beaupré, Bellegarde, Bligny, Bourlemont, Bretagne, Carignan, Champagne, Chartres, Chevalier Due's troopers, Chevalier Dumiers' troopers, Cologne forces, Comagny, Corsican Italian, Crillon, Crussol, Dauphin, Dubordage, Erlach, Faressay cavalry, La Ferté, Flanders, French Guards, Furstenberg, Greder, Harcourt, Humières, Kerman, Konigsmarck, Limousin, La Marine, Navarre, Normandy, Picardy, Piedmont, Plessy Praslin, Rouvigny, Royal fusiliers, Royal Roussillon, Salis, Saulx, Stouppa, Swiss free company, Turenne and Vendome; British and Irish regiments in the pay of France in which they served included: de Bely, de Bleyny, Charlemont, Dongan, Douglas, English Gendarmes, Hamilton,

Hogan (*Ougan*), Inchiquin, Jones, Monmouth, Muskerry, O'Brien, Preston, Reynolds, Royal anglois, Scots Gendarmes, and York.

These men's regiment would have changed name when the colonel-proprietor died or when he sold his regiment. The story of the Furstenberg regiment illustrates that of many other formations. It was raised in 1668 by William-Hégon, landgrave of Fürstenberg, and arrived in France in 1670 uniformly clad in blue and yellow. The formation was immediately admitted to the French army and was paid at the rate applying to foreign troops. It participated with distinction in the Dutch War (1672–8) and saw action in the Pyrenees and notably in the siege of Puycerda (1678). The regiment remained the property of the Fürstenberg family until 1686. Afterwards, it became, successively, Greder, Sparre, Saxe, Bentheim, Anhalt and Salm-Salm. Many of its veterans held certificates[29] showing that they had served in several regiments.

The following is the entry in respect of such an individual:

> **Eugene MacAuley**; Irish; sergeant in Windich's company, Koinigsmarck regiment, where he said he served four years; previously, he had served four years in Furstenberg, ten years in Hamilton, five years in the Royal, eight years in Douglas, five years in O'Brien and three years in Muskerry regiments; his certificates showed that 'he had served the king of France for thirty years'. He was admitted to the Invalides on 26 April 1686; on 4 July 1686, he renounced his rights to a place in the hôtel and was given 15 *livres* to help him on his way.

The institution was very reluctant to admit men who suffered from scrofula (the King's evil), a contagious form of tuberculosis. The following man found this out the hard way:

> **Charles Cassidy**; Irish by nation; aged 35 years; soldier in captain Anthony Hamilton's company in the regiment of the same name where he served for five years; he could no longer continue in active service, having received two musket balls, one to the body and the other to the throat, on the day that Turenne was killed; he held a certificate from his captain dated 29 August 1676; he was married in his own country. On 5 September 1676, he was refused admission because he was not disabled and because he suffered from scrofula; he was given six *livres* to travel to his own country.

king. 29 Commentators have remarked on the fact that these old soldiers, who had experienced many ordeals, had usually managed to hold on to the regimental documents that established their rights to a place in the *Invalides*. There are instances of men who were turned away for not being able to produce written proof, and then seeking out a former commander and getting him to sign a text.

Generally, the administration was quite tolerant in its attitude to postulants. Some of these men would have spent their last evening with their old comrades and could arrive at the Council under the weather. The following was such a case:

> **William Shaughnessy** (*Guillaume Scharnesy*), known as *La Vallée*; aged 46 years; *brigadier* (sergeant) in the quartermaster's troop of the De Bleyny regiment where he had served for seventeen years as trooper and as sergeant, according to the certificate signed by his quartermaster on 7 March 1677; he claimed that he had served for a further eight years in the Reynolds (*Renel*) regiment; he could no longer continue in service, because of his infirmities; unmarried. On 20 March 1677, having presented himself to the Council in a drunken state, he was placed in custody for a week; on 27 March 1677, he was admitted as a trooper. On 22 February 1679, he died.

Many of their captains were Irish – Burke, Butler (*Boutilier*), Byrne (*Borne*), Callaghan, Carmody, Carney, Doherty, Fitzgerald (*Geraldin*), Grace, Hannigan (*Anogan*), Kelleher, Kennedy, Lacy, Macarty, MacNamara, McHugh (*Machieu*), Mellows (*Malissy*), O'Conor – but the majority of the pre-1690 Irish soldiers would have served under French, English, German or Swiss captains. Many of these men, often in poor physical condition, were able to get former non-Irish commanding officers – some of very high rank – to give them the required documentation.

These veterans had sustained wounds in the following sieges and battles: Altenheim, Ath, Besançon, Cambrai, Cassel, Douai, Entsheim, Fleurus, Fribourg, Gerona, Gigery, Gravelines, Haguenau, Lens, Lille, near Luxembourg, Montmedy, Ponnsalbruck, the crossing of the Rhine, Maastricht, Molsheim, between Nancy and Toul, Perpignan, Puycerda, Saint-Denis, Saverne, near Seneffe, Singem, Strasbourg, Tarding village in Germany, Valenciennes, Villefranche en Roussillon and Walcourt. Yet other Irish soldiers sustained crippling injuries whilst working on the construction of fortifications at Montlouis and Saarlouis. During the pre-1690 period, specific regions of origin in Ireland were given for only ten of the Irish veterans of which six counties are readily identifiable: Clare, Cork, Dublin, Kilkenny, Limerick and Roscommon; Grand Plac and Hartiriork are not easy to place.

Incidentally, the entry for the Kilkenny man is interesting because of the fact that he was accepted in the Invalides though very young, and because of the wealth of information contained in his file which tells the tragic story of an Irishman man who had suffered much:

> **Thomas Commerfort**; native of Kilquenin in Ireland; aged 21 years; soldier in captain Macnamara's company, Hamilton regiment, where he

served since the regiment was created; he was entirely unfit to continue his service as a result of a musket shot to the body received at the combat on the Rhine the previous year; since then, he had been in the hospital at Toul, according to the certificate dated 9 June 1676 which was issued by the doctor (*medecin*) of that hospital; on the foot of that certificate was a discharge given by the major of his regiment which instructed him to return to his regiment when he had recovered from his wound; he also had a certificate issued by the surgeon of the hospital at Chalons dated 24 June 1676; unmarried: Catholic; he was admitted on 4 July 1676. On 24 December 1676, he died, obviously being in a bad way when admitted.

Some Irish veterans, who had given long and honourable service but were not disabled, approached the authorities in the hope of being admitted to the *Hôtel*, but were to be disappointed, as was this man:

Robert Biota[30] (Betagh?); Irish; aged 65 years; gendarme in the English gendarmes company where he served eight years according to the certificate of his captain issued at Lille on 25 January 1679; previously, he had served in the duke of York regiment for 18 years as volunteer, ensign and lieutenant as was shown in the certificate issued by his lieutenant-colonel on 9 September 1678; his age, his wounds and his disabilities did not allow him to continue in service any longer; Catholic; unmarried. On 29 April 1679, he was refused admission on the grounds that he was not disabled. He was given 15 *livres* to return to his home country (*pour gagner son pays*).

The years 1691 and 1692 were marked by the arrival at the Invalides of French soldiers who had been wounded at the battle of the Boyne. There were:

Dominique Bressant, known as *Dominique*; aged 38 years; native of Clinsa near Cony in Piedmont; soldier in the chevalier Neuville's company, Zurlauben regiment, where he served six years as shown on his certificate; he said he had served previously for seven years in the Lalemand regiment and a further seven years in the Maglotty regiment; his left hand[31] was amputated (*coupé*) following a musket shot he had received at the Boyne in Ireland. He was admitted on 18 August 1691.

30 The Betaghs were an eminent family who had estates at Moynalty in Co. Meath; they were stripped of their property through fraud and perjury to the benefit of Cromwellians and other English land-adventurers. Members of the family gave distinguished service in France for several generations in the cavalry regiment of Fitzjames and in the infantry regiment of Clare (O'Callaghan, op. cit., pp 45f). 31 Musket shot wounds to the left side

Edme Dyotte, known as *La Verdure*; aged 28 years native of Auxerre; soldier in the Conflans company, Forest regiment, where he said he had served for two years; his left arm was amputated following a musket shot he received at the battle of Drogheda[32] in Ireland: he was a draper by trade. He was admitted on 3 November 1691. On 7 December 1728, he died there, 37 years later, aged 65.

The year 1692 saw other young men who had been wounded in Ireland present themselves at the Invalides. Theirs was a forlorn hope, because they had suffered serious injuries in the Williamite War but not, strictly speaking, in the service of a French formation. Here are two such stories, which showed a better awareness of Irish counties by the registrars. Both young men were sent away with a sum of 13 *livres* each. One wonders what became of them subsequently.

Hugh Farrell; Irish; native of Co. Longford: dragoon in the colonel's troop, Luttrell's Irish dragoons, where he says he served three years; his right leg was amputated following the effects of a cannon shot which he received at the siege of Limerick according to his certificate.

Anthony Kennedy; Irish; native of Co. Tipperary; soldier in the colonel's company, Charlemont Irish regiment, where he claimed he had served for three years ; his right arm was amputated following the effects of a cannon shot at the siege of Limerick, according to his certificate.

THEIR PLACE OF ORIGIN IN IRELAND

From 1692 onwards, the numbers of Irish seeking admission to the Invalides increased significantly, and the county of origin was entered in virtually all cases. The Wild Geese came from every Irish county; even counties with small populations such as Carlow, Leitrim and Wicklow supplied a cohort.

The largest number came from Co. Cork; there followed, in order of magnitude, the groups of men from Limerick, Dublin, Tipperary, Kerry, Galway, Armagh and Clare. With the exceptions of Armagh and Tipperary, all these were seaboard counties and leaving for France in small boats was relatively easy to organize. Interestingly, these were also the counties which provided most resistance to the occupier in the eighteenth and nineteenth centuries and were most active in the War of Independence. Moreover, they

were held to reflect being struck by an enemy musket ball whilst in the act of aiming.
32 The French tended to name battles after the nearest town.

were the counties in which – much later – the Gaelic Athletic Association was to become most highly organized.

SOME HISTORIC REGIMENTS

Each of the regiments in the service of France had a history and involved considerable investment by their colonel and by their captains. Following are summary histories[33] of a few of the formations which had Irish involvement:

Champagne: first mentioned on 29 May 1569, this regiment was renowned for its continuity of service, excellent discipline, remarkable esprit de corps; it was the most popular of all the regiments of France.

Duke of York: raised on 31 October 1652 by James Stuart, duke of York and commanded by Lord Muskerry: Lorraine, siege of Bar, Arras: 1654; Catalonia: 1655; given the title 'Royal Irish' with the chevalier James d'Arcy as lieutenant-colonel on 22 March 1657; Italy, siege of Marsal, 1663; stood down: 24 February 1664.

Forest (*recte* Forez): 1690: at the Boyne and Limerick: given on 4 April 1693 to Count Montmorency-Fosseuse who was killed at Marsaglia.

Zurlauben: given on 14 April 1688 to Béat-Jacques de la Tour-Châtillon, Count Zurlauben; Ireland, cut to pieces at the Boyne: 1690 ; Flanders: 1691; Mons, Namur, Steenkerque: 1692; Neerwinden, Charleroi: 1693; stood down on 21 September 1704.

Inchiquin: raised on 20 June 1653 by Lord Inchiquin; Naples expedition, 1654; Catalonia, 1655; incorporated in the Royal Irish regiment on 6 January 1657.

Muskerry: raised on 18 June 1647; Flanders, Landrecies: 1655; Condé: 1656; stood down: 1662.

Preston: raised on 18 June 1647 by James Viscount Preston; Flanders, Italy: 1653; Valenza: 1656; stood down: 1662.

Royal Irish regiment: formed on 6 January 1657 at La Charité-sur-Loire by James, the Chevalier d'Arcy from the debris of several Irish, English and Scottish regiments; incorporated into the duke of York regiment on 22 March 1657.

33 Susane, *Histoire de l'Infanterie française* (Paris, 1876).

A LONG MILITARY CAREER

The longest seventeenth-century military career of an Irishman in the French service was probably that of a Galway man who had been in six French and two Irish formations – Simon Kelly; Irish; aged 85 years; native of Galway; soldier in Conor's company, Dublin regiment where he served five years; earlier, he had served in Navarre for seven years, in Brittany for two years, Plessis Praslin for three years, La Ferté[34] for seven years; the French Guards for seven years, three years in the free company of Fabert when he was in Crete and was enslaved, and nine years in Flaharty's regiment; he had sixty years' certified service. His great age together with his wounds and the inconvenience caused by a large hernia made him unfit for service; Catholic; admitted on 30 May 1697. On 13 October 1697, he renounced his rights to a place in the Invalides and was given 15 *livres*.

This veteran joined a French regiment in 1637. Later, he was one of the Irishmen who, at the behest of the pope, had volunteered to serve against the Turks in Crete as part of one of the two expeditions sent to aid the Venetians in their struggle to repel a concentrated Muslim drive to capture the island. Fabert's force was overwhelmed when they made a sortie from the last fortified place in Christian possession. Fabert was killed – and the Turks took few prisoners. Kelly is the only one who seems to have survived captivity and to have been ransomed. He spent a relatively short while in the old soldiers' home before he found that he could stand it no longer. This man was very old and no doubt in poor physical shape and in no condition to fend for himself.

At times, the Irish mercenaries were considered to have little loyalty – particularly when pay was in arrears. There was a point at which Cardinal Mazarin discouraged the French ambassador to London from recruiting further in Ireland for he had noted that the Irish had a tendency to desert, 'due to their natural flightiness.'[35]

RELATIONS WITH OTHER IRISH, AFTER CEASING ACTIVE SERVICE

Many of the Wild Geese had been recruited for service in France through contacts with sergeants and corporals from their home areas. Even after they became unfit for further service, friendly contacts continued.

34 This regiment fought beside Hamilton and two other regiments during the rearguard actions in late July 1675. A measure of the ferocity of the fighting in the Rhineland can be had from the fact that the La Ferté regiment lost 15 of its 16 captains. 35 Affaires Etrangères, Paris, Correspondance Politique, Angleterre, vol. 66, fol. 272, quoted in: P. Gouhier, 'Mercenaires Irlandais en France', *Revue d'Histoire Moderne et Contemporaine*, 24 (Oct.-Dec. 1968), 672–90 at 688.

This bond is touchingly illustrated by the case of Cornelius Ronan of Drogheda 'in Co. Meath', a trooper in Fitzgerald's troop, Irish cavalry regiment of Fitzjames. On 15 June 1747, aged 60 years, married, he was admitted to the Invalides having been pronounced 'worn out' by 35 years' service. Almost fourteen years later, on 1 April 1761, Madame la Marquise de Nugent – whose family had been proprietors of the regiment prior to disposing of it to Fitzjames – went in person to the governor of the Invalides and declared that Ronan had died and handed in his discharge document. The declaration was accepted and Ronan's discharge was lodged with the institution's secretariat.

Other veterans were employed by Irish industrialists and others in France as supervisors, coachmen, watchmen and blacksmiths.[36] One such case was that of James Russell of Carlingford, Co. Louth. A farrier in Offarrell's troop of the Fitzjames cavalry regiment, he was aged 54 when the regiment was stood down in 1763 and he had served for 24 years. He had received wounds to the head and to several parts of the body from sabre slashes at the battle of Rossbach in Germany in 1757, an encounter in which his regiment had suffered heavily. Ten years later, he was buried in the parish of La Chevrollière in Brittany, having been on invalidity pension. The parish records show that, when he was buried on 21 June 1773, the burial register was signed by William Murphy and John Byrne, who were invalided veterans on detachment to the castle of Nantes. The register also indicates that Russell was employed at the castle of La Frudière as a supervisor.

Thomas Kelly, aged 53, had been quartermaster in Geoghegan's troop of the Fitzjames regiment when he, too, was recognized as being unfit for active service on 12 May 1763. He had 29 years' service of which four years were as quartermaster. The doctors recorded that he still suffered from the effects of an old wound received at the battle of Fontenoy. He was buried at Chapelle Saint Sauveur, about 30 kms north-east of Nantes, on 24 September 1771. The burial register described him as a former quartermaster of the Fitzjames regiment and inspector of the stables attached to the mines at Ceans. The record further states that the burial was witnessed by Irish (*Hirlandois*) officials of the mines. Eight signatures are appended but only one, 'Magrath', is decipherable. The mines at Ceans were under the direction of Stephen Misset – son of John-Baptist Misset – born at Mouzon in Champagne, diocese of Reims, about 1740; the Misset family was of Irish origin, according to M. Loncle de Forville.

36 M. Alain Loncle de Forville of Nantes kindly provided much information on the Irish veterans in Brittany.

DISABLED OFFICERS

Not only troopers and infantrymen suffered; so too, of course, did officers who led from the front and took great risks, as the following cases illustrate:

> **Christopher Plunkett** from Clinkatt in Co. Meath was 49 years old when his medical condition was recognized as being sufficiently grave as to render him unfit for further service. A half-pay lieutenant who had served nine years with the Fitzgerald (formerly Albemarle) regiment, he had previously been an ensign in the Dublin regiment for seven years and had been captain and lieutenant in Ireland for five years. His right hand had been crippled by a gunshot wound received at Cassano on 16 August 1705 and his right leg was disabled by a gunshot wound received at Calsanato. He was admitted to the Invalides on 19 August 1707 where he lived for a further thirty years, passing away on 6 September 1737, aged 79 years.

A second member of this Meath family was Oliver Plunkett of Killinne. He had served for fourteen years in lieutenant-colonel Marnegia's company, Aunay (formerly Croüy and Poitiers) regiment, as lieutenant, sub-lieutenant and cadet; previously, he had served for eight years as half-pay lieutenant and cadet in the Dublin regiment and had been lieutenant for six years in Gormanston in Ireland. His left arm was crippled as a result of it having been broken by a blow of a stone; he also had a chest condition which resulted from another blow of a stone to the stomach during the defence of Béthune, and by a gunshot wound to the head received at the siege of Verceil for which he had been trepanned. He also suffered from many other wounds. Withdrawn from active service on 1 July 1712, he was sent to Arras on detachment with the rank of lieutenant; he died there on 17 December 1713.

At the end of 1746, Michael Davoren was admitted to the Invalides. He belonged to the great Brehon laws custodian family of Co. Clare. He was a lieutenant in captain Terence O'Brien's company where he had served for eighteen years and nine months, of which three years were as lieutenant, two years and nine months as half-pay lieutenant and thirteen years as cadet. His right leg had been amputated as a result of a gunshot wound received at Fontenoy. He is one of the better-documented of the Wild Geese. The control list of the Clare regiment for 1737 describes him as having enlisted in the colonel's company on 15 December 1728, aged 20 years. He was 5 feet, 9 inches tall with blond hair and blue eyes. His face was said to be *beau et unis* which would indicate that he had not smallpox marks. The colonel's company was the unit from which promotions to the rank of officer were made. Only one man in thirteen could be in the colonel's company. Thus from the outset,

Davoren had support and favour. Nevertheless, promotion to the rank of lieutenant did not come until 26 April 1741.

On 30 December 1746, the French minister for war, d'Argenson, wrote from Versailles to the authorities at the Invalides stating that Davoren had been amputated at the thigh because of a cannon ball wound sustained at Fontenoy (11 May 1745) and the king (Louis XV) had granted him a place in the Invalides.[37] The ledgers show that he was admitted immediately to the veterans' home. In the lists of the dead and wounded at Fontenoy, he is shown as having had his leg broken in that battle. Because he was not hopelessly disabled, he was assigned to a military function at the citadel of Amiens where he served for 24 years. The records at Amiens show that he died in the Hôtel-Dieu hospital there on 2 March 1771 at the age of 63 years. On the following day, he was interred in the adjoining cemetery; the parish priest and two other clergy officiated at his funeral services. Although Amiens suffered heavily during the Second World War and lost most of its archives, the registration of Michael Davoren's demise is still extant there. Five separate references to this man of the Irish Brigade can be seen today; had he remained in Ireland, it is probable that he would have vanished without a trace. At an Irish hosting in Amiens on 19–21 June 1998, this officer was commemorated.[38]

INTEGRATION

Many of the Irish veterans opted to take a pension and remain with their families. The following two cases illustrate such men. Captain Henry Barnewall Plunkett declared that he was of Rathmore near Athboy in Co. Meath. (The Rathmore reference was to the Plunkett's hereditary estate that had been forfeited in the previous century, but to which the family still made claim.) Born on 1 January 1717, he joined Berwick regiment as a cadet on 12 January 1737, was appointed lieutenant on half-pay on 1 September 1741, full lieutenant on 8 February 1745, captain *en second* on 14 July 1747 and full captain on 22 April 1756. He married Catherine Ley of Kilkenny and they had four daughters.[39] At the battle of Lawfelt on 2 July 1747, he received a wound to the neck from which he never recovered and was obliged to retired on invalidity pension of 1,000 *livres* in September 1763. The family remained in their house in the town of Aire-sur-la-Lys where he died on 16 September 1765. He was interred in the chapel of the Poor Clares in that town; at least two of his daughters were in that religious order.

37 This letter is conserved in the de la Ponce papers at the Royal Irish Academy, Dublin. 38 Tim Pat Coogan, *Wherever green is worn* (London, 2000), p. 3. 39 A detailed account of their story is in 'Captain Henry Barnewall Plunkett of Rathmore's widow and four daughters', *Ríocht na Midhe* (2006), 144–72.

After his death, the only source of income for his widow and children was the reduced pension they received. When the French Revolution broke out, one of his daughters took up a petition in the town, gathering more than 1,200 signatures, commending Louis XVI for having refused to sign two laws which disfavoured the Catholic clergy. For this, she was arrested and, despite her spirited defence, was guillotined at Arras on 13 June 1794, following a second trial. At her first trial, she convinced the jury of her innocence. In the interim between the two trials, Robespierre had seen to the adoption of a new law which provided that the accusation was sufficient in itself and silenced the defence.

Two of her sisters most courageously defended the Poor Clares' convent in Aire-sur-la-Lys from being sold by the revolutionaries. These two remarkable Franco–Irish ladies died in the 1820s at their home in Aire-sur-la-Lys: Marie Elisabeth Plunkett on 24 March 1825, aged 71, and Marguerite Catherine Joseph Plunkett on 8 December 1826, aged 72. Prior to their deaths, both women had arranged to have their deaths recorded in the civic registers with a specific mention to Rathmore – a place that they had never seen. Thus, at least for one generation, this family was loyal to the faith which had led to their losing their property and to their heritage. They illustrated what Philippe Loupès has written of Irish Jacobite families in eighteenth-century France: 'These immigrants were profoundly attached to France but did not forget their country of origin. Sometimes, they made reference to ancient domains and they appeared to believe in the possibility of recuperating them.'[40]

DRUNK AND DISORDERLY

Many veterans were racked by continuous pain because of old wounds; others regretted lost opportunities; some took refuge in drink – in the following instances, to excess.

Philip Reddan, a Limerick man, was one of eleven Irishmen who were recognized as being unfit for further service, on 22 March 1715 at the end of the War of the Spanish Succession. Aged 59 years, he had served for 24 years in Kelly's company of the Berwick regiment. He suffered from hernia, from the effects of a sabre slash to his left ear, and other wounds.

Being mobile, he was sent on detachment to the garrison of invalids on the Iles Ste. Marguerites off the Mediterranean coast of France. There, he was found to be 'much given to wine, mutinous and uncontrollable', despite having been imprisoned for a time. On 3 August 1717, his commanding officer had him brought out of prison and ordered him to take his place in the ranks. Reddan refused and threw down his arms. A court martial was held on

40 Patrick Clarke de Dromantin, *Les réfugiés jacobites dans la France du XVIIIe siècle* (Bordeaux, 2005), p. 10.

him and on four other Irish invalids in the same company who were also mutinous drunkards. A report was sent to the authorities in Paris and Reddan was expelled. He was thus used to serve as an example to the others and lost all his privileges.

William Power, a native of Waterford, aged 30 years, was admitted to the Invalides on 2 December 1694. He had seven years' service as sergeant in Brian O'Rourke's company, the Queen of England's[41] regiment, and the extremities of both his feet had been carried off by a cannon ball at the battle of Marsaglia. He was lodged in the Invalides in Paris which was a favour, in view of his relatively young age and his short service. On 9 January 1703, he was sentenced to three years in Bicêtre, a tough prison for the unruly and the unstable, because he had assaulted a priest and insulted many other persons in the rue de Grenelle which is close to the old soldiers' home. However, he did not serve the full sentence, being brought back to the Invalides on 20 December 1704 having been pardoned by Chamillart, the minister for defence. We do not know who made representations on his behalf. In any event, he learned a lesson for he spent the remainder of his life – over 25 years – quietly in the Invalides, passing away on 12 September 1730.

Other men also drank and got into debt. One of these was Charles MacShane (*Maxanne* as he was entered phonetically in the records). Aged 40 years and a native of the Fews in Co. Armagh, he was a soldier in O'Neill's company, the Irish regiment of O'Brien (formerly Clare), where he had served for twenty years according to his certificate. His right leg had been wounded by a bomb splinter at Valence in Italy; the wound had not healed and he suffered from an ulcer and he had other wounds, too. On 5 July 1715, he was recognized as being unfit for further service and was sent on detachment to Armand Dupuy's invalids company in garrison at Entrevaux. (This is an ancient town in the Var valley on the old route to Italy. It is still walled, and no four-wheeled traffic is allowed within its gates. The old fortifications are in excellent condition.) The records state that, on 30 November 1716, Charles Maxanne was on guard at the Savoy gate, a place with great strategic importance that is still very evident today. Between the hours of one and two after midnight, he took a greatcoat and all the reserve of arms and deserted. The entry says that he left more than 15 *livres* of debts in the town, and the governor was informed by letter dated 12 December 1716.

DUELLING

Some of the Irish veterans were given to violence against other old soldiers. On 20 August 1693, Francis Higgins of Offaly (*Comté du Roy*), a carbineer in

41 She was Mary of Modena, the wife of James II.

Tobin's company, the Royal English regiment, was recognized as being unfit for active service because his left hand was crippled by a sabre slash he had received whilst on duty between Marche (en Famenne) and Liège; he also suffered from other wounds. He was sent on detachment to Captain Pierlé's company in garrison at Amiens. He deserted from there on 5 May 1700, after having killed Jean Gauthier, a native of Champagne.

William Dwyer, aged 31, a native of Tipperary and a sergeant in MacMahon's company, Fitzgerald (formerly Albemarle and Dublin) regiment – where he had served for eight years and claimed to have had four years previous service - was recognized on 29 June 1703 as being unfit for further active service because his left hand was disabled by a musket shot received at the battle of Luzzara. On 4 June 1705, he deserted from Captain Lucenoy's company on detachment to Tournai, after having killed his comrade, David Condon, an Irishman from Co. Cork. This Condon had been recognized as being unfit for further services on 24 November 1702; a soldier with twelve years' certified service in the Burke (formerly the Queen of England's) regiment, his left arm had been crippled by a musket shot received at Cremona on 1 February 1702.

Duelling could also have led to the death of Patrick Cantelon of Kilkenny. Aged 50, he was an NCO in Betagh's troop, Nugent cavalry, when recognized as being unfit for active service on 11 June 1716. He was on detachment at Colmar when he was killed by a comrade on 20 January 1718.

INTERMARRIAGE

Captain Henry Barnewall Plunkett married an Irishwoman and his children became integrated, as we have seen above. Other Wild Geese married French women.

Brigadier Edmund Finnegan of the Fitzjames cavalry shows how rapidly such families became integrated into French society. Finnegan was a native of Castlebellingham, 'Drogheda, Ulster'. He enlisted as a trooper in the Fitzjames cavalry on 2 April 1750 at the age of 26 years; his mother's name was Rose Reiley and his father's first name was James. He was described as being 5' 10" tall, with chestnut hair, a ruddy face and well-built. Eight years later, he was stated to be a *brigadier* (sergeant) in Offarrell's troop. On 16 April 1763, he was given an honourable discharge with permission to go the Invalides and was stated to be an honest man and a good subject who had given proof of courage on many occasions. This certificate was signed by Captain Offarrell of Ballintober, by MacDermott, the regimental aide-major, and was given the visa of d'Argenson. At the foot of the document[42] there is a

42 Conserved in the Royal Irish Academy, Dublin: de la Ponce papers, MSS 12 N, pp. 13/4,

special written permission allowing him to go to Louvain to collect some personal effects before going to the Invalides. Troopers and soldiers could carry with them on active service only a limited amount of baggage, and it would seem that Finnegan stored some of his possessions in Louvain.

The files of the Invalides show that he was received there with 24 other Irishmen on 12 May 1763. This was the largest ever intake of Irish on a single day. The registers indicate further that he had been shot through the neck at the battle of Graebenstein near Cassel in Germany on 24 June 1762; that was the encounter in which the Fitzjames regiment suffered very serious losses which led to its being stood down. Finnegan did not remain in the Invalides; he retired on pension to Faoüet in Morbihan, Brittany where on 10 July 1753 he had married Marie Josephe Talhouarn who belonged to a family of jurists. Three officers of the Fitzjames regiment signed the act registering the marriage, which reflected well on the relations between officers and NCOs.

Edmund Finnegan died on 17 November 1777, aged 55 years, and was interred in the presence of his widow, his son Thomas and his brother-in-law, *maître* Jean-Marie Talhouarn, Royal notary. On 16 October 1785, Thomas Marie Fenigan, *avocat à la Cour*, Edmund's son, married Marie Anne Mancel, daughter of the late *maître* Yves Jean Mancel, *procureur, notaire et receveur du domaine royal de Locronan*, at Châteaulin, Finistère.

On 28 August 1825, Edouard Esprit Marie Finegan, *avoué près le tribunal de* Châteaulin, born at Quimper on 27 May 1798, son of Thomas Marie Finegan, married Mélanie Foucault, daughter of François Robert Foucault, *receveur de l'enregistrement* at Châteaulin. Among the witnesses who signed the entry in the register were 31-year-old Lilas-Thomas Finegan, doctor of law and the bridegroom's brother, and 25-year-old Thomas Alain O'Finegan, principal clerk to the notary at Rennes. Five Finegans appended their signatures to the entry.

A daughter born to this couple on 14 June 1826, Melanie Finegan, described as 'proprietor', married Raimond Bernard, *contrôleur des contributions*, on 31 August 1846. Among those present at the ceremony were Lilas-Thomas Finegan, aged 52, councillor at the Royal Court at Rennes and a chevalier of the Legion of Honour, aged 52, who was the bride's uncle, and Thomas Alain Finegan, aged 46, advocate at Rennes.

A RESOURCEFUL VETERAN

At the age of 28 years, Daniel Morphy was admitted to the Invalides on 8 March 1696.[43] The entry in the register is relatively brief; it states that he was

item 40. 43 On that same day, six other Irish veterans were admitted to the Invalides.

from Harmanacq which is almost certainly Fermanagh, pronounced in the Irish manner.

He had five years certified service in Captain Reilly's company, the Dublin regiment and claimed that he had served for seven years previously in Ireland. He was unfit for service because his right leg had been crippled by a gunshot wound received at Speyer in Germany. The register also shows that he died at his home in Gros Caillou, Paris, over sixty years later on 8 April 1756, aged 88 years.

However, the minutes of the management Council of the old soldiers' home give us a most interesting insight into Morphy's resourcefulness. The Council discussed on 7 December 1747, in the presence of count d'Argenson, a concession that Morphy had obtained from Le Blanc, who was a senior administrator of the *hôtel*, on 29 December 1718, which entitled him to build a small house on an *arpent*[44] of land that belonged to the Invalides and which was close to the institution. Further, Morphy was given permission to live in the house for the rest of his life with his wife, Annable Maguy (*Nábla Nic Aoidh*). He had previously got permission from the authorities to marry Nábla – which was not easy to obtain for the institution did not allow disabled veterans to bring a spouse into the institution, because of the potential effect on unmarried inmates, and of the problems that the support of children could cause. Furthermore, Morphy got the authorities to agree that his wife could remain on in the little house, if she should outlive him. But Nábla died and, in 1725, Morphy was given permission to marry Marie Joseph Heslin.

In December 1747, Morphy asked the Council for approval of Marie Joseph staying on in the house, should she outlive him; he was now 79 years of age and his second wife was 73. The HRI Council agreed to his request. On 16 March 1752, the Council had before it a petition from a labourer named Etienne Dutfoy who asked that he and his wife Marie Anne Charpentier be allowed to continue to occupy a house that Dutfoy's father, a master gardener, had built on a half-*arpent* of land following an agreement between Dutfoy senior and Morphy and his wife Marie Joseph Heslin. The agreement was made before a notary in Paris on 15 September 1732 and prescribed the rent to be paid on an annual basis. The *hôtel's* Council agreed that Etienne Dutfoy and his wife could live out their days in the house on the half-*arpent* of land, but stipulated that the terrain would then revert to the *hôtel*. Morphy had entrepreneurial skills and also the nerve to risk his little house by going for further gain.

Other men with lesser powers of persuasion were refused permission to marry and remain in the institution. One such was Bernard McGurk, native of Armagh, soldier in Cahan's company, Galmoy regiment, where he had served for eleven years and claimed to have served for three years in Ireland.

44 An *arpent* was between 350 and 500 square metres of land.

His left arm was crippled by a musket shot received at Chiari and he suffered from other wounds also. Aged 32, he was recognized as being unfit for active service on 28 July 1702. On 19 November 1707, he resigned voluntarily so as to be free to marry.

Other Irish veterans consorted with prostitutes – which was strongly disapproved of by the authorities. One such was Theodore Macarty, aged 51, from Co. Clare, who had served in Lieutenant-colonel Colgrane's company, Lee (formerly Mountcashel's) regiment. He claimed to have 24 years' service, in all. He had lost his left eye when a piece of wood fell on him during the siege of Kehl and he had very poor sight in his right eye. Admitted on 4 July 1704, he came before the Council on 17 October 1704 for having been found lying with a woman of ill-repute in full view of the *hôtel* and was sent to Bicêtre for six months. On his return, he lived uneventfully in the institution for over nine years until his death on 3 December 1714.

Still other Irish veterans, longing for female companionship, became enamoured of unsuitable partners. Nicholas Locke, a native of Dublin, was one of these. Aged 58, a trooper in Francis Nugent's troop, Nugent 'Foreign Cavalry' regiment, where he had 23 years' certified service, he had poor sight and other infirmities when he was judged unfit for active service on 5 April 1715 and sent on detachment to Fort Barraux. On 2 March 1722, it was recorded that, now aged 65 years, he had deserted from Captain de Mouchy's company taking with him an official-issue belt, a bayonet and a quantity of wheat, and had gone off to marry a trollop (*une gueuse*) whom he consorted with, despite his captain having strongly objected. In fact, his captain had kept him locked up for two months and only released him when he promised not even to think of eloping – but he ran away the very next day. The Council struck Locke from the register.

LONGEVITY IN THE INSTITUTION

Daniel Morphy described above spent just over sixty years as a beneficiary of the invalidity system. Many other Irish also enjoyed quite long stays in the HRI. The following were typical of seriously disabled men who survived for decades in the institution and who could not have eked out an existence outside:

> **Henry Lucas**, aged 66 years, native of Ormond, Co. Tipperary, was admitted on 25 April 1710. He was then a sub-lieutenant in Callaghan's company, Dorrington (formerly the King of England's Guards) regiment, where he had served for 19 years in that capacity and as half-pay lieutenant and enseign; previously, he had served for three years as lieutenant, cornette and quartermaster in Lucan's cavalry regiment; his

weak sight and problems with his legs made him unfit for further service. On 19 March 1741, he died.

John Doherty, aged 25 years, was a Derry man and a soldier who had served for four years in Brian O'Neill's company of the Charlemont regiment. Married in Paris, his right leg had been carried off by a cannon ball in the assault on Heidelberg. He was admitted on 17 November 1695 and died there over 37 years later on 18 February 1733.

Edmond Condon, aged 34 years, a native of Cork and a cadet in the colonel's company of the Lee regiment, where he claimed to have served for seven years, had his right leg badly injured by a musket shot received at the siege of Rinfels. On 2 May 1697, he was admitted until such time as he had fully recovered. He remained on for more than 34 years, passing away on 28 January 1732.

Several Irish were badly cut up at Ramillies on 23 May 1706 where, in a rearguard action, the Irish regiments saved the French from total disintegration – at great cost. Thomas Reilly of Sixmilebridge was one of those wounded there. Aged 30 years, he was a trooper in Marshall's troop, Nugent (formerly Sheldon) cavalry, where he had eight years certified service. His left arm was crippled by a sabre slash received at Ramillies and he had received a further seven sabre slashes to the head. He was recognized as being no longer fit for active service, on 8 October 1706, and placed on detached service. On 15 August 1752, he died at Caen, aged 76 years.

ACCIDENTAL INJURIES

Some troopers and soldiers were disabled as a result of accidents that occurred whilst on duty or in training.

Alexander Macswiney,[45] aged 50, of Tireban in Donegal was recognized as being unfit for further service on 20 November 1732. A trooper in the Brissac (formerly Cossée) regiment, where he had served for six and a half years; he had previously served for three and a half years in Vaudemont cavalry and twelve years in the English *gendarmes*. His right hand had been injured by a gunshot at the siege of Fribourg and was crippled by a sabre slash which he had received accidentally in March 1732; his wife resided at Chalons-en-Champagne. He died in the Invalides on 2 March 1744.

Daniel Macmolin, aged 19, a native of Blarney, Co. Cork, had an unusual story. A draper by trade, he had only two years' service in Captain Kennedy's

45 He belonged to the famous clan who lost everything with the destruction of the old order.

company, Bulkeley regiment, when he became the victim of a 'work-related accident'; he was on sentry duty on the ramparts of Avesnes-sur-Helpe when he fired on persons who were in the fosse; his musket exploded, his left hand was shattered and had to be amputated at the wrist. On 8 August 1737, he was considered to be unfit for further active service. However, he was too young and had too short a period of service to be kept in Paris and was sent to the Château d'If,[46] an island fortress off Marseilles. The records state that he was suspected of deserting on 25 March 1738 and was struck from the registers on 1 May 1738 but, with only one good hand, it is distinctly possible that he fell into the sea from this windswept rocky island.

Another Irishman to have suffered a similar injury was Arthur Machanally (Mac An Fhailí). Aged 26, he had served for four years in Macnamara's company, the Furstenberg (formerly Hamilton) regiment. His left hand had to be amputated after his musket blew up at the changing of the guard at Nancy. Unmarried, he was admitted to the Invalides on 15 July 1679, but on 7 April 1682, he voluntarily renounced his rights to a place in the institution and was given 15 *livres* to enable him to return to Ireland.

A third Irishman to experience a similar mishap was Fiacre Morphy, known as *St. Fiacre*; aged 45 years, he was a native of Rochefort and a dragoon in the chevalier de Bresseillac's troop, Rohan (formerly St Hermine) regiment, where he served eight years; previously, he had served for four years in the Queen of England's dragoons and four years in the Clare infantry regiment. He had 16 years' certified service. His left hand was crippled when his musket blew up in his hands at the battle of Blenheim; his wife was at Morlaix in Lower Brittany. When he applied to the Invalides on 20 March 1708, he was refused on a technicality; he had left the army two years earlier; he was given 13 *sols* to help him travel to rejoin his family.

There were many other similar accidents attributable to the uneven quality of gunpowder or to poor metal in gun barrels. In any event, it was worrying for the infantrymen – and may have contributed to the French tendency to favour the bayonet charge rather than the exchange of volleys at a distance, particularly as the English muskets were of superior quality.

French cavalry men had also grounds for complaint about the tactics they were required to apply. They charged the enemy cavalry with their officers advancing in front of the troopers. If the initial shock failed to break the enemy ranks, the officers were very vulnerable on colliding with the wall of enemy cavalry and had great difficulty in disengaging their men and reforming for a further charge. Furthermore, their sabres were inferior. The Marquis de Castries complained that in combat with the Prussians the

46 Anthony Fitzmaurice, a native of Connacht, who had had his left leg amputated, was also sent on detachment to the château d'If . He was drowned there on 9 April 1719, aged

troopers showed great courage but their blades were very light; they often broke on impact[47] and were also shorter than the enemy weapons. The wounds that the Prussians received were relatively light. As well, the guards to cover the hands of French troopers did not give adequate protection from enemy slashes in hand-to-hand encounters. The marquis strongly recommended that the cavalry be equipped with swords about eight inches longer than the existing issue, and that the weapons should be pointed so as to enable the trooper to thrust as well as slash.

The horses tended to be overloaded, because troopers carried on their horse – in addition to their weapons and harness – a large range of reserve supplies and provisions which exhausted the animals.[48]

The records show that some of the veterans suffered injuries due to less warlike incidents. **Hugh Cullen**, sergeant in the grenadiers' company, Berwick regiment, and a native of Rathfarnham, Co. Dublin, married, was aged 38 when admitted on 4 April 1776. He had fallen down the stairs in the barracks at Port Louis Lorient in 1774 and had been at a spa which failed to cure his condition.

TREPANNING

Trepanning was a painful operation that many of the Irish underwent in the hope of relieving an excruciating pressure on the brain; the surgery, without anesthetics, entailed boring a hole in the skull to release fluid.

The survival rate was about 50% at the time, but Martin Mongon, known as *Saint Martin*, survived the ordeal. He had been struck on the head by a musket ball, had nineteen years' service in a French unit under the command of Captain de Maumont; married in Paris; he was admitted on 10 March 1685, then aged 67 years; he died in the Invalides on 8 October 1696.

INTENSE COLD

Cornelius Callaghan, aged 39 years, a native of Co. Cork, a grenadier in MacMahon's company, O'Donnell (formerly Fitzgerald, Albemarle and the Irish Marine) regiment, where he had twenty years' service, had his left arm crippled and all his toes amputated as a result of the extreme cold of the 1709 winter[49] experienced whilst being conveyed by boat from Gand to Nieuport. Admitted on 11 July 1710, he died on 5 January 1716.

44 years. 47 Captain Warren wrote that, at Fontenoy, he slashed a fleeing enemy with his sword which broke off at the hilt; he pursued others for a considerable distance before he realized his own exposed situation. 48 Édouard Desbière and Captain Maurice Sautai, *La cavalerie de 1740 à 1789* (Paris, 1906), pp 32ff. 49 Marcel Lachiver, *Les Années de Misère*

FAR DISTANT FIELDS

Some of the Irish invalids had served France in remote places. Daniel Canavan, aged 32, a native of Tyrone with three years' service as grenadier in Misset's company, Berwick regiment, and five years with O'Neill in Ireland, married at Arras, had his left hand crippled by a gunshot wound received at the siege of Gibraltar. He had also been struck by a stone between the shoulders at the same siege; he was admitted on 9 October 1705. He was one of the small force which was given the impossible task of trying to recover the Rock, after it had been captured rather fortuitously by the Landgrave George of Hesse-Darmstadt who had been on the way back from an otherwise unsuccessful cruise in the Mediterranean under admiral Sir George Rooke. On 20 March 1705, count Villiers wrote to Chamillart, the French minister for war, from 'before Gibraltar', seeking reinforcements and stating that he had only 650 men under arms.[50]

Edward Clarke, aged 60, of Milltown, Co. Dublin, was admitted on 3 April 1749. His left leg had been crippled by a gunshot wound at Culloden in Scotland and he had other wounds also. He died in the Invalides on 18 January 1752. Michael Shee of Kilkenny was admitted on 27 October 1757. He had received several wounds at Culloden where he had been 'with Prince Edward' (Stuart). On 21 April 1758, he died on detachment at Fort Médoc, Blaye.

William Swan, aged 38, a native of Slivstenn, Co. Tyrone, was among the last of the Irish to be admitted to the Invalides – on 8 June 1786. A Protestant,[51] part of his right leg had been amputated after 'he was struck by a bullet at Tobago in the Americas'.

Luke Lynott of Ballina, aged 40, was admitted on 31 March 1763. Married, he had 17 years' service and had been a sergeant in Lally's regiment. A gunshot had inflicted serious injuries to his left foot at the siege of Madras in India.

HARD DECISIONS

Some of those who applied for admission, only to be refused, would seem to have been rather harshly treated. Timothy Macartie was one such unfortunate case. A native of Cork and a 67-year-old half-pay officer with Magrath's

describes the conditions. 50 Saint-Simon, *Mémoirs* (Paris, 1986), iv, 1286, n. 5. 51 There was a restriction on the admission of non-Catholics to the Invalides. However, as we have seen above, several Irish Protestants were admitted. The file concerning trooper Whitehead who served in Wall's troop of the Fitzjames Irish cavalry is interesting. When that regiment was stood down at Haguenau, it was stated that his head had been split (*partagée*) by a sabre slash at the battle of Graebenstein and he was accepted as an invalid. Very soon, he enlisted in the Dillon regiment where he was a sergeant for four years and from which company, the Irish Marine regiment, he had six years' certified service and claimed to have also served earlier for two years with Milord Macarty in Flanders. His age, his weak eyesight and other unspecified disabilities made him unfit for active service. The record states that he had brought his wife to Lille. However, he was turned away and was given the sum of 15 *livres* on 11 November 1694.

One could not but feel some sympathy for Brian McGeeny of Co. Armagh, aged 80 years, a soldier in Magennis' company, Galmoy (formerly Charlemont) regiment, where he had served for fourteen years; because of wounds and disabilities, he was no longer fit for active service. But when he applied for admission on 22 May 1705, he was turned away with four *livres*.

TRIPLE DESERTION

Another Irishman, Arthur Magennis (*Macquenis*) illustrated very well the hazards of accepting very young men in the institution. A native of Co. Down, a soldier in the colonel's company, the Carra regiment, where he said he had served for four years, his right shoulder was crippled by a musket shot he received at the battle of Aughrim; he was admitted on 13 March 1692. Sent on detachment to Bapaume, he deserted on 11 March 1696. However, he obtained a royal pardon on 28 August 1706 *de la propre bouche du Roy*.

On 15 March 1708, he was arrested in Paris for causing a disturbance, was imprisoned in the Châtelet on the orders of d'Argenson and from there taken to Bicêtre, another place of confinement. On 11 July 1709, he was allowed back to the Invalides. On 29 January 1718, whilst under arrest in the institution, he deserted; the record states that he did not steal anything but that four or five years previously he had lost his chamber pot which was the property of the *hôtel* – which shows the detailed recording that operated within the place. He was taken back again but, on 6 July 1722, he deserted for the third time and was then definitively struck off the records.

A CABALIST

The *hôtel* records very rarely refer to the personal interests of individual veterans. Maurice Honan (*Huonin*) is a remarkable exception to this general rule. A native of Limerick, aged 64, he was a sergeant in O'Shaughnessy's company, Clare regiment, where he had served for eight years and he claimed

he later got an honourable discharge. He stated that he was an Anglican, not a Catholic, and was told that he could not remain in the *hôtel* unless he abjured. His comrades in two regiments would have known of his religious persuasion, but they never made an issue of it.

to have served previously in the Languedoc regiment for nine years, in Hamilton for twenty two years and in Royal Roussillon for three years; on 27 March 1698, it was acknowledged that his wounds and disabilities made him unfit for further active service.

He was not admitted to the old soldiers' home in Paris but was sent on detachment to La Rougerieau's company at the fort of Scarpe. He deserted from his company there on 30 July 1713. His movements were known for the records state that, on 16 August 1713, he went to Arras without permission. The records mention that he was much given to writing (*grand escrivain*) and a Cabalist. However, he must have been able to use considerable influence for, on 19 August 1718, he received a royal pardon. He died on 20 October 1721, aged 87 years.

THE IRISH LANGUAGE

Officers in the Irish regiments in the service of France were obliged to work through the medium of French, in their communications with their paymasters and in maintaining the regimental records. They owed loyalty to the Stuarts whose language was English and a disproportionately high number of officers were of Scots or English origin,[52] for James II and, later, Bonnie Prince Charlie tried to maintain the fiction that they were heirs to the three kingdoms. But the vast majority of the soldiers in the Wild Geese regiments were of Irish origin and many would have been fluent in the Irish language.

This cultural dimension shows up in the records on many occasions, notably in family names and in place names. Eugene Reilly of Cavan, registered on 6 May 1700 stated that he was from Cawain = *Cábháin*. The entry in the records for another man of the same clan is very relevant. He was recorded as Flany Rilly, which is almost certainly *Flann Uí Raghallaigh*; the text is as follows: Flany Rilly, aged 40 years, native of Cuynanore, diocese of Kilmore in Ireland, trooper in the quartermaster's troop, Irish regiment of Nugent, where he served 18 years according to his certificate; his left hand was crippled by a sabre stroke which he received at the battle of Ramillies; on 5 June 1711, he was attached to a company of invalids at Arras; on 10 March 1712, he died in the town of Arras. A Monaghan man, Thomas Connolly, gave his place of origin as Carikmaharache, which is a phonetic rendering of *Carraig Machaire Rois*. A county Waterford man, James Lamy, admitted to the Invalides on 15 November 1736, gave his place of origin as Wnngerwan. This is the closest that a clerk could come to noting down '*ó Dhún Garbhán*', that is, Dungarvan. Men from the north Tipperary area gave their place of

52 For evidence of the friction between these nations, see Nathalie Genet-Rouffiac in E. Cruickshanks and E. Corp (eds), *The Stuart court in exile and the Jacobites* (London, 1995).

origin as Enagharoon, that is, *Aonach Urmhumhan* or Nenagh in English. An O'Driscoll from Baltimore stated, when asked where he was from, that he came from *Grand Maison* which was a French translation of *Tí Mór*.

Nevertheless, reading the names, first names and places of origin of the Wild Geese listed in the archives of the Invalides, one gets the impression that those who travelled to France were from the more Anglicized regions of Ireland; whilst Gaelic Ireland may have dreamt of help coming from abroad, the hillside men did not feel a sufficiently strong loyalty to a foreign king as to take the virtually irreversible step of going to serve overseas. Nevertheless, those who did enlist in the Irish Brigade were highly motivated. The risks of being captured and hanged – as was the fate of hundreds - were very great.[53]

For the battered veterans who got to the Invalides, the quality of care was enviable – even by modern standards. The audited accounts of the expenditures of the institution from the 1670s onwards, are still conserved in the château de Vincennes. Officers fared better than soldiers, but the food provided for all was varied and excellent. Cod, salmon, eggs, salted and unsalted butter, oil and vinegar, rice, lentils, prunes, peas, beans, cheese, soups, salads, beef and dried fruits (almonds, figs, nuts and raisins) were on the menus. Wine of good quality was also supplied, and any complaints were examined. There was a good pharmacy, and the medical care was ahead of that available elsewhere, at the time. Some twenty priests looked after the spiritual needs of the invalids – although this number was reduced to eight priests and four brothers in 1770. Married men were allowed to spend three nights a week outside the institution. A dead veteran was accorded a guard of honour and three priests at his funeral.

FALSIFICATION OF DOCUMENTS

The excellent fare available in the institution prompted some persons to gain admittance by way of falsified documents. The unravelling of this fraud came about when the English ambassador to the court of Louis XIV, Matthew Prior,[54] protested vigorously about the activities of Cornelius O'Driscoll, a privateer turned pirate. Many of O'Driscoll's clansmen[55] and all his hereditary territories had been lost in the Williamite Wars and he obtained a privateer's permit during the War of the League of Augsburg. With the peace of Utrecht, his permit lapsed, but he continued to attack English merchant shipping and allegedly put the captured crews to death. One of O'Driscoll's

53 See Eamon Ó Ciardha, *Ireland and the Jacobite cause, 1615–1766* (Dublin, 2001) and Breandán Ó Buachalla, *Aisling Ghéar* (Dublin, 1996). 54 For more on this remarkable diplomat, spymaster and friend of Dean Swift see Charles Kenneth Eves, *Matthew Prior, poet and diplomatist* (New York, 1973). 55 John D'Alton, *King James's Irish army list*

team of pirates, William Burke of Galway, became distressed at the fate of unfortunates thrown overboard. To hush him up, O'Driscoll obtained falsified papers which enabled Burke to gain admission to the Invalides, but inside, Burke still talked of what he had been done.

Burke was arrested and placed in the Bastille on 19 March 1699; he died there on 3 November 1704. O'Driscoll learned that he was sought and got out of France. But he returned in August 1701, was denounced by a French priest and thrown in the Bastille with his wife and mother-in-law. His wife gave birth to a daughter there and this was the sole instance of three generations of the same family being in the Bastille at one time.[56]

Following the scandal that arose from William Burke's papers, the authorities began to look more closely at the documents produced by the Irish who sought admission. On 25 November 1700, Cornelius Sheehan of Limerick presented himself. He had papers claiming to be a soldier in Macarty's company, Lee (formerly Greder alemand, Furstenberg and Hamilton) regiment, where he had served 24 years and he suffered from a chest condition since the siege of Rinfelds. He was placed under arrest for having produced a false certificate. With effect from 23 December 1700, he was expelled after having been placed on the wooden horse (*après avoir esté sur le cheval de Troie*), a humiliating and uncomfortable punishment. For others expelled at this time, see the Appendix to this chapter.

RECRUITERS

The Irish regiments had constant need for new recruits. The large army that followed Patrick Sarsfield into exile in France was consumed in the battles and sieges of the closing years of the War of the League of Augsburg. For the following seventy years, each colonel and captain had a constant struggle to find replacements for casualties. As mentioned above, recruiters in Ireland risked hanging. Yet, some went to great lengths to earn the bounties that were given for new recruits. After the battle of Fontenoy, Daniel White[57] slipped into the English camp and endeavored to get men to desert. He was detected and sentenced to 1,000 lashes – a slow death sentence. A few years later, Irish and Scottish agents went to Dover and London on recruiting missions. Several, including a sergeant of the Rothe regiment, were caught and hanged.

Irish recruiters also feature in the records of the *Invalides*, as the following entry shows:

(Limerick, 1997, reprint), pp 904f. **56** During her confinement, Mrs O'Driscoll was assisted by the royal obstetrician. A full account of the O'Driscoll family – and the Vauban connection – is given in Ó hAnnracháin, 'A Cork family in the Bastille', *Journal of the Cork Historical and Archaeological Society*, 107 (2002), 177–94. **57** Remarkably, the last man to be flogged to death in the British army, in 1843, was also named White and the colonel

John Offarrell, Irish, aged 63 years, native of Monasteroray in Ireland, province of Leinster, formerly a trooper in the Fitzjames cavalry where he served for about ten years, and then he entered the Berwick Irish infantry where he served for 35 years as a recruiting officer, according to a Ministerial letter dated 21 June 1769 which admitted him to the 'intermediate class of invalid'. He was unable to produce any documentary proof of his length of service, having been obliged to burn his papers when he was being pursued by the English whilst recruiting for the Irish regiments in the French service; this resulted in his being deprived of all his possessions in his homeland since 1765 and he could no longer return to his own country; he was married and a Catholic. He was recognized as being unfit for further active service, on 22 June 1769. On 29 November 1782, he died in the hospital at Montreuil sur Mer on detachment with captain de la Haye's company.[58]

DAILY LIFE IN THE *INVALIDES*

Great attention was paid to ensuring that the invalids attended religious services, for Louis XIV attached importance to preparing the veterans for eternity. They were still soldiers and were expected to drill. Louvois and the king wished to keep them occupied and set up workshops for the manufacture of shoes, stockings, uniforms, matches and tapestries.

There was also a calligraphy centre and an illuminated manuscripts workshop which produced some fine items. The veterans received modest payments, but the workshops eventually fell into disuse. Hence, some of the old soldiers, as we saw above, got into trouble outside the institution.[59]

CONCLUSION

Probably some 250,000 Irishmen served in the armies of France prior to the French Revolution. About 1% of these Irishmen feature in the registers of the *Invalides*. The vast majority of the Irish who were paid by the *ancien régime* died in the many sieges and battles or perished in the epidemics that swept their barracks in winter time. Others, especially those married or with a trade, preferred to remain outside the institution, even when in a handicapped state. Of the invalided, it was truly written:

responsible was named Whyte. 58 Registers of the Invalides, vol. 60, matricule 295. 59 François Lagrange and Jean-Pierre Reverseau, *Les Invalides* (Paris, 2007), pp 31f.

In Austria and France they roved
Through ways as sad as death;
In alien paths the tired feet bled,
The laurel crowns that decked the head
Were thorn-set underneath.[60]

They suffered much but for those who were received into the *Invalides*, there was repose and good medical attention at the end of their days. Those sent on detachment were far less well-off, but still had a minimum support. Influence played a part in deciding whether the veteran was kept in the HRI or detached. The brief, representative biographies given above tell us a good deal about seventeenth and eighteenth century men of our nation in the service of France, of their military organisation and of the injuries they sustained.

APPENDIX

List of 27 Irish expelled from the Invalides, *after the major review, giving their names; county of origin; age; date of admission and marital status, where stated.*

Maurice Curran: Kilkenny; 25; 21 Jan. 1700; married in Pont-à-Mousson.
Thomas Nisbette; Sligo; 50; 28 Jan. 1700.
Nicholas Barry; Cork; 35; 28 Jan. 1700; married in Paris.
Patrick Rogers; Tyrone; 30; 04 Feb. 1700; married at Paris.
Terence MacDermott; Roscommon; 40; 04 Feb. 1700; married at Paris.
Brien O'Brien; Monaghan; 48; 11 Feb. 1700; married in Paris.
Niall Hagan; Kerry; 36; 11 Feb. 1700; married at Paris.
Peter Butler; Waterford; 40; 11 Feb. 1700.
Germain MacLaughlin; Galway; 30; 18 Feb. 1700; barber-surgeon.
Oine Brady; Cavan; 27; 18 Feb. 1700; married at Paris.
Germain O'Sullivan; Cork; 26; 25 Feb. 1700.
Felix Quinn; Wicklow; 24; 25 Feb. 1700.
John Callaghan; Cork; 39; 11 Mar. 1700; married in Paris.
Bernard Mahony; Armagh; 40; 25 Mar. 1700.
William Fitzgerald; Limerick; 65; 25 Mar. 1700.
Patrick Maguire; Fermanagh; 45; 01 Apr. 1700.
Charles Hughes; Armagh; 36; 22 Apr. 1700; married at Paris.
David Barry; Cork; 50; 22 Apr. 1700.
(Another) Niall Hagan; Cork; 50; 22 Apr. 1700.
Eugene Reilly; Cavan; 34; 06 May. 1700.

60 Katharine Tynan, *The Flight of the Wild Geese*.

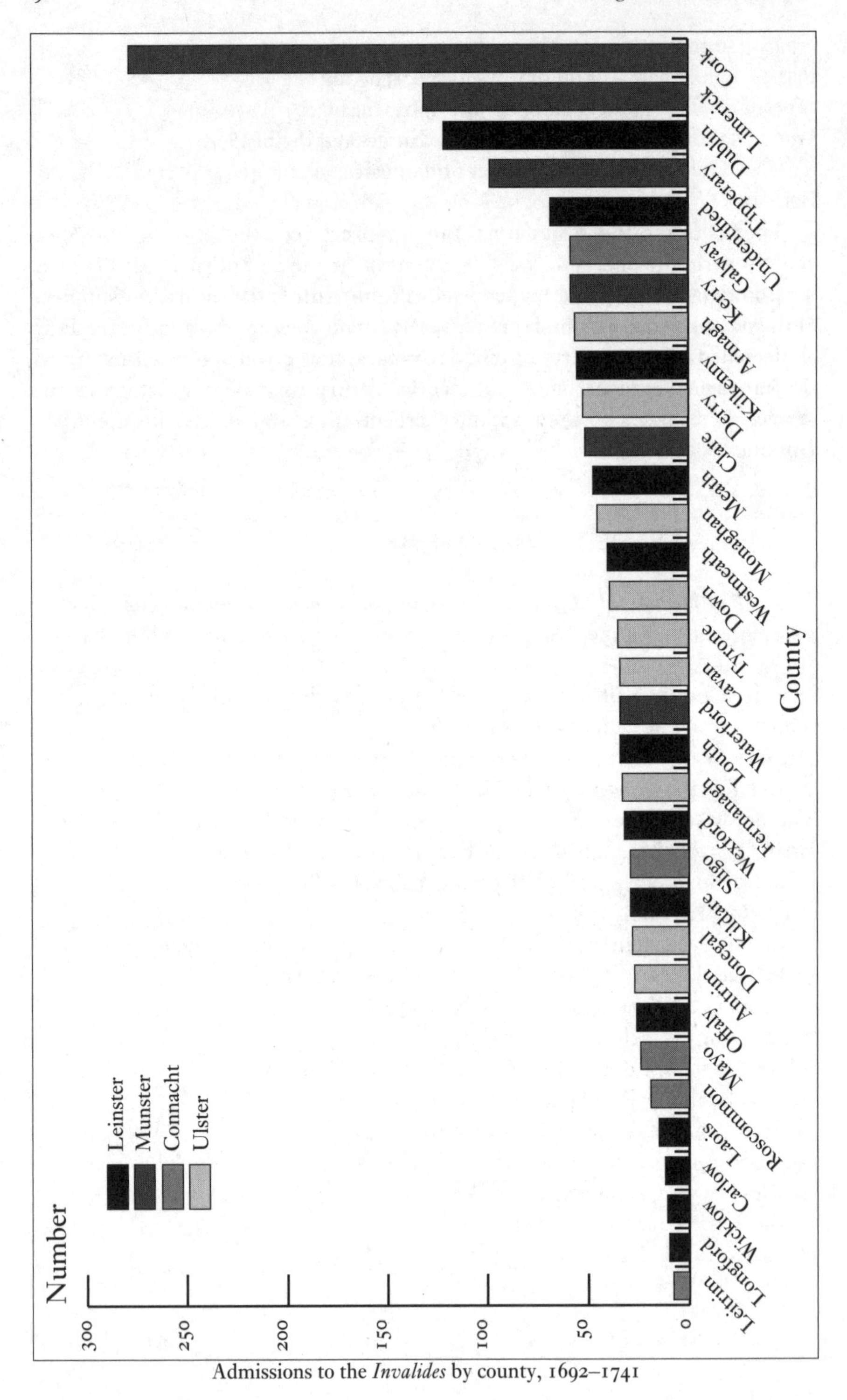

Admissions to the *Invalides* by county, 1692–1741

Thady Cunningham; Cork; 55; June 1700; married at Paris.
Patrick Faulkener; Louth; 63; June 1700; married at Paris.
Sibert Devet(t); Donegal; 42; 08 July 1700; married at Paris.
James Mahony; Cork; 57; 22 July 1700; married at Saint Germain en Laye.
Terence Gohine; Sligo; 35; 19 Aug. 1700; married at Paris; gardener.
Bernard O'Moan; Cavan; 50; 04 Nov. 1700; married at Strasbourg.
Terence O'Brien; Wexford; 60; 18 Nov. 1700.

These men would appear to have spent some time in the service of France. However, a remarkably high proportion of them were married in Paris. They all presented a document that purported to have been signed by the Marquis de Barbesieux. Apart from expulsion, no other punishment was inflicted on them, Sheehan excepted. Indeed, three of them were re-admitted. Felix Quinn, was taken back on 31 August 1703, on the grounds that he was blind (and thus could not know he was presenting a fake document!). Yet the same Quinn re-enlisted in a cavalry regiment on 18 April 1713, for one year and then returned to the Invalides where he died on 1 March 1721. Charles Hughes was re-admitted on 9 October 1707, having been on active service again. He died on 15 August 1709. Eugene Reilly was re-admitted on 3 July 1711, after serving a further ten years; on 30 September 1721, he died at Le Havre, on detachment. Patrick Faulkener tried to get back on 14 November 1710, but was refused.

'Irlandois de nation': Irish soldiers in Angers as an illustration of Franco-Irish relationships in the seventeenth and eighteenth centuries

PIERRE-LOUIS COUDRAY

> '[...] le gros du peuple est toujours le même: vivant sans instruction dans la campagne, et n'ayant que de fort légères teintures du Christianisme, ils ignorent les premiers éléments de la charité chrétienne. Ils sont vindicatifs, implacables, et si enclins au brigandage et à la volerie [*sic*] qu'elle ne passent point parmi eux pour un métier infâme.'
>
> James Beeverell[1]

If the 1600s marked the highest point of the links between Ireland and Spain culminating with the battle of Kinsale, the rise of Louis XIV as the most powerful Catholic king and the progressive decline of the Spanish monarchy throughout the seventeenth century provoked a watershed in Irish alliances. Irish people loosened their allegiance to the Spanish crown and began to migrate to France, creating networks of friends and families all along the western borders. Tradesmen, sailors and priests found refuge and prosperity mainly in Brittany before spreading further afield. As the island found itself at the centre of high politics in the last decade of the seventeenth century, those links became even more important. By the 1690s, France had become the most influential state in mainland Europe and, being involved in what was to be called later the Nine Years War, the French government saw Ireland as a valuable vantage point from which they could easily divert their enemies' attention for minimal cost and maximum effect. Likewise, for James II, who had fled to France after the Glorious Revolution, Ireland was the perfect stepping stone towards a recovery of his English crown. Finally, the English, still apprehensive about their Irish neighbours after the rebellion of 1641, considered Ireland as the backdoor through which Catholic absolutism and brutality could reappear. War ensued, and the Irish partisans of James II, the Jacobites, were finally defeated and sent into exile to France after the signature of the treaty of Limerick in October 1691.

1 J. Beeverell, P. Vander Aa, *Les Délices de la Grand' Bretagne, & de l'Irlande* (La Haye, 1707, vii, 1441). Both the spelling and the grammar of this extract have been modernized for this article. '[...] the greatest part of the people is still the same: living without any

Being such a bone of contention for monarchs, Ireland often appears in official records, from secret correspondence between kings and ambassadors to state papers and national gazettes. But official archives in current-day Paris only give us a very general idea about the way those Irishmen and women were treated as they arrived in France, while nineteenth-century Irish authors like John Cornelius O'Callaghan or Thomas Davis brought us a very romantic view of those exiles. To better understand the daily lives of ordinary Irish people living in France at the time, it is necessary to delve deep into local archives in western cities, where clerics, magistrates and mayors tackled the sometimes unsavoury task of receiving those defeated allies.

One would probably expect Brest or Nantes to be most logical places to look for relevant evidence, but a town like Angers holds in its local archives some material elements. The city itself is quite far inland and isn't particularly famous for its military past in French history, unlike northern or eastern cities like Lille or Strasbourg. Yet Angers is a remarkable place, for it was one of the major stopping-off points for cavalry and infantry regiments enroute to and from Brittany and Paris. Troops came in and out through its gates on a regular basis during the period under study, offering us a unique insight into Irish military presence in France. Other sources, such as municipal decisions, birth certificates or hospital registers, can also be used to better understand the evolution of the relations between the local *bourgeoisie* and the Irish soldiers during the seventeenth and eighteenth centuries.

IRISH SOLDIERS IN ANGERS IN THE SEVENTEENTH CENTURY

Very early on, relations between 'Angevins', the local inhabitants, and Irish people were rather tense.[2] Irish people appear in Angers as early as the beginning of the seventeenth century, when the Flight of the Earls, one of the first massive exoduses of Irish people accompanying their nobles fleeing for the Continent, brought many refugees to the western shores of France. Some of them travelled as far as Angers and had to be expelled from the city by municipal officers specially appointed for that purpose.

As the century went on, Irish military presence in Europe grew more important. Most interestingly for us, recruitment of Irishmen into the ranks of the French army became more frequent when Spain's influence as a military power began to wane. Thus, in the 1650s, no less than four foreign regiments in the service of the French king were said to be *Hibernois de nation*, that is, of Irish nationality.[3] There are some traces of those Irish units in the

instruction in the countryside, having but a slight tincture of Christianity, they are ignorant of the basic elements of Christian charity. They are vindictive, merciless, and so inclined to banditry and robbery that they don't regard as infamous trades.' 2 Municipal Archives of Angers (MAA) Série BB 52, fol. 146. 3 Corvisier, André & Contamine Philippe (eds),

local archives of Angers. The first instance dates back to 1667,[4] when two regiments of Irish infantry passed through the town. The document, a municipal decision concerning their lodgement, helps us understand how billeting was organized in France and more specifically in a town which wasn't used to frequent military presence. The troops were to report their arrival to one municipal clerk, the *greffier*, who then determined the number of billets that were to be distributed among the troops for their stay, which normally only lasted for a night. The clerk's orders were to be followed to avoid any disorder with the civilian population, and, apparently, nothing of notice happened during that first stay.

Bearing this in mind, two other documents found in Angers' archives throw some new light on the way Irish soldiers and officers were considered by the local population. The first example, taken from the municipal archives, dates back to 1674 when one Irish regiment of no less than 1,200 men passed through the city for a day on its way to Nantes.[5] Angers' mayor had to gather his officers and aldermen for an 'special meeting' to notify his town that no exceptions to the general rule of billeting would be accepted unless the 'bourgeois' were to pay a predefined sum to the local magistrate. Since billeting prevailed at the time (barracks were very expensive and were only avaliable in the late 1670s in Paris for the Gardes Françaises), French local archives are always laden with rich citizens' pleas for exemptions on the grounds of nobility, personal influence or religion. Though we do not have explicit terms used in the text suggesting that Irish soldiers were particularly resented, the perspective of having several soldiers in their own home was obviously not a pleasing one for the Angers inhabitants, and it appears that Irish soldiers were even less welcomed as guests considering the rumours that spread in town and these resulted in the mayor's exceptional reaction. What is peculiar about this first encounter is the fact that the mayor insisted on relieving the poorest inhabitants in whose homes visiting troops were usually taken in, alongside local taverns.

The other example of the tensions that could arise between French magistrates and Irish soldiers can also be found in Angers departmental archives, but occurred in the nearby town of Saumur.[6] On 8 May 1676, a major, who claimed to be an Irish officer, burst into the local magistrate's house and demanded in his *mauvais françois* (note that the ending in 'ais' in modern French was written 'ois' in the seventeenth century) to be given *billets de logement*, the official documents that were issued for the officers' lodgings by the local authorities. The man himself was alone, riding in town ahead of his fellow countrymen in order to get the best house available. The magistrates explained 'with all civility' that the normal procedure for billeting wasn't

Histoire Militaire de la France (Paris, 1992), p. 394. 4 MAA., Série BB 89, fol. 15. 5 MAA., Série BB 94, fol. 47. 6 Archives of the Maine-et-Loire Department (AMLD).

being followed and that he would have to wait for the other soldiers to catch up with him so the French could organize their overnight stay more easily. But the Irish officer began to lose patience and ended up abusing the chief magistrate and his assistants. Soon enough, the senior officer of the regiment joined the already heated conversation and fuelled it with even more *blasphèmes et injures*, going as far as to threaten their lives. The local justices of the peace wrote an eight-page report about the whole affair and were adamant in their intention to send it to Louvois, France's minister of war. The Angevins conceded that it was always more difficult to understand what foreign troops or officers wanted, since they couldn't speak or understand French, and they clearly explained that they refrained from sending the two Irishmen to prison only because they a had 'a deep respect for his British Majesty', that is Charles II, the two officers' official master who was stopped to furnish troops to the French army after the treaty of Dover signed with France six years earlier.

Those two cases already give us an idea of the problems faced by Angers authorities when it came to the Irish. As France was becoming a highly centralized country, the military had to obey orders coming from above, even if it meant from a mere civil servant.[7] Such a situation was already difficult for French nobles, but was particularly harsh for foreigners probably more used to the rough ways of war than to French civility.

IRISH SOLDIERS IN ANGERS: 1690–1692

The real turning point for the relations between French people and Irish soldiers came with the end of the Jacobite war in Ireland. James II, the Stuart king of England, had been defeated initially at the battle of the Boyne in July 1690. His supporters went on fighting until they finally left Ireland after the signature of the treaty of Limerick at the end of the following year. The fate of many an Irish soldier was linked to James' court in France. A first wave of volunteers came to France to be organized as separate units in the service of Louis XIV when the war still seemed to favour the Jacobite cause. Those soldiers were said to be healthy, strong men according to Irish officials. O'Callaghan in his *History of the Irish Brigade in the service of France* reaffirmed this idea at the end of the nineteenth century,[8] something that was to be repeated by other authors until the late twentieth century. But the archives of Angers tell quite a different story. Thanks to hospital registers taken from the Hôtel-Dieu Saint Jean l'Evangéliste, one of the oldest hospitals

Série 2 B article 498. 7 To get more information about the way French noble officers had to accept discipline from outside their own circles, see Drévillon, Hervé, *l'Impôt du sang* (Paris, 2005). 8 John Cornelius O'Callaghan, *History of the Irish Brigades in the service of France* (Eastbourne, 2005, new ed.), p. 8: 'The Irish soldiers, on their landing in France,

in France and incidentally one of the best equipped at the time, we get a better idea of the health and lives of those who came as members of Lord Mountcashel's brigade.

The passage of these troops through Angers in the summer months of 1690 is of great value for us since we very rarely get first-hand accounts on Jacobite soldiers. The dates here are important since we know that after springtime the Irish units brought over to France were reorganized. If O'Callaghan explains that three regiments were created from the five original units, he doesn't give any reason for such a new state of affairs. It is no stretch of the imagination that the poor state of health of these 'well-made men' could well be the real cause of such a decision.

More often than not, books from the nineteenth century dwell on the officers' lineage and forget about the rank and file, most probably because their authors didn't have good material sources or ignored it. Angers' archives partially fill that very gap. The Hôtel-Dieu had a very well planned system of 'entrées et sorties de malades', that is to say, that diseased people had their names noted down with their birthplace, the date of their arrival to the Hôtel-Dieu and the date of departure, or sometimes of death. Those lists were arranged both alphabetically and in a chronological order, which helps us get a clearer picture of the presence of Irish people in Angers. Both soldiers and civilians were received in that hospital. One large building, which is still in existence, was separated into two rows of beds and was kept in good order by priests and nuns in charge of the patients coming from and around Angers.

While we do not have contemporary accounts of the troops' behaviour towards the local population, we know that the Hôtel-Dieu only had 100 beds or so to receive patients, and the figures show that some 70% of those beds were taken by Irish soldiers on their way to the northern frontiers of the realm (of course, we must bear in mind that beds were very usually used by several people at the same time; this allowed more patients to be accommodated, but it also meant that diseases were spread more rapidly). They were most probably still sick from the sea journey, or maybe they suffered from the so-called 'Irish plague' that had decimated troops in both camps during the fighting in Ireland the previous year. We know that some of them actually died in Angers – which proves that the troops brought over to France were far from being exceptional in their physical aptitudes as was claimed later on. Other entries in those registers did include soldiers from other units, either from French or foreign regiments; but even if some of them were sick, they very rarely died. The year 1690 stands alone as counting no less than 105 soldiers on its register, most of them admitted between May and June of that year.[9]

having been remarked as "tous gens bien faits", or "all well-made men".' 9 All the following data about Irish soldiers admitted to Angers' hospital in 1690 and 1692 are

When it comes to the Jacobite period itself, that is, after 1691, the same hospital registers also provide information about Irish people, mainly the presence of women and children alongside soldiers. What is striking about this second wave during the first months of 1692 is that, compared to the first one, the figures for sick soldiers and civilians are far less impressive. Though the roll calls prove that all sorts of people – wives, daughters, children, grand parents were in Angers – the Hôtel-Dieu wasn't invaded the way it had been a few months earlier.

The presence of women and children is clearly stated in those archives and contradicts the traditional idea that families had stayed behind. The entries included numerous mentions of children, some of whom were actually born in France, accompanying their parents during their journey. But most of all, other sources attest that those families were far from being well-off and welcomed in Angers. The mayor had to gather funds from his citizens to get rid of some Irish families, the number of which isn't specified but who were said to be 'more numerous everyday'.[10] Unlike what happened almost a century earlier, those Irishmen and women weren't expelled from the city; they were paid to be 'dispersed throughout the realm'. Being both allies and fellow Catholics, they couldn't be treated as enemies. We should bear in mind the dire situation of France at the time, the early 1690s being plagued by poor crops and dreadful weather. Irish refugees were just too numerous and had to compete with the local poor for relief, and obviously the mayor gave priority to his fellow citizens in need.

Another relevant document for our study is an account written by an alderman about the passage of James II through Angers in 1692.[11] His story gives us two major points: firstly, he alludes to the number of Irish people who fled to France to join their rightful king and secondly he underlines the fact that both James II and his Irish supporters believed their exile was to be short-lived, the soldiers being still considered as British subjects, obeying British rules and customs on French soil.

This comes as an illustration of what O'Callaghan explained in his book. When he quotes the different Irish regiments reorganized in Vannes as belonging to James II, he makes the point that Louis XIV recognized Irish refugees as his own subjects.[12] But Irish soldiers were in fact under British law and this was to have a major consequence during one particular incident that shocked the population of Angers.[13] In 1692, as James II and Louis XIV were organizing a second attempt to reinstall the Stuart King on his English throne, Irish troops numbering 8,000 men were mustered in Angers before leaving for Brittany. Out of those 8,000 men, 1,400 stayed within the walls of Angers itself.

available under file GG 338 at the Municipal archives of Angers. **10** MAA. BB 99 fol. 103, 31 January 1693. **11** MAA. class mark BB99 fos 119, 120 et 121. **12** O'Callaghan, p. 31. **13** MAA. BB 94 fol. 178.

Thanks to the diary of an Angevin lawyer,[14] we know that contrary to what usually happened in the seventeenth century, they didn't stay for just one day, but were garrisoned on the outskirts of the city for eighteen days. One can imagine the difficulties local authorities encountered when they were faced with such a huge number of men, since Angers was only used to accommodating a hundred militia men from time to time. One small village in the vicinity of Angers called Champtocé fell prey to one of these Irish regiments: the soldiers were so ill disciplined that the local magistrates went to see the mayor of Angers, who was also the king's lieutenant, to denounce one particular non-commissioned officer who had sexually assaulted a local girl. Had the culprit been a Frenchman, he would have been sentenced to death or sent to the galleys, but considering that he hadn't actually raped the girl, and particularly the defendant was Irish, that is a British subject, he was simply demoted in front of the regiment and condemned to have his halberd 'shattered on his head'. The local magistrates considered the sentence not to be severe enough, but the culprit simply was outside of their jurisdiction. We have only one case of such a sexual assault being committed by a Irish soldier. Despite the fact that there were over 8,000 troops gathered around Angers, the reputation of the Irish must have been tarnished in some way. Yet an absence of local newspapers or other memoirs written at the time makes this hypothesis difficult to corroborate. Nonetheless, the jurisprudence could have been disastrous for the French population nationwide, and this is probably the reason why it survived in local archives, which abound with reports of violence against or by soldiers.

Tracking the presence of Irish people in local archives, and especially that of Irish soldiers, can be quite puzzling, as one can easily imagine that French officials, be they clerks or priests, had huge difficulties when it came to noting down the names of Irish refugees in their registers. Contrary to what happened in Brittany during the same period, they could not resort to the Breton language as an intermediary, and so Angers gives us a pretty good idea of how Irish names were modified by French ears and quills. It also explains why it is sometimes difficult to find traces of those exiles.

For instance, the Hôtel-Dieu's registers already mentioned provide an impressive list of the Christian names and surnames given by these soldiers on their arrival. The results of our study allow us to complete Eoghan Ó hAnnrachàin's work in the military archives of the Hôtel Royal des Invalides,[15] corroborating his findings – for instance, about the overwhelming presence of Christianized names while Gaelic ones were by then the

14 *Mémoire de Me Estienne Toysonnier Avocat au Présidial d'Angers* (AMLD, class mark BIB 3368). 15 Eoghan Ó hAnnrachàin 'Irish veterans et the Hôtel Royal des Invalides (1692–1769)', *Irish Sword*, 21:83 (1998), 5–42, and 'Some early Wild Geese at the Invalides', *Irish Sword*, 22:89 (2001), 249–63.

exception. French officers, aldermen and priests constantly translated names – which explains why it is sometimes hard to discern an Irish person behind a Frenchified name. It appears that names like Thadée or Daniel or Denis were in 1690 far more popular than Guillaume (William) or Corneille (Cornelius).

Out of the 105 names that are to be found in those hospital archives for the year 1690 alone, one can just have an idea of the changes that were brought to Irish names by French people. Very common or famous names like Mac Carthy were transformed into Macarti, Makarty or Mackarty; soldiers called Brien became Brin, Brein or Brain. Fielding's regiment became Philding, Filding or even Filing. Names of officers or soldiers were no exceptions. Plunkets slowly mutated into Plonquet, Butlers were changed into Bouteler, and Murphys into Morphy. Only the expression *soldat irlandois* provides us with a certain degree of certainty when it comes to the origin of the names given. Yet even the term *irlandois*, which is mainly used to describe the place of birth of those soldiers, is not always the one appearing in the entries though listed on the same page. Words like *Hibernois* were also in use, and one case of *natif d'Hibernie* was found. Finally, there is an indiscriminate series of spellings *Yrlandois, Hirlandois* or even abbreviations, *Sol. Irl.* completing the sequence. According to the series of names one can distil from this evidence, we can assume that the vast majority of the men who were admitted at the Hôtel-Dieu were from Co. Cork, while Monaghan, Donegal, Wexford and Offaly were far less represented.

The registers also offer evidence of the strong, closely united bonds that linked men and officers in Irish regiments. One could consider Denis Boulder, 25-year-old soldier of the regiment and company of Captain Brian (O Brien) as a good example. Caytbery Berny's own case as a 's. [soldat] hibernois regiment Dilon [sic] comp. [company] Dilon' is another. Those two examples prove that members of the same family were both senior and junior officers in Irish regiments during that period, a recurrent feature in Irish units in French service until their official disbandment in 1792.

They also have details about their age, and more rarely, about their origins in Ireland. Most of the men were very young, the majority being in their early twenties, but the presence of more experienced or at least older soldiers should also be underlined in the first contingent of soldiers in 1690. The fact that a man's age is not given in most of the cases dating back to 1692 leaves us unable to determine any recurring pattern when it comes to the age of the Jacobite soldiers *per se*, but we may suppose that it would have amounted to the same results.

On the whole, the military presence of Irish soldiers in Angers petered out at the end of the seventeenth century. Only some examples of Irish soldiers seem to appear here and there, and they rarely are linked to Irish regiments after 1697, simply because those original units had been disbanded at the end of the war (many soldiers being urged to find another nation ready to employ them, or else turning to robbery). Some of them did manage to become soldiers in other non-Irish regiments after the signature of the peace of Ryswick, as the registers show – which proves that a certain form of integration was possible in late seventeenth-century France.

IRISH SOLDIERS IN EIGHTEENTH-CENTURY ANGERS

Thus, the Hôtel-Dieu registers provide evidence of Irish integration into French society through baptisms and conversions at the very beginning of the eighteenth century. Most obviously, the only persons who could help the exiles in their new lives were priests. Early in the century, Irish students were sent to different colleges on the Continent, mostly to France and Spain, to become priests. Since they mastered both Gaelic and French, alongside Latin, they were surely the ones who translated names and facilitated hospital admissions for their countrymen. The archives furnish different examples throughout the seventeenth and eighteenth centuries of Irish priests who were present at the Hôtel-Dieu and who would sign official documents in favour of Irish soldiers.

The number of Irish people who actually lived in Angers was probably too small for one to be able to talk about an Irish community as such, but there are traces of Irishmen and women, connected with the military, who lived in the very last years of the seventeenth century in or around Angers. The example of the MacLoughlin and Burgo families is interesting in so far as Jean (most probably John) MacLoughlin, clearly defined as an Irish soldier (*cy-devant soldat hirlandois*), was married to Mary Higgins and had two children in 1698 and 1700. The Burgo (or Bourgo as once again the spelling changed over time) family provided godparents for the children – with Thomas and Elionore Burgo, two siblings who were themselves the children of an Irish captain, signing the parish record.[16] The presence of French people at the first baptism in that case the daughter of a French officer could lead us to reconsider our perceived ideas about Irish people staying in a closely knit community away from French society.

One can find another example of a baptism showing Franco-Irish links in the Hôtel-Dieu archives, in which a certain Guillaume Louis Butler, born on 12 May 1693, the son of Pierre Butler (also a soldier serving 'his Majesty' though the text does not say whether it was James II or Louis XIV) and of Marguerite Fithgerald [*sic*], had for godparents a man called Louis Runier and a woman called Françoise Hureau.[17] Does that imply that integration was

16 Municipal Library of Angers. *Registre paroissial de St Jacques à Angers* n° XI (tome 2) class mark C91838. 17 MAA class mark GG 226 (5 Mi 61) baptisms for 13 May 1693.

possible at the time? We cannot ignore this particular example, but we must remember that the death toll in childbirth was extremely high during that period and that baptism was all the more important. Maybe those two French persons just happened to be there, but of course we don't really know.

Since in France charity was offered only to soldiers who professed the Catholic faith, the links between Irish soldiers and priests was all the more important. The Angevin archives provide an example in this connection, with the life of Geoffrey O'Connell summarily described in a short text written by the priest officiating at Geoffrey's abjuration of his Protestant faith[18]. Aged 36 and born in Ballydeloughy, Co. Cork, he accepted the Catholic faith in August 1716 and was given the absolution by several priests, one of which was called Macarthy. It is probably thanks to this cleric that we get so many details about O'Connell's life, down to his parents' names and address in Ireland. Incidentally, this document provides evidence that Irish soldiers at the time did not always have a Catholic background.

Encounters between French and Irish people occurred later in the century, as illustrated by the presence of Irishmen held prisoners in Angers in 1779, when France decided to help the American insurgents. English prisoners, some of them of Irish origin, were sent to Angers and held in the chateau that still dominates the city today. During their forced stay in Angers, these prisoners tried to escape on 16 December 1779, troubling the public peace.[19] Some of those prisoners actually died during that winter, and the parish records give some interesting ideas about the way French people conceived Irishness, since one of them, George Flower, is said to be 'irlandois de nation' but was born, according to the French priest, in 'Bristol, Ireland.'[20] French people at the time often had confused ideas about the geography of the British Isles.

According to what the town hall proceedings claim, it seems that an Irishman named O'Sullivan[21] sounded the alarm. It is quite difficult to know exactly who this man was, but one can find traces of a citizen called Louis Jacques O'Sullivan. Married to a French woman, Félicité Cesbron, and with a daughter named Charlotte, he embodied the second or third generation of Irish people who were by then no more limited to their own exiled community when it came to family affairs and could also seek openings in society other than serving as soldiers or officers.

By 1779, such an Irish presence was Irish only in name. But Irishmen came to Angers directly from the British Isles, and, contrary to what we may think, Angers was probably a rather familiar placename in France for some Irish gentlemen of the period. Many of them, wanting to enhance their skills as cavalry officers, looked forward to studying at Angers' *académie d'équitation,* a

18 MAA class mark 5 Mi 72. **19** MAA class mark BB 129 fol. 8 end of December 1779. **20** 6E7/22 microfilm 5 Mi 0071 (A.M.L.D.). **21** Ibid., fol. 4, 3 December 1779.

riding school that had been founded at the turn of the seventeenth century to offer nobles from all over Europe a proper education in the arts of war. The academy was to attract quite a lot of students during the eighteenth century, mainly boarders who came from as far as North America or Denmark. The school really became successful thanks to the Avril de Pignerolles family who ran it from 1679 to the very eve of the French Revolution. From 1755 to 1790, no less than 449 students attended lessons in riding, fencing and military tactics.[22] Interestingly enough, out of those 449 pupils, 261 were foreigners. Irish youngsters, numbering 51 individuals, actually came second in the ranks after their English counterparts. Incidentally, the most famous Irish gentleman to attend military lessons in the mid-1780s here was Arthur Wellesley, the future duke of Wellington.

Alongside local archives relating to Irish personal lives in eighteenth-century Angers, contemporary French institutions – especially local military archives – also furnish their own crop of dates that should be considered.

The town hall holds military archives dating back to 1727,[23] the year in which the *étapes* were re-established after a few years' elapse. The *registres d'étapes* or registers of journeys are invaluable in so far as they provide proof of what recruitment for French Irish regiment actually looked like. The passage of those recruits through Angers corresponds to the evolution of the military relations between Ireland and France in the eighteenth century. If one looks at the official reports of the passage of troops in Angers from 1728 to 1747, three distinct periods emerge. The first one, in 1728, saw no less than 53 recruits, both cavalry and infantry, registered in Angers, and was probably the result of the political détente that marked Franco-English relations at the end of the 1720s. The second period corresponds to the years 1734–6 when Irish regiments were heavily involved in the War of the Polish Succession and badly needed recruits to fill the gaps in their ranks. Those recruits belonged to three different infantry regiments in the service of France – Dillon, Clare and Rothe. The only Irish cavalry regiment left at the time, Fitzjames, also had its recruits marching through Angers. The last period (1741–5) coincides with Ireland's last great contribution to the French army as well as to the Jacobite cause. Fontenoy on 11 May 1745 coincided with the high point of this recruiting trend.

The mayor, as a local officer of the French king, was in charge of organizing these men's stay in his city, providing food and shelter for the night. He was also in charge of counting the number of recruits or soldiers passing through the gates, and his official report stated both the date of arrival and of departure of the men and also their destination, the 'route' they had to

22 Uzureau, Chanoine, 'l'Académie d'équitation d'Angers', Annales Historiques, 24 (1924), 198–204. 23 These archives can be found under the code EE 10 to 17 for the years 1728–65 (MAA). 24 MAA, class mark EE 11, p. 112, 115 et 116 (14 May 1744, 14

follow, thus leaving us some valuable documents about the lives of Irish soldiers in eighteenth-century France.

The circuits that Irish recruits had to follow was always the same. They were marched from Brittany (either starting from Rennes or Nantes) to join their nation's units up north, in towns like Lille, Saint Omer, Amiens, or Béthune. This confirms the idea that, up until 1745, most if not all the recruits of Irish units were actually coming from Ireland, a tendency that was to disappear away as the Jacobite cause faded and English coercion got more prominent. The men generally had to follow one or two officers and a few non-commissioned officers. Their numbers varied from 12 to 47. The local archives are also proof that these trips from the western to the northern borders were organized by seasoned people who knew both Ireland and France. The name of one particular sergeant, a man called Heynes,[24] who worked as a recruiter for the infantry regiment of Clare, appears several times during the first half of the eighteenth century. He was able to sign his own name at the bottom of the document.

The mayor's report gives us another element about Irish recruits compared to their French counterparts. The Compagnie des Indes Orientales also had recruits coming through Angers en route to Breton harbours, notably Port-Royal and Lorient, to embark for India. If we compare these two kinds of unit, the French were far more prone to desertion than the Irish. Only one example of Irish soldiers deserting can be found in the municipal archives. On 25 February 1735[25] one lieutenant and a sergeant from the Regiment of Clare arrived in Angers with ten men, but left the town the day after with only four of them; these men were not sick, since the mayor would have mentioned such an important detail in his report. Irish soldiers were probably less tempted by desertion since they were in a foreign country where no one spoke their language (unlike Germans in the east or Italians in the south-east who could easily mingle with the local population).

Another document, though not directly linked with Angers since it is a copy of the names of soldiers left behind in Nantes hospital as their regiment stayed for a few days in Anjou in 1747, reveals that after 1745 Irish regiments were so desperately in need of men that they recruited from either the Low countries or Germany and gradually lost their national character.[26] Only the names of the officers were still Irish, but most of them by that time were only descendants of Jacobite refugees.

The archives illustrate in details the organization of Irish regiments in the eighteenth century, offering us two examples of Irish units as they were mustered in Angers on their way to either Brittany or Flanders. The cavalry

January 1745). **25** MAA, EE 10, p. 156. **26** MAA, EE 10, p. 176, *Régiment de Rothe Irlandois*, 6 March 1747.

regiment of Fitzjames is the first one being mentioned in the *registres d'étapes* in 1739. It numbered some twelve companies totalling 300 horsemen and some thirty junior officers. A few years later, in 1744, when the regiment came back to Angers, the same report was made, but with a slight change:

> Arrivé à Angers le onze mai mille sept cent quarante quatre pour en partir le treize du dit mois le régiment de cavalerie de Fitzjames composé d'un lieutenant colonel, un major, un aide-major, un aumônier, un chirurgien major, onze lieutenants, cinq cornettes, douze maréchaux des logis, trois cent soixante trois cavaliers et quatre cent un chevaux.[27]

The absence of captains is interesting, though it is not known whether this is a mistake in the aldermen's calculations or if they were casualties from a previous battle.

The other example given is the Regiment of Rothe precisely described as it was reviewed by the mayor in 1747. It had 13 companies and numbered exactly 317 men (figures which correspond to the general rule of 25 men per company). The senior officer mentioned in the mayor's report was as in the case of Fitzjames a lieutenant-colonel – a clear sign that the men who gave their names to Irish regiments were not necessarily in charge and that most of the time the units were under the command of another senior officer, himself Irish or of Irish origin.

Angers' local archives even offer us the possibility of imagining what these men looked like. Since some soldiers were admitted in hospitals in the eighteenth century, documents detailing their physical appearance had to be completed in order to avoid desertion. This is why we get some examples of convalescence certificates throughout the *registres d'étapes*:

> Certificat de convalescence
>
> Nous soussigné certifions à tous ceux qu'il appartiendra que le nommé Jean Roc cavalier de la compagnie de Mylord Tyrconnell au régiment de Fitzjames natif de Bellefast [*sic*] en Irlande en la province d'Ultonie [*sic*] généralité de Belefast [*sic*] âgé de vingt quatre ans de la taille de cinq pieds quatre pouces six lignes cheveux bruns les yeux gris visage [illegible] a été à l'hôpital de Nantes le vingt neuf de septembre mille sept cent trente neuf et que [l'étape ?] et le logement [pourront ?] lui être fournis conformément a l'ordonnance du roy 13 juillet 1727 fait a

27 MAA EE 11, 11 May 1744. 'Arrived in Angers on the 11th of May 1744 to leave on the 13th of the said month the cavalry regiment of Fitzjames composed of one lieutenant-colonel, one major, one deputy major, one chaplain, one surgeon, eleven lieutenants, five ensigns, twelve sergeant majors, three hundred and sixty three horsemen and four hundred

> Nantes le vingt neuvième jour du mois de septembre mil sept cent trente neuf signé L. Callaghan lieutenant commandant ladite compagnie.[28]

The document once again confirms the fact that French official systematically translated foreign names since that particular cavalryman signed as 'John Roc'. Another interesting feature is the way Ulster is renamed in the Spanish fashion, *Ultonia* being the name of one of Spain's Irish regiments at the time. A different case underlines a recurring fact about eighteenth-century soldiers, with another soldier from the same regiment having traces of small-pox on his face.

But Angers was not just a temporary stopping-off point for regiments as they journeyed towards Flanders or to face English threats along the coasts of Brittany. The Hôtel Royal des Invalides could also send old soldiers to Nantes which, at the time, was one of the cities where wounded soldiers guarded the town, and some Irishmen did exactly that.[29] The 'companies d'invalides' were in charge of the fortifications, a far less demanding task compared to a soldier's life in the field.

After 1745, the flow of Irish soldiers passing through Angers dwindled and finally disappeared. The last example of recruits marching north dates back to 1763.[30] The only Irish soldiers evoked after that date were no longer in the service of France, but of the British crown. The Revolution with its defiance towards foreign troops in the service of the king and the crisis of 1793 that emphasized the importance of a national, that is, entirely French, army brought the Irish military tradition to an end but certainly didn't sound the knell of Irish military presence in France.

Even though Angers at first may appear to be just a stopping-place for Irish refugees, local archives reveal quite a different picture. Even if the presence of Irish people there cannot be compared to that prevailing in towns like Nantes, Rennes or Bordeaux, there being no Irish community as such, it is still possible to find traces of their passage and their lives in Anjou. The way they managed, even on a very small scale, to organize supportive networks not only within the military community but also between Irish people at large is a testimony to their resourcefulness. This of course implies that Irish integration into French society was more complex than expected. Maybe we

and one horses.' **28** MAA EE 10, autumn 1739. The French text has been modernized. 'Medical certificate of recovery, we, undersigned, declare to whoever it may concern that the named Jean Roc horseman in Mylord Tyrconnell's company of the Fitzjames [regiment of] cavalry born in Belfast, Ireland, in the province of Ulster, district of Belfast, aged 24, height 5 feet 4 inches and 6 fractions, brown hair, grey eyes, face [illegible] was at Nantes hospital on the 29th of September 1739 and that he was given board and lodging as prescribed by the king's edict of July 13th 1727. Made at Nantes, on the 29th day of September, 1739, signed L. Callaghan lieutenant in charge of the said company.' **29** MAA EE 10, 7 September 1739. **30** MAA EE 12, July 1763.

are not faced with isolated pockets of Irishness that gradually opened up to French society as a whole, though this was true when it came to nobles and trades people, but with a lot more flexible population which managed to survive through hardships and defiance. It also reveals the way French people conceived Ireland, with a mixture of sympathy and dread. It is hard to tell whether the collection of documents one can consult in local Angevin archives, be they municipal or departmental, are a mere fragment of what can be found there or whether they represent the whole record. This essay must be viewed as an incentive to look for Irish people where they were not supposed to be. Even if the findings in Angers do not really re-define what we know about the subject, they offer a perspective which is very rarely seen. Information about the lives and actions of Irish soldiers in French exile often comes from the authorities in Versailles, but it was the local administrations that had to deal on a day-to-day basis with the thousands of refugees the treaty of Limerick brought to western France.

The life of Thomas Lally

LAVINIA GREACEN

Military historians and biographers do not always see eye to eye. The meticulous analysis of individual battles may clash with the emphasis on human interest and dramatic potential which is the biographer's approach. Yet the life and, ultimately, the tragic fate of Thomas Arthur Lally has room for both. Each speciality illuminates the other.

At first glance, Lally was a consummate tragic hero who suffered an unjust death at the hands of corrupt accusers and a weak king: on that, both military historian and biographer agree. But the broad sweep of his life reveals a man with a complex personality which would help to shape historical events – a brilliant soldier of great intellectual talent, but a man with the vulnerable emotions of a contemporary outsider. It is his character, as well as the impressive number of his great military contributions, that reaches forward across the 250 years since his death, and speaks directly to us today. Lally was a child of his time, of course, but in many ways he was a twenty-first-century man, and for that he was to pay a high eighteenth-century price.

Despite a successful campaign by Voltaire which finally won Lally a posthumous re-opening of his case in 1778, negativity still surrounds him. But it was the dramatic potential of his life that inspired Robert Louis Stevenson's novel *The Master of Ballantrae*, as well as Stevenson's choice of compassionate adjective when he came to describe him. 'It began to occur to me', Stevenson once wrote,[1] explaining the genesis of his book, 'it would be like my Master to curry favour with the Prince's Irishmen, and that an Irish refugee would have a particular reason to find himself in India with his countryman, the unfortunate Lally. Irish, therefore, I decided he should be.' The unfortunate Lally … his own true story is as strange as any fiction.

Thomas Lally, who is disguised so often as a Frenchman, masked by his title Comte de Lally-Tollendal, was born in 1702. He was the only child of Gerard Lally, an Irishman who had fled to France with his cousin Arthur Dillon after the rout of the Stuarts, and who had continued the fight against William of Orange's successor, Queen Anne, and the English by joining his cousin's regiment in the Irish brigade in the service of France.

The years between Gerard Lally's arrival in France and the birth of his son had swung from carnage to tedium and back again. His older brother James

1 R.L. Stevenson, *The art of writing and other essays* (London, 1905).

had died in battle within a year, and his own luck had almost run out in 1694, when in the acting rank of lieutenant-colonel he had been badly wounded while leading his men in a futile attempt to break the siege of the Spanish city of Gerona. By 1697, 48 officers and 520 men from the original 800 had been killed or wounded, but the Irish Brigade in the service of France fought on. Soldiering was addictive, with the adrenalin rush of the attack, the camaraderie, and the fatalism that developed from the randomness of death. During a night attack at the siege of Barcelona that year, for instance, 15 officers and 100 men of the Regiment of Dillon were lost, one of whom was Gerard's younger brother, William. 'To the honour of Mr Lally, seeing the disorder he rallied [those] retiring and halted the onward rush of the enemy,' wrote a regimental observer. 'Mr Dillon would have perished, but for an officer who killed a German whose sabre was raised to split his head.'[2]

The Irish troops were described by one of their commanders, Marshal de Vendôme, as 'the butchers of the army',[3] and they took it as a compliment, as was intended. Each year brought a fresh campaign in extremes of weather, frequently about territory that was all too familiar because it had been fought over before. In the spring, summer and early autumn they marched to take on the enemy; when temperatures fell, they marched off to winter quarters instead; these were claustrophobic and unhealthy barracks that for the uncommissioned became a dismal home from home.

Thomas, the son whose presence brought Gerard a happier purpose, had been conceived in a year of international upheaval. In 1701, while the regiment was wintering in Dauphiné, a threat was building up of a realignment of power caused by the death of Charles II of Spain, whose successor, Philip V, was a grandson of Louis XIV. The prospect of such rapid French expansion created a Grand Alliance of England, Austria, Bavaria and Holland, and led to a war that would last for eleven years. That same year James Stuart died at Saint-Germain-en-Laye, outside Paris, the royal château lent to him by Louis XIV who backed his claim to the English throne. Jacobite hopes fastened onto his son, now referred to by them as James III.

Meanwhile, in the south-eastern town of Romans-sur-Isène, in the mountainous region of Dauphiné, Gerard Lally had met Anne Marie de Bressac, the comfortably-off widow of Philipe de Faventine du Vivier, and allegedly they were married on 18 April, a convenient nine months before their son was born. The Dillon family, who had an apartment far away at Saint-Germain-en-Laye, would have their doubts about that ceremony. 'This first bastard distinguished himself in the wars of James II,' a descendant would relate a century later, giving public vent to entrenched family opinion.

2 Chevalier Gayden, *Memoir of the Regiment of Dillon* (1738), translated and edited by Professor Liam O Briain, and published in the *Irish Sword*. 3 Thomas Bartlett and Keith Jeffery (eds), *A military history of Ireland* (Cambridge, 1996).

'Although he never married, he [too] left a natural son …'[4], and the story would be picked up by historians. 'Like his father, Gerard Lally liked to relax with women,' speculated a twentieth-century version.[5] 'That is how he had one day seduced Anne Marie […] who was burning with the fires of a still hot and ardent setting sun. He rapidly gave her a son whom he legitimized later through marriage.' Anne Marie's early death may have been hastened by the birth.

Thomas Lally's childhood, in consequence, was heavily influenced by his father, who cut an authoritative and vividly masculine figure as a high-ranking officer in his cousin Arthur Dillon's regiment, the Regiment of Dillon. Frequent absences in annual campaigns in the War of Spanish Succession emphasized the impact of his all-too-brief presence. As a boy, Thomas absorbed his father's Jacobite loyalties, his family pride, his tales of a distant home, and his longing for Ireland. It was the increasingly nostalgic view of an exile, shot through with the idealistic goal of restoring the Stuarts to the throne – a gallant dream, tailor-made to appeal to an imaginative child.

Gradually – partly history lesson, partly confided personal experience – Thomas was admitted to his father's world. He was taught that throughout the civil wars of the seventeenth century, despite confiscation of their Galway castle of Tollendal and the risk of exile, the Catholic Lallys had fought for the Stuart crown. Under Charles II's Catholic brother, James II, there had been a brief change of fortune when his oldest uncle, the ill-fated James, who was to be killed in battle within a year of reaching France, had been elected the member of the Irish parliament for Tuam. But, as Thomas' father explained, a regime that advanced Catholics was a threat to Protestant power, and resulted in an unconstitutional invitation to James II's Protestant son-in-law, William of Orange, to become king. The journey from the west of Ireland to the remote French province of Dauphiné was portrayed as a logical progression, driven by the two qualities Thomas was encouraged to hold most dear: loyalty and honour.

At the age of seven, commissioned as a half-pay captain, he experienced his first siege by his father's side at Gerona, a decision taken in the belief that it was high time that he 'smelled powder'.[6] The child expected soldiers to be tall, but they appeared colossal, with grenadiers and pikemen looming over the musketeers. Drums beat, shouted commands were in Irish, few men washed, and the stained, sun-bleached uniforms stank of stale sweat. Under fire, the tumult was deafening, its impact heightened by the acrid smell of cannon-fire that burned Thomas' throat and made his eyes sting. But he was determined to live up to his father's expectations. Overcoming the desire to

4 Felice Harcourt (trans. and ed.), *Memoirs of Madame de la Tour du Pin* (London, 1985). 5 Pierre Antoine Perrod, *L'Affaire Lally-Tolendal – Le Journal d'un Juge* (Paris, 1976). 6 John C. O'Callaghan, *History of the Irish brigades in the service of France* (London, 1870).

choke, he inhaled deeply, and disguised his revulsion at the sight of blood and the cries of wounded men. Afterwards, he was judged tough enough to join the regiment for a longer period and in more terrifying circumstances; before he was twelve he was mounting his first guard in the trenches at the siege of Barcelona, where many years earlier, during a previous encounter, one of his uncles had been killed. By his teens, on his own admission, Thomas's sole ambition was to join the army and win a marshal's baton.

Indoctrination to war was balanced by a privileged classical education, and in 1714 he was sent away to college. Physically well-built and athletic, he also excelled at the useful military subjects of mathematics, geography and languages, but his quick wit, unusually, was literary, and his favourite subjects were history and the Classics. One of his favourite books was *The Courtier* [*L'Homme de Cour*] by Balthazar Gracian, and he paid keen attention to the Spanish Jesuit's maxims for achieving personal and professional success, consciously preparing himself for opportunities ahead. The emphasis on honour and idealism, which underlay Gracian's advice for impressing superiors, confounding rivals and making the best use of subordinates, appealed to him, unlike the more popular cunning of Machiavelli's *The Prince*. At Court, where he gained an introduction through his late mother's contacts, he was able to put Gracian's advice into practice; soon the powerful duke of Orleans, the regent, became his patron.

At the age of nineteen he was not the only one who saw himself at the head of a company in the Irish Brigade. He was widely considered to be eminently suitable, possessing unquestionable courage, an excellent understanding of the arts of war, and already showing signs of that martial air said to have great influence in a commander. Eminently suitable, he was soon to discover, by everyone except his own father, who without explanation blocked him from joining the Regiment of Dillon. In public, Lally had to demonstrate filial obedience, but suppressing his strong feelings would impose a long-term price. At the age of sixty, when adolescent resentment, usually, has long mellowed, he would still blame his father bitterly for holding him back. His behaviour at the time, however, during the lengthy peace between France and England which marked those years, demonstrated the wisdom of the decision.

With the short-cut of royal favour ruled out, the sole option for an outsider like Lally was to harness his aptitude for concentration and analysis, which he had honed at school. Military skills could be studied in books. Battles and the reasons for their outcome, he soon reasoned, were essentially no harder than history lessons, and the more practical arts – siegecraft, the setting up of camps and the movement of armies across unfamiliar ground – were not unlike geography. His short periods of earlier experience proved invaluable in shedding light on complicated descriptions, and enabled him to comprehend cause and effect. For several years the enforced inactivity was put to good account as he studied textbooks of past battles without interruption, analysing why and how victory or defeat had come about. It was a self-taught schedule of military technique that can be likened to Higher Command and Staff College training today.

In 1728, Lally's patience paid off. With his father's agreement he was appointed to the rank of full captain in the Regiment of Dillon on 15 February that year, and by the time the War of Polish Succession broke out five years later he had been promoted to the future high flyer's rank of regimental adjutant-general (*aide major*). His knowledge was to prove crucial at the siege of Kehl in 1733. As his extensive reading had taught him, siege warfare against a resolute governor and well-stocked garrison had a strict international choreography. A half-century earlier, the French engineer Vauban had brought fortress-building to a fine art by designing a classic star-shaped pattern which had since been adopted throughout Europe. Low outer defences of earth, revetted with brick or stone, fronted deep ditches which could be flooded. Behind these loomed the main defences, a 'curtain' wall with arrow-headed bastions jutting out that held guns to provide extra cover by flanking fire, and the reinforcement of further works – ravelins or *demi-lunes*. From the outside all that could be seen was a gentle slope, the *glacis*, falling away into a shallow walkway, the covered way, which in turn was protected by sharpened stakes called the *palisade*.

Theory and diagrams had shown Lally that, if there was sufficient time, attackers could dig a tunnel underneath, culminating in a chamber packed with explosives which could be blown up. But more often a physical assault would have to be launched to gain possession of the covered way, which meant rushing it, sword in hand, after lobbying grenades over the *palisade*. Once the covered way was secured, and the guns beyond it silenced, breaching batteries of heavy guns could be dragged up to pound the base of the curtain wall until a long groove was opened, the *cannelure*. Then the entire rampart would slide into the ditch, leaving an indefensible breach, and at that point the governor would have to surrender or accept that no quarter would be given.

Under Lally's instructions, engineers laboured day and night to open a line of trenches parallel with the fort, just out of range of the defender's guns. They then 'sapped', driving zig-zag trenches forward to open a second parallel wide enough to take batteries of guns and four battalions of men armed with mortars; this provided cover for the digging of a third, nearer the bastion under attack. Continuous fire was kept up to undermine the enemy's defence. By the second day, buildings in the town had been reduced to rubble, and within hours the covered way had been stormed without loss of life, aided by the blaze of a large store of wood inside the fort, on fire from a direct hit. Berwick, the French commander, was heard to remark to the Prince de Conti, who accompanied him, that the Irish Brigade might be slaughtered, but they would never be beaten.

The siege provided Lally, additionally, with a glimpse of the psychological aspect omitted from textbooks. Although regimental casualties were only sixty killed and wounded, compared to the 150 who were about to be lost from illness on a march to Cambrai, it was apparent that the drawn-out stress of siegework was liable to prove greater than confronting the enemy in the crimson heat of battle; men trained for action were the least fitted of all to endure life in slow motion, pinned down under fire. His authority was bolstered by his own aptitude for the cut and thrust. During fighting at Ettlingen in 1734, it was only his quick reactions in the smoke and chaos that saved his father's life.

'Grievously wounded, [Gerard Lally] was upon the point of falling into the enemy's' hands,' a regimental account would relate,[7] 'when his son threw himself between them and his father, covered him with his own body, and, by prodigies of valour, succeeded in disengaging him; thus preserving at once the life and the liberty of the author of his existence!' A shift in the generational balance of power was evident as soon as his father recovered: from then on, proudly but sometimes, surely, with tongue in cheek, he referred to his son as 'My protector'.[8] His death in 1737 left Lally in emotional limbo between Ireland and France.

The short War of Polish Succession laid the groundwork for Lally's steadily growing reputation, although, to his frustration, further promotion was withheld. Kehl and the siege of Philipsbourg – '[Lally was] as much distinguished there by his brilliant valour as by his uncommon military knowledge,'[9] – had brought France significant strategical gains, and the famous soldiers who were coming to rely on his advice at moments of crisis included the marshals Belleisle and de Saxe, as well as Berwick. In 1744 Marshal de Noailles insisted upon having him as his chief of staff for the entire campaign. That year England and France were again at war, and George II's ruthless son, William Augustus, duke of Cumberland, arrived in Flanders to lead the Allies. The 120,000 strong French army, in which the Irish Brigade was attached as a single unit, was led by Saxe, another of Lally's patrons, and the fighting swept as far afield as Alsace, after the capture of Menin, Ypres and Furnes. On 1 October 1744 Lally's expertise was rewarded. He was commissioned to head a new Irish infantry regiment bearing his own name, the Regiment of Lally, and promoted to colonel. Short of his ultimate ambition of a marshal's baton – but indicating that in time such an honour might well be achieved – it was a very public accolade.

By the spring of 1745 Lally was satisfied with recruitment and training, and the timing was to be propitious. The Regiment of Lally's first deployment was at Tournai, where it won praise, and in May that year, under the overall

7 *Biographie Universelle*, tome 23 (Paris, 1819). 8 Ibid. 9 Ibid.

command of Marshal de Saxe, it approached the critical battlefield of Fontenoy, eighty kilometres due west of Brussels. At the head of the Anglo-Dutch, who faced them, was Cumberland, and intelligence revealed that his army was greater, but had less artillery. Lally, conscious of his father's lifelong cause, relished the prospect of directly combating Hanoverian power.

De Saxe, too bloated with dropsy to ride and with the added responsibility of the presence of Louis XV and the dauphin, had chosen his ground along an advantageous slope, with fortifications dug for defence. Over 50,000 men and 100 cannon guarded the position, which ran from the river Scheldt at the town of Antoing, on the right of the French army, due north-east for three-quarters of a mile to Fontenoy. There the line turned a right-angle, towards the dense wood of Barri. Both Antoing and Fontenoy were faced with trenches and linked by three redoubts; two more protected the ground from Fontenoy to Barri wood, with a field in between left unfortified because two redoubts already covered the flanks. The six regiments of the Irish Brigade took up position in reserve at the rear of Barri wood.

On 10 May, on the eve of battle, Lally reverted to self-reliance. 'He wanted to reconnoitre the battlefield for himself,' ran one account,[10] '[and] discovered a lane from Antoing to Fontenoy which had been wrongly judged impassable, and by means of which the French army would for certain be circumvented.' Going straight to de Saxe, he reported that a strong enemy force there would be able to turn the French position, and orders were issued for the lane to be secured with a further three redoubts and sixteen cannon. Lally rode back to wait until dawn with the Irish Brigade in the rear of the wood. Drawn up alongside the Regiment of Lally, 'in a little plain of 100 paces',[11] were the infantry regiments of Dillon, Clare, Bulkeley, Rothe and Berwick, and the cavalry regiment of Fitzjames.

On 11 May, while it was still dark, de Saxe awoke; with his nervous companion, the Marquis d'Argenson, the foreign secretary, he waited in silence for the break of day. At 4 am, attended by his aides and principal officers, he went to visit all the posts. 'Gentlemen,' de Saxe told them, 'there will be occasion for your lives today.'[12] At dawn he inspected the artillery: a hundred cannon ought to have been present and only sixty could be counted. The rest were hastily brought up, but in 'the tumult and hurry unavoidable on such an occasion' the cannonballs were forgotten. Eighteen thousand of the army were still pinned down investing Tournai. Cumberland had 21,000 British, 22,000 Dutch, 8,000 Hanoverians and 2,000 Austrians; with last-minute reinforcements, the number was estimated to be '56,000 of the finest troops in Europe and all their corps complete'.[13] The total French force at Fontenoy would subsequently be put at 52,000.

10 Ibid. 11 *The age of Louis XV by Voltaire* (translation of *Précis de Siècle de Louis* XV) MXCCLVIX. 12 Ibid. 13 O'Callaghan, op. cit.

At 5 am the battle began. There was a confusing similarity of red uniforms on both sides and the fighting was brutal, with little medical aid in prospect for the wounded of either side. Quickly the Allies gained the upper hand. 'The English made a running fire,' opened a French account.[14] 'The line of French infantry did not fire […] their eyes must have been surprised at the depth of the English corps and their ears stunned […]'. The first rank was swept away, leaving many killed and wounded and others disorientated and fleeing; so many officers had become casualties that there was little leadership to re-motivate them. 'The English […] advanced gradually as if they were performing their exercise; one might see the majors holding their canes upon the soldiers' muskets, to make them fire low and straight. Thus the English pierced beyond Fontenoy and the redoubt'.[15] By 7 am, the English 'encompassed the whole ground of the village of Fontenoy' and were attacking on every side. 'The animosity, the blows, the cries, the reciprocal menaces of above 100,000 combatants armed for mutual destruction […] the flashes and reports of 100,000 muskets and of 200 pieces of cannon, the terrible thunder of which was 1000 and 1000 times reverberated along the [rivers] Escaut and Scheldt as well as by all the forests about it […]'.[16]

Now the forward thrust of the massed Anglo-Dutch column appeared unbreakable. 'Infantry battalion after battalion […] and cavalry squadron after squadron,' one witness described the slaughter, 'gave way to that moving citadel of gallant men'.[17] Louis XV's once-proud army had become scarcely recognizable, 'entrenched behind the heaps of their comrades who were either killed or wounded',[18] and Marquis D'Argenson, unused to the saddle and a civilian 'looker on', as he would admit, felt giddy and sick. 'There was one dreadful hour,' he wrote afterwards, 'in which we expected nothing less than a renewal of […] Dettingen; our Frenchmen being awed by the steadiness of the English, and by their rolling fire, which is really infernal […]. Then it was that we began to despair of our cause.'[19]

The Irish Brigade, impatiently waiting for the call, were forced to watch and curse. Lally was mounted at the head of his regiment, concentrating upon move and counter-move. He had noticed that the Allies were employing cannon and musketry against French musketry alone, and as soon as he remembered four cannon lying idle in reserve, the idea of using close-range artillery on the approaching column became feasible. Spotting the king's *aide de camp* and acting lieutenant-general, the Duc de Richelieu, in the distance, he spurred his horse and galloped up. In a rapid conversation he suggested a combined manoeuvre, and with Richelieu's support and de Saxe's approval the cannon were brought up and trained on the front of the oncoming column,

14 *The Age of Louis XV by Voltaire*. 15 Ibid. 16 O'Callaghan, op. cit. 17 Ibid. 18 *The Age of Louis XV by Voltaire*. 19 O'Callaghan, op. cit.

with both flanks to be attacked simultaneously. At that stage the battle was hopelessly lost, according to the French historian Michelet.

'The more the English column advanced,' Voltaire would describe the aftermath,[20] 'the deeper it became, and of course the better able to repair the continual losses which it must have sustained. It still marched on close and compact, over the bodies of the dead and wounded on both sides, seeming to form one single corps of about sixteen thousand men.' On the left flank the infantry regiments were brigaded together, ready for a united assault under the joint command of Charles O'Brien, 6th Viscount Clare, and Marshal de Thomond.

Lally, back with the Irish Brigade, understood the psychological strain of waiting for so long, and turned to rouse his men. 'Remember that those you fight are not only France's enemies, but your own,' he shouted, beginning a longer speech that would become legendary. 'March without firing until you have the points of your bayonets upon their bellies!'[21] He was the first to charge the enemy column, and the Regiment of Lally was on his heels, roaring, '*Cuimhnigí ar Luimneach agus feall na Sasanach!*'[22] (Remember Limerick and English treachery!) The four cannon pounded the centre, and, in Thomas Carlyle's words,[23] 'slit up like a ribbon the terrible English column'.

As the tumult died and everywhere cold, shock and pain set in, de Saxe found Lally to congratulate him, in sombre circumstances despite the victory. The ground beneath the English column was strewn with dead and wounded – the figure would later be reckoned to be fourteen thousand – and French relief was mixed with admiration for enemy bravery. Lally, 'with his own hand',[24] had taken several English officers prisoner to spare their lives, and when the king and dauphin sought him out they found him sitting on a regimental drum, covered in blood from wounds sustained in the recent fighting and surrounded by the 'mutilated remnants' of his regiment. In 1819, the usually impersonal *Biographie Universelle* broke with precedent to describe the scene.

'Beside him were a few English officers to whom he had given help after wounding them himself. The dauphin ran up to him and expressed in advance the gratitude of the king. 'My lord,' replied Lally, 'Gratitude is like that in the Gospel; it falls on the one-eyed and the lame.' As he said this, he pointed to his lieutenant-colonel who had received a bayonet-thrust in his eye, and his major whose knee had been shattered by bullets. At that very moment the king summoned him to the front of the army and made him a brigadier on the battlefield.'[25] The self-serving Duc de Richelieu had taken full credit for the

20 *The age of Louis XV by Voltaire.* **21** Pierre la Maziere, *Lally Tolendal* (Paris, 1931). **22** Bartlett and Jeffery (eds), *A military history of Ireland.* **23** Richard Hayes, *Irish swordsmen of France* (Dublin, 1934). **24** *The age of Louis XV by Voltaire.* **25** *Biographie Universelle*, tome 23 (Paris, 1819).

crucial advice, having relayed it to de Saxe and the king at the height of battle, but according to Voltaire's account, Lally's courage and humanity, combined with his ability to fashion a witticism at such a moment, made a deep impression on the king. Twenty years later that memory, conveniently, would be suppressed.

The decisive attack by the Irish Brigade was widely held to be the turning point in the French victory of Fontenoy. But it was Lally's defensive precaution of sealing the lane the previous night which had prevented the Dutch coming to the aid of the English, and his placement of the four cannon, which were credited by de Saxe as being the decisive factors. 'It was unquestionably due to this,' he wrote in his official report, 'that the battle was won.'[26] He made no secret of his trust and approbation, and when faced with growing public hysteria at a time of sudden conflict three years afterwards, before peace was once again restored, he spoke out. 'We can sleep calmly in our beds,' he stated then, 'because we know that Lally is with us'.[27]

Off duty, Lally by now had a townhouse in Paris with a well-stocked library, and the open support of Madame de Pompadour, the king's favourite; unlike a conventional soldier, his friendships had expanded to include intellectuals like Voltaire and Montesquieu. He had a titled French mistress who canvassed loyally for further promotion on his behalf, and he was loved by Arthur Dillon's daughter, Mary, with whom he felt a growing but unacknowledged rapport. His regiment was the nearest he had to a family, and he had ruled out marriage until he reached the peak of his profession, when he could pick and choose among the rich and elite. In 1751 a son was born and fostered out; according to contemporary gossip, the child was the result of an affair with the wife of an aristocrat. The years of peace following de Saxe's pronouncement passed quickly, spiced by intrigue, because the failure of the '45 had left him with deep disappointment, not lasting despair. He continued to advise the prince, and draw up plans for potential Stuart campaigns.

As has often been pointed out, if Lally had lost his life at any time before the outbreak of the Seven Years War in 1756, his reputation today would be a glorious one. Up until then, he could do no wrong; from that time onwards nothing would go right. Circumstances would be against him, but also that personality which had built his high career and now, exacerbated by events beyond his control, would begin to undermine it. Concentration in this brief sketch upon his military abilities, rather than his character, has been to show the scale of his achievements; it is time now to look more closely at the man himself.

Lally had advanced through his personal qualities of extreme professionalism in an age of well-connected amateurs, based on iron self-discipline,

26 Hayes, *Irish swordsmen of France*. **27** Ibid.

perfectionism and high principles. As a soldier, he had mastered his brief by concentration, patience and self-denial, refusing to be sidetracked from his goal even in his hedonistic late teens and twenties, when temptations were readily available. Confident and highly motivated, Lally had won admiration for his acerbic wit, and his humour had been deft and to the point. He had felt at ease in the diverse worlds that he inhabited: that of aristocratic French friends and acquaintances; that of the dedicated Jacobite plotters who could count on him as one of their own; and that of the regiment where he belonged and – as we would say today – drew upon heavily for emotional support and his sense of identity.

But had he ever blended into the French establishment as well as he believed? Even his good friend Voltaire, who would eventually take up Lally's cause and win, always described him as an Irishman, and, proud as he was of his Irish blood and dual allegiance, Lally was beginning to resent being categorized as a foreigner. French vacillation over Stuart support intensified that irritation. Since childhood, honour had been his guiding light, and though money, possessions and power certainly interested him, he had a soldier's disdain for political equivocation or mendacity, for sharp practice of any sort, and for civilian priorities. The higher up the social scale he rose, the less he was finding in common with worldly, manipulative men like the Comte de Stainville (the future Duc de Choiseul) with whom he had to deal, and the more super-sensitive and judgemental he was becoming. His wit was razorsharp now, and it could cut.

In 1756, however, when war against England was renewed on a global scale, all boded well for the fulfilment of his lifelong ambition to reach marshal's rank. Louis XV summoned him privately to Versailles to ask for his strategic advice, and Lally's reply was unhesitating. 'Invade England with Prince Charles. Destroy English power in India. Attack and conquer their American colonies'.[28] Because war was his speciality, the sole reason for his presence on that occasion, he pressed home all three points in the brusque manner of a soldier confronting a civilian, not a subject facing his king, 'But you *must* do all three together,' he emphasized, adding, 'if not, you will fail.'

He hoped to be involved at senior level with the invasion of England, on which Jacobite planning had secretly continued, and his appointment, when it came, took him by surprise. He was to lead an army to India drawn up to his own specifications, with full naval support, and his objective – as he himself had advised the king – was to expel the English and monopolize trade and influence for the French. He began planning at once. Filled with a successful first-generation immigrant's veneration for his adopted king and country, he intended to repay that trust with a swift victory, based on meticulous

28 *Biographie Universelle*, tome 23.

preparation. On a personal level, command of that army would bring general's rank and the reward of the ultimate insider's decoration, the Cordon Rouge. If he was successful, as he fully expected to be, there would be a marshal's baton on his triumphant return.

As preparations at Brest took impressive shape, Lally's estimate of three million *livres* for a war chest was agreed without question. He kept his requirements to a professional minimum, however, planning for flexibility and superiority in battle; the English army on the spot would be taken by surprise, and the local French forces in India would swell his numbers as soon as he landed. He chose French as well as Irish regiments, and ambitious aristocratic volunteers at officer level were eager to share the glory. Upkeep of the army on arrival was to be paid by the Compagnie des Indes, the long-established trading company in which he himself held shares, because its profits were guaranteed to soar as soon as the rival East India Company was routed.

In the late spring of 1757, shortly before embarkation, one-third of the men, munitions and money that his tight planning considered essential was cut back, for fighting on another front; his protests were met with a promise that replacements would follow on. And, without warning, an alien new element was added. Widespread corruption and tax evasion among traders in the Compagnie were suspected in Paris, and at the last minute a senior taxation official was detailed to accompany Lally. A thorough investigation into Compagnie accounts and practices was made Lally's equal responsibility, and he was briefed in advance, to his disgust, on the culpability of individual officials.

The sea passage to India should have taken six months, but the elderly, timid Admiral d'Aché, in command of the naval fleet, stretched it out to a year. Lally's double mission was common knowledge in Pondicherry, headquarters of the Compagnie, when in the early summer of 1758 they eventually reached the French city on India's Coromandel Coast. Little help or money, ominously, was forthcoming. Taut with impatience, he refused to pause for the *Te Deum* sung in his honour in the chapel, and he told the civilian governor that he was going to bring war straight to the English. Shortly afterwards D'Aché unaccountably sailed away, at a stroke removing naval assistance from the diminished and heat-stricken army.

The loss of French India, with which Lally is historically linked, is quickly told, despite being played out in slow motion for almost three interminable years. He began as robustly as he promised, taking three English forts in quick succession, and, after a lengthy siege, was poised to take the prize of their capital of Madras the following spring when fresh English reinforcements arrived there by sea. From then onwards Lally was doomed, as, with his strategic mind, he was well aware. Pushed back inexorably towards Pondicherry, he fought every inch of the way, often paying his dwindling army from his own pocket, and faced with illness, deserters and mutiny. A quarrel had soon broken out with Count de Bussy, the longstanding commander of the Compagnie's army, an irregular force which he considered vastly inferior to the king's army and corrupt to a man. In January 1760 the reinforced English army, led now by Eyre Coote from Co. Limerick, won a decisive victory at Wandewash, and Lally and his remaining troops fell back to Pondicherry.

Under siege by Coote's army there, at bay among hostile traders he openly despised and whose illicit profiteering had been proved, Lally held out for a year. Balthazar Gracian's subtle arts were discounted. In vain the horizon was searched for the promised return of d'Aché's ships, and when general provisions ran out, accelerated by the rapacious private stockpiling of certain officials, all animals within the city's walls – from elephants to rats – began to be eaten. In mid-January 1761 the imminent starvation of Pondicherry's poorer inmates and the hopelessness of the military situation forced Lally to surrender. By now his previous certainties mocked him as naïve illusions. No promised assistance had been sent, a reminder of the alacrity with which Louis XV had dropped Charles Edward Stuart fifteen years earlier, and the conclusion was inescapable: Lally and his army, which included his own precious regiment, had been considered expendable.

In London, under house arrest as a favoured prisoner of the English, rumours reached him that returning Compagnie officials to Paris were saying that he had been in league with the enemy all along. He requested and was granted Prime Minister Pitt's permission to return to clear his name. But the political landscape he found on arrival had changed; he had left as a hero, and returned as a scapegoat. The English were winning on every front of the war, and with India lost, as well as Canada, the popular Compagnie shares had plunged at home, wiping out investments. The tax official, who had kept him fully abreast of investigations, had been murdered before leaving Pondicherry and his dossier of evidence destroyed; Lally knew that he, too, would have to be silenced if the reputations and fortunes of the guilty were to be saved. Vociferous coloured versions of events had already swung public opinion against him.

With rumours flying, his assessment was realistic. Although D'Aché had failed to provide naval support, he had close family connections at Court, and Bussy had used his ill-gained India fortune to marry a cousin of the Duc de Choiseul, who had risen to become France's most powerful minister. A formal enquiry by his military peers into all the reasons for the loss of Pondicherry was Lally's by right, but previous supporters would be no help. Noailles was in his eighties, Belleisle was dying and de Saxe was dead. Richelieu, unsurprisingly, remained silent, Voltaire was out of touch and unaware, on the other side of the Swiss border at Ferney, and Pompadour's influence over the king had waned. Bussy ran a theatrically effective smear campaign – 'Either Lally's head must fall or mine!' – and raised a mob to 'insult' him in the street, but everything would be properly investigated in due course, Lally was assured, if in the meantime he kept quiet.

He wrapped himself in his honour, embittered and proud, and scorned Choiseul's man-to-man advice to leave France until the situation cooled. In Pondicherry he had been unable to fraternize with men he despised, however much it might have helped him. He would not – could not – do so now. A year of frustration passed, to no effect, and in November 1762 he took his carriage out to Fontainebleu, where the court moved in winter, for a showdown with Choiseul. His request for a meeting to discuss the military enquiry was refused, and on return he learned that Choiseul had already persuaded the king to sign a *lettre de cachet* for his imprisonment in the Bastille. To prove his innocence, he went there voluntarily, rather than wait for arrest.

Anxiety was deepened by not knowing the precise charges that he faced, and his continued request for his conduct in India to be judged by his army peers was ignored. Weeks in solitary confinement became months. In 1764 a formal *arrêt* was issued, and he learned that instead of a military enquiry to examine his decisions as army commander, the investigation was sanctioned into events in Pondicherry alone, and would be carried out by Parlement, headed by its feared chief prosecuter, Denis Louis Pasquier. He was charged with oppression of the king's subjects, extortion and high treason, punishable by the forfeiture of his property and death, and was denied counsel for a defence. Months in the damp, dark depths of the Bastille became years.

In May 1766 he was brought to the Conciergerie for a secret trial, at which the written defence he had drawn up himself was not read. Accused of having betrayed the king, he shouted out, 'That is not true. Never, never!'[29] He was sentenced to death for treason, and on 9 May 1766 – almost twenty-one years to the day since the victorious battle of Fontenoy – he was beheaded on the Place de Grève.

'All the places, all the windows, were rented at crazy prices for the occasion, and tiles were taken off roofs to make temporary stands,' wrote one anonymous eyewitness.[30] 'People were hanging off chimneys, because it was decided to carry out the execution by daylight instead of at night. The count's arms were tied so he couldn't strangle himself, and when he continued to rant against the king and the judges he was gagged. Instead of being placed on a state carriage draped in black, he was brought on a common criminal's cart. All the way from the Conciegerie to the site of execution he regarded the crowds impassively. At the Place de Grève he was placed with his back to the

29 Lecture on the career of Count Lally by G.B. Malleson, 1865. **30** Quoted by Ric Erikson, editor of *Metropole Paris*.

Hôtel de Ville. The eldest son of the old executioner, on the left facing his father, took the Damascus sword handed to him and, without measuring his stroke, chopped too high. The old man instantly took the sword and did the job with the second stroke.'

The crude wooden and rope gag, according to other witnesses, had been tied on in prison to prevent him committing suicide *en route* by swallowing his tongue, and was taken off at the scaffold in front of the representatives of Parlement, after he was blindfolded. 'Tell my judges', he said then, 'that God has given me grace to pardon them; if I were to see them again, I might no longer have the forbearance to do it.'[31]

And one public epitaph would run counter to majority French opinion. 'No one has a higher opinion of Lally than myself,' mourned Eyre Coote in England, who had taken his surrender. 'He has fought against obstacles which I believed invincible, and he has conquered them. There is not another man in all India who could have kept on foot for the same length of time an army without pay and receiving no assistance from any quarter.'[32] It was the highest praise that one soldier could pay another. The bitter irony – as Lally, above all, would have appreciated – was that the tribute came from the enemy commander.

31 Pierre la Maziere, *Lally Tolendal* (Paris, 1931). 32 E.W. Sheppard, *Coote Bahadur* (London, 1956).

The Jacobites in the American War of Independence

PATRICK CLARKE DE DROMANTIN

One has to agree with the Baron de Contenson that the American War of Independence has never found in France the place that it deserved in French history.[1] The explanation is simple and is to be found in the short period between the Versailles treaty of 1783 and the beginning of the French Revolution, which left no time for a huge sympathy movement to begin in French opinion. Very quickly indeed, the French had other subjects to preoccupy them, which in turn has hidden this aspect of French history. If French involvement in this war has not yet been given prominence, the role played by the Jacobite troops in these events has received even less attention. The Jacobite involvement, however, was not restricted to the participation of two Irish regiments but also included the original and especially effective diplomatic actions of a certain Jacobite, who is largely unknown in American Revolutionary historiography.

THE INTERVENTION OF FRANCE IN THE REVOLUTION AND THE IRISH REGIMENTS IN AMERICA

When George Washington was appointed in 1775 as the commander of the Continental Army he understood that he would have to find an ally among the western sea powers. Almost immediately, France became the obvious choice if only because of the desire for revenge that the Bourbon kingdom was harbouring following the humiliations received from England through the centuries. The American victory at Saratoga in 1777 and the successful diplomatic mission of Benjamin Franklin to Paris led to French intervention. Through the Alliance Treaty of 6 February 1778, France and the American Revolutionary government committed themselves not to sign separate peace treaties with England. In spite of Turgot's and Necker's reluctance, Louis XVI and his foreign affairs minister, Vergennes, who had always supported the revolutionaries, had perfectly understood that a victory by England might soon result in the defeat of France.[2]

Military operations took place at land and sea but naval operations were prominent. The French navy had to fight against the powerful British fleet

1 Baron Ludovic de Contenson, *La société des Cincinnati de France et la guerre d'Amérique, 1778–83* (Paris, 1934), p. 1. 2 André Lasseray, *Les Français sous les treize étoiles, 1775–1783*

while also transporting the French expeditionary force under the command of the Comte de Rochambeau. Less than two months after the Franco-American treaty was signed, the French squadron under Admiral d'Estaing, which included twelve ships-of-the line and four frigates, left Toulon on 13 April 1778 bound for the mouth of the Delaware, which they reached on the 8 July. In July 1779 D'Estaing, having been reinforced by thirteen ships under the command of the Chevalier de la Motte Piquet, attacked the English fleet in Grenadian waters and forced it to retire. A few months later, Admiral d'Estaing left the Antilles at the head of a fleet of more than 30 ships, carrying over 4,000 troops with the intention of joining the American army and assiting in an attempt to retake the town of Savannah, which had been occupied by the English for over a year.[3] This expedition was a failure and resulted in the loss of 400 American soldiers and 700 French. Among the French casualties, the 1st Battalion of the Regiment of Dillon had 147 casualties out of a strength of 434, all ranks.[4] But at least the Savannah expedition put an end to the English offensive against the southern provinces.

George Washington quickly understood that French military aid would have to be extended to include land operations: 'If we don't have money and soldiers from France, our cause is lost.'[5] As a result, an expeditionary army had to be created, mainly out of infantry. It numbered almost 10,000 men, including 6,000 among the initial force under the command of General Rochambeau and a force of 3,000 men coming from the Antilles under the command of the Marquis de Saint Simon. On the whole, the French army had engaged more than 20,000 men in this conflict, if one also includes the naval contingent made up of ships' crew and marines. This French military aid proved to be decisive in the campaign.[6]

THE INTERVENTION OF THE IRISH REGIMENTS[7]

One has first to remember the conditions in which the Irish regiments came to be involved in the American War of Independence. A famous memorandum written by General Arthur Dillon to try and ensure the continued existence of

(Paris, 1935), p. 55. 3 Eoghan Ó hAnnracháin, 'Réfexions sur la vie du general Jacques O'Moran', *Valentiana*, 7 (June 1991), 30–41. 4 Mark G. McLaughlin, *The Wild Geese: the Irish brigades of France and Spain* (London, 1980), p. 20. 5 Ministère des Affaires Etrangères, *Les combattants Française de la guerre Américaine, 1778–1783* (Paris, 1903), p. vi. 6 André Corvisier, *L'armée Française de la fin du XVII siècle au ministère de Choiseul* (2 volumes, Paris, 1964). 7 Pierre Carles, 'Troupes Irlandaises au service de la France, 1635–1815', *Etudes Irlandaises*, 8 (December 1983), pp 193–212. See the PhD thesis of Nathalie Genet-Rouffiac, 'La première génération de l'exil jacobite à Paris and Saint Germain-en-Laye, 1688–1715' (Paris, 1995). See also Nathalie Genet-Rouffiac, 'The Wild Geese: les regiments Irlandais au service de Louis XIV', *Revue Historique des Armées*, 222

these Irish regiments which 'have always claimed the privilege to be the first ones to march against the English in whatever conditions France came to fight against them and it is according to that principle that the Regiment of Dillon asked and was given the leave to go to America at the beginning of 1779'.[8]

It is clear that the Irish regiments were deliberately chosen to take part in the American War of Independence, unlike most of the other French units, as it gave the Irish a chance to fight against the hereditary enemy of their nation. As Gilbert Bodinier has stated, most of the soldiers were not drawn by the will to fight for liberty or egality or to settle democracy in America as they only had a very tenuous idea of what democracy actually was.[9] Most of them simply went because the regiment that they belonged to had been ordered to go. In the case of gentlemen such as Lafayette, Noailles, Ségur or Lauzun, these young men often were looking to gain military experience abroad and 'it was even more appealing to do so in that case in that it meant fighting England'.[10] This war in America obviously offered these young aristocrats an unexpected career opportunity that lived up to their taste for glory and which also included a chance to take revenge on England.

Two Irish regiments were involved in the American War of Independence – the regiments of Dillon and Walsh. The first battalion of the Regiment of Dillon, which numbered 1,400 men, was embarked at Brest on 1 May 1779 in the ships of the squadron of the Chevalier de la Motte Piquet and joined the army of Comte d'Estaing who had thus far not been able to confront the English for lack of infantry.[11] With these re-inforcements d'Estaing decided to act and he made an attempt to take Grenadian Islands on 1 July 1779. At the head of the grenadier company of the Regiment of Dillon and assisted by the regiments of Champagne and Auxurrois, d'Estaing succeeded in taking the hospital buildings where the English had entrenched themselves. In September, the Regiment of Dillon, having left part of its men at La Grenade, was embarked for Savannah, capital of Georgia. General Arthur Dillon in his memorandum recorded the difficult, not to say awful, conditions in which the troops were disembarked in Savannah:

> Because of the distance from the coast that the squadron was anchored and the because of perpetual gales, during several days communications with the disembarked army and the ships were stopped, so that it was not possible to gather enough troops to take the place. It left time for

(March, 2001), 35–48. 8 SHD, Vincennes, General Arthur Dillon, 'Observations historiques sur l'origine, les services and l'état civil des officers irlandois au service de la France, addressées à l'Assemblée Nationale en 1790'. 9 Gilbert Bodinier, *Les officers de l'armée royale combattants de la guerre d'Indépendance des Etats-Unis d'Amérique* (Vincennes, 1983), p. 539. 10 Henri Doniol, *Histoire de la participation de la France à l'établissement des Etats-Unis d'Amérique* (2 volumes, Paris, 1886), i, 647. 11 SHD, Vincennes, Sub-series

> 800 Scottish from the 71st Regiment to reinforce the British positions. On 9 October, Monsieur le Comte d'Estaing, after several days of entrenching, made an unsuccessful but costly attack at the head of the troops. The French army was repelled with many casualties, the Regiment of Dillon lost M. Browne, their major, a very distinguished officer and had 63 grenadiers killed or wounded, not counting fusiliers.[12]

Actually, the French casualties were even more numerous and the regiment lost three more officers on that day – Georges Taffe, *lieutenant-en-second*; Jean Lambert, *cadet gentilhomme*; and N. d'Elloy, *cadet gentilhomme*.[13] The siege of Savannah obviously was one of the high points of the campaign in America. We have a detailed record of this painful event thanks to the diary of the Antoine O'Connor, *engineer du roi*, in which the Comte d'Estaing himself had written a preamble.[14]

> It was an awful shedding of blood [...]. General Putasky is dangerously wounded, M. le Comte de Bethysy was wounded by two shots, M. le Vicomte de Fontanges, major-general, is dangerously wounded, M. Browne, major of the Regiment of Dillon, was killed, Le Baron de Steding was wounded and soon after that the general [d'Estaing] himself was wounded in the arm ... all this spread the worst confusion. The front ranks charged for a second time without success, the disorder increased, most of the leaders being wounded [...]; the front ranks charged for the third time with the support of the troops that the general had just gathered. The advance was long and the guns of the enemy made great devastation [...]. M. le General was wounded for the second time by a bullet through his leg. Witnessing the disorder that started again, he gave M. le Vicomte de Noailles the order to retreat [...]; the whole action lasted no more than one hour but was extremely fierce [...]; it is almost certain that two American deserters had warned the enemy the day before of the place where the attack would take place.[15]

This first-hand account of O'Connor is annoted by personal comments by Admiral d'Estaing, which are no less interesting when referring to the conduct of the French soldiers:

> The enemy was not surprised while surprising them was the main point. Everything had depended on it in my eyes [...]. M. le Comte Dillon, in spite of conduct of the highest bravery, was not followed by more than

Y6. 12 Arthur Dillon, 'Observations historiques', op. cit. 13 SHD, Vincennes, Subseries Y6. 14 SHD, Vincennes, Antoine O'Connor, 'Journal du siege de Savannah', Mémoires Historiques, MR248/2. 15 Ibid.

> 80 men. The enemy having advanced on him in large numbers, it became unavoidable to retreat, which led to an infinite number of casualties. This fierce moment would have been decisive if he [Dillon] had been sustained – it was extremely fierce [...]; all the regimental commanders behaved with the greatest courage and with firmness, intelligence and will – looking for and finding remedies for everything. M le Comte Dillon, M. le Vicomte Noailles, M. le Baron de Steding, the colonels, M. le Marquis de Pondevaux, M. le Comte O'Dune [*sic*], the lieutenant-colonels, all highly distinguished themselves either during the siege and in the command of the siege works or during the day of the attack [...] three times M. le Vicomte de Fontanges rallied the troops and led them back in an attack. All the officers of the etat-major have behaved themselves perfectly, they have helped with the coolest courage to reform the troops under heavy fire [...]; the loss of M. Browne is real indeed for the service. After a long discussion with him, he had promised me to renounce forever, every time on active service, that in which he had lost himself twice. It is with the highest respect for his memory that I am content to pay homage to the military qualities that he had.[16]

The first battalion of the Regiment of Walsh was also involved in America in 1779 and one of its companies took part in the conquest of Senegal. The second battalion of the same regiment reached Martinique in early 1780 and took part with the Regiment of Dillon in the three naval engagements led by the Comte de Guichen against Admiral Rodney. On 29 April 1781, the French and British fleets, commanded by the Comte de Grasse and Admiral Hood respectively, were engaged. On 21 June the Tobago Islands were taken by a French force commanded by the Marquis de Bouillé. These battalions of the Regiment of Walsh took part in the conquest of St Eustache in conditions that were later reported by General Arthur Dillon:

> General de Bouillé, wishing to take advantage of the absence of the English squadron, embarked on 15 November 1781 at Martinique, with a body of 1,200 men taken from the regiments of Dillon, Walsh, Auxerrois and Royal-Comtois, and arrived at St Eustache Island on the night of 25–26 November. His measures had been perfectly well-taken but the tide could not allow the frigates to get close to the coast and he could disembark only 377 men, without any hope of re-embarking them or being reinforced by the rest of his troops, without being discovered

16 Ibid. It is unclear just what it was that Browne had decided to 'renounce forever'. This was perhaps a habit of heavy drinking.

> by the enemy. In this thorny situation the cool courage and determination of the general [Bouillé], that he shared with the men of his small force, overcame their lack of numbers. He marched on the enemy. The Irish were at the head of the column. The surprise was complete. 840 men of the English garrison had to put down their arms and were taken prisoner by half their number.[17]

While some authors have been played-down the severity of warfare during the eighteenth century, referring to it as 'la guerre en dentelles' ('the war in lace'), these accounts illustrate the fact that during the war in America many of these men highly distinguished themselves and their actions deserve not to be forgotten.

THE IRISH JACOBITE OFFICERS DURING THE AMERICAN WAR OF INDEPENDENCE

The commitment and courage of the Irish Jacobite officers during the American War of Independence never faltered as shown clearly by some witnesses. The first of these reports concerns Jean O'Brien, one of the many officers without any financial means who were totally devoted to the service of the king. In spring 1779 he went to America in very difficult conditions.

> This officer, having suffered from an attack of esquinqancie (a respiratory illness) in April 1779 while aboard ship in Brest, would not suffer to be disembarked and would have rather died on board than leave his regiment, which was being sent against the enemy. During the voyage, a stroke of paralysis deprived him of the use of the left side of his body, from his head to his feet. In spite of this unhappy situation he has been present everywhere with his ship, especially at the battles at La Grenade, Savannah and at Monte Christi. Not wishing to be taken to the hospital to be cured, he had, on 13 August 1779, while at Cape Town, the regimental paymaster pay him an advance in the amount of 900 *livres* tournois and again 3,600 *livres* in March 1780 … thanks to that help … he was partly relieved … despite the fact that he was born of an old and illustrious family, he had nothing for his living but his wages and if the two amounts aforementioned were taken from his wages, he would rather have died 1,000 times during in the war than have had to live 10 more years in misery to pay back this money.[18]

17 Arthur Dillon, 'Observations historiques', op. cit. 18 Centre des Archives d'Outre-Mer, Colonie E325, 'Mémoire du sieur O'Brein au maréchal de Castries du 25 Septembre 1781'.

Jean O'Brien finished his memoir by asking to be granted a gratuity to compensate the heavy expenses that he incurred.

> The bad situation of his health, his long and harsh services like those of his family that has almost become extinguished in His Majesty's service. The zeal with which he always has shown in his military duties, makes him hope that the well-known and famous humanity of M. de Castries, will result in him taking account of his situation and will give an order that the two sums will not be taken from his wages but rather given to him as a gratuity.[19]

The behaviour of another Jacobite officer named Jacques O'Moran was nonetheless interesting. He entered the service on 15 November 1752 as a *cadet gentilhomme*, while not yet being 14 years of age. He reached the first officer ranks and served in the artillery and distinguished himself during the battle of Marpurg and appears as a major during the attack on Grenada on 4 July 1779. On 6 July he took part in the naval engagement aboard the *Dauphin Royal*.[20] A few days before the siege of Savannah he was very badly wounded during a violent encounter with English troops and had to be evacuated. After several months in a local hospital, he was sent back to France to finish his convalescence. He landed Rochefort in March 1780 and didn't lose any opportunity to report to the administration how precarious his financial situation was. On 7 April 1780 he wrote two memoirs – one for the *ministre de guerre*, the Prince de Montbarey, and the other to the *ministre de la marine*, the Comte de Sartine. In the first, he asked for a colonel's commission and gave a precise report of his campaign in America:

> He dares to hope that he has deserved the justice due to his conduct during the campaign in the American colonies under the orders of the Comte D'Estaing. He has been dangerously wounded, being shot in the thigh, breaking a bone and tearing apart the ligaments while charging the English who were making an attempt, in the number of 300–400 men, on a French entrenchment on 24 October 1779.[21]

Comte Arthur Dillon had supported the memoir and stated: 'I want to believe that M. O'Moran deserves to be granted his demand because of the way he has always served during the last campaign in America.'[22] O'Moran's attempt was successful because on 24 June 1780 the Prince de Montbarey notified him that 'on the account that I have made to the king of your services, especially of your behaviour during the last campaign in America under the orders of M. le

19 Ibid. **20** SHD, Vincennes, 7Yd2/GD/2, Jacques O'Moran, dossier personnel, Mémoire du 7 Avril 1780. **21** Ibid. **22** Ibid.

Comte d'Estaing, His Majesty has granted you a commission as *maître de camp de infantry*'.[23]

In his second request to the minister de la marine, O'Moran asked for 'a pension or gratuity; which would allow him the means to go to Barèges'.[24] This Pyrenean town was renowned for the healing of wounded soldiers and Lord Mountcashel himself and died from his wounds there. O'Moran went to Barèges but having not received a gratuity, he sent a further letter on 1 January 1781 to the new *ministre de la guerre*, the Marquis de Ségur.

> Sir O'Moran has the honour to present you the fact that he has made 600 miles in a coach to go and come back from Barèges and that he has spent three and a half months there, that he still is very anxious that the state of his wound, which is still open and from which not long ago a splinter of bone has been drawn, will force him to return there. Having no other income than his wages and having spent 3,000 *livres* in advance since 15 March, when he landed in Rochefort. On this consideration and because of the extraordinary and daily expenses that he is still subject to, he begs M. le Marquis de Ségur to take interest in his condition and to help him by requesting a gratuity on his behalf in proportion to his expenses and the state of his position.[25]

Once again, he had the support of General Arthur Dillon who estimated that O'Moran needed at least 800 *livres*: 'the memoir of M. O'Moran is true in every aspect and it seems to me that it is fair not to expose an officer of his distinction to be ruined by the expenses made necessary by the cruel position he has been placed in for the last fifteen months because of his wound'.[26] However, O'Moran recovered and went back to America in 1781 and took part in the Jamaica expedition. Louis XVI gave him a sign of recognition by giving him a pension of 400 *livres* on 8 July 1784 as a gratuity in connection with O'Moran's Croix de Saint Louis, which he had been awarded on 28 January 1784. But the recognition seemed to be modest in comparison to the commitment of O'Moran, who expressed his surprise in a new memoir on 10 September 1789, in which he solicits an appointed as the colonel of a regiment of foreign infantry.

> Sir O'Moran has been given the commission of colonel without protection or favour but only in consideration for his services on 24 June 1780. Full of confidence in the bounty and justice of M. le Comte de la Tour du Pin, he dares to flatter himself that he [the comte] will wish to

23 Ibid., Mémoire du 10 Septembre 1789. 24 C.A.O.M., Colonie E325bis. 25 SHD, Vincennes, 7Yd2/GD/2, Jacques O'Moran, dossier personnel, Mémoire du 1 Janvier 1781. 26 Ibid.

> obtain from the king, that he [O'Moran] will be put in service as the colonel in whatever foreign infantry regiment to which His Majesty is pleased to appoint him.[27]

He was appointed as the colonel of the 88e Regiment d'Infanterie on 25 July 1791 and then of the 87e Regiment d'Infanterie one month later. He was later made *maréchal de camp* on 6 February 1792 and lieutenant-general on 9 October, but the French Revolution resulted in his death and he was guillotined in 1794, like the two other Jacobite heroes of the American War of Independence, General Arthur Dillon and General Thomas Ward.

One of the other interesting characteristics of General O'Moran was that he had been admitted to the Order of the Cincinnati Society on 28 August 1784. After his death, some even claimed that this decoration was the only one that he ever wore.[28] This hereditary decoration had been created on 10 May 1783 by George Washington to perpetuate the commemoration of the American War of Independence. It first was only given to general officers and colonels and their heirs but was later on extended to heirs of officers of all ranks who had served for more than three years and eventually it was awarded to all descendants, in direct or collateral line, of all those who had been wounded or killed during the war. Among all the French families that had been awarded this decoration, the Dillon family was the only one, as far as we know, to be granted three seats in the Cincinnati Society.

In effect, seven members of the Dillon family fought in America. These included General Comte Arthur Dillon himself and the four sons of Robert Dillon and Mary Dicconson - Comte Edouard Dillon, Guillaume Henri Dillon (also called Billy), Robert Guillaume Dillon and François Theobald Dillon (also called Frank). Comte Edouard Dillon was wounded in the action at Grenada and was a founding member of the Cincinnati Society. Billy Dillon was wounded at Yorktown. Other members of the Dillon family included the cousins of the above – Theobald Hyacinthe Dillon, wounded at Savannah, and Barthélémy Dillon.

The best-known of all of the Dillons was Comte Arthur Dillon. This was due to his prestigious career and also his death at the hands of the Revolutionaries. In the spring of 1779 he landed in America at the head of his regiment, still not being 30 years of age. He fought with great bravery at the taking of Grenada and during the naval operation of 3 and 5 July 1779. In October, during the siege of Savannah, he was put in command of more than 10,000 troops after Admiral d'Estaing was wounded. On 29 April 1781 he took part in a naval engagement under Admiral de Grasse at Martinique. He

27 SHD, Vincennes, 7Yd2/GD/2, Jacques O'Moran, dossier personnel, Mémoire du 10 Septembre 1789. 28 Eoghan Ó hAnnracháin, 'Réfexions sur la vie du general Jacques O'Moran', *Valentiana*, 7 (June 1991), 30–41.

was also present at the engagement in St Lucia on 13 May and the taking of Tobago. In November 1781 he led the advance guard in the seizing of the island of St Eustache, surprising the English garrison that was superior in numbers. He led the advance guard of the army during the operations in the island of St Christophe and Montserrat and was at the head of the expedition to Jamaica in April 1782. Dillon distinguished himself during the unsuccessful engagement of Admiral de Grasse on 12 April 1782. In July 1786 he was made governor of the island of Tobago.

Eduoard Dillon, the eldest of the four Dillon brothers who were involved in the American War of Independence, was embarked for the Antilles on 27 March 1779 aboard *Le Diadème*, one of the ships of the fleet commanded by the Comte de la Motte Picquet. Their arrival at Fort Royal on 17 June gave a chance for Admiral d'Estaing to take the initiative in the campaign. D'Estaing was very enthusiastic in his letter to Sartine, the *ministre de la marine*. On 29 June 1779, he wrote: 'The convoy has arrived. These four words say everything. The royal army now gathered is something never seen before in America. This masterpiece of your ministry will bring it eternal fame. We are now equal to the English.'[29] On 3 July 1779, the young officer was assigned to take command of one of the three assault columns that were designated to take the island of Grenada. The official chronicler reported that 'the three columns debouched at the same time, rushing into the (English) entrenchments, each in its own direction and forced the palisades as the drums beat the charge. Dillon's regiment was mauled by the fire from the battlements of the port but never slowed its pace for a moment'.[30] On 4 July, Lord Macartney and the garrison and colony surrendered and 700 prisoners, three regimental colours, 102 cannon and 16 mortars were taken. Eduoard Dillon had been slightly wounded in the attack but took part in the naval operations of 6 July against the fleet of Admiral Byron. He was wounded again and his arm was broken in two places and he had to return to France for his recovery. This wound caused him pain for years. Even if he had never actually landed on American ground, by special request of Admiral d'Estaing, on 19 August 1784 he was admitted as a founding member of the Cincinnati Society and was also granted the Order of St Louis.[31]

Robert Guillaume Dillon, the second brother, embarked to America with Rochambeau's army on 2 May 1780 with is two younger brothers under his command, Guillaume Henri (Billy) and François Theobald (Frank), who was not yet sixteen. The three of them took part in the battle of Yorktown. After the arrival of the French reinforcements under the command of Admiral de Grasse in Chesapeake Bay, Rochambeau decided to attack the positions of

29 Papers of Comte Jacques Dillon in the private collection of la Comtesse de Scoraille.
30 Ibid. 31 Ibid.

Lord Cornwallis at Yorktown. At the beginning of the siege, the Hussards des Volontaires Etranger of the Duc de Lauzun, in which the Dillons served, covered themselves in glory in a battle with Tarleton's Dragoons. The English cavalry was three times more numerous. Three charges were led before the English dragoons were pushed back in spite of the fire support of their infantry. A few days later, Robert Guillaume Dillon distinguished himself again during the attack on the redoubt. Despite several sorties by Cornwallis' troops, the French artillery succeeded in reducing the position. On 19 October 1781, 8,000 English soldiers surrendered, leaving 22 colours, 190 cannon, 8 mortars, one frigate, 2 corvettes and 60 transport ships. After the departure of Lauzun in October 1781, Colonel Robert Guillaume Dillon was put in command of the Legion de Volontaires Etranger. He was in command of the flank column of the expeditionary army during the winter of 1781–2 and accomplished this task to the great satisfaction of Rochambeau. In June 1783, Rochambeau himself presented him with the Order of St Louis – 'for his good behaviour during the siege of Yorktown' and Dillon was admitted to the Society of Cincinnati on 7 January 1784.[32]

In the case of the cousin of the Dillon brothers, Chevalier Theobald Hyacinthe Dillon, he took part in the Grenada action and in the siege of Savannah, where he distinguished himself and received wounds he would suffer from for the rest of his life. In spite of his wounds, he refused to be evacuated and remained to take the command of the Regiment of Dillon when his cousin, Arthur Dillon, was put in overall command when D'Estaing was wounded. Theobald Dillon brought back the remains of the battalion, which had suffered heavy losses. He then fell very ill and had to return to France in November 1779.[33] The importance of the Dillon family is not surprising when one considers that it gave thirty-five officers to the service of France in the eighteenth century, including several colonels who were killed at the head of their regiments, for example at Fontenoy and Laffeldt.[34]

Other Irish families have also given several officers to the service of France. For example, Charles Laure de MacMahon, the Marquis de Viange, had three sons who commanded regiments in the royal service.[35] Having taken part in the American War of Independence, Charles Laure MacMahon had solicited for the rank of colonel and was given several testimonies of satisfaction. The Marquis de Vioménil, *commissaire des guerre*, wrote to Ségur, *ministre de guerre*; the following:

32 François William Vanbrock, 'Le lieutenant general Robert Dillon', *Revue Historique des Armées*, 158 (March 1985), 27. 33 SHD, Vincennes, 3Yd3816/MC/1, Arthur Dillon, Dossier personnel. 34 The Chevalier Jacques Dillon was killed in action at Fontenoy in 1745 while commanding the regiment. In 1747, the Comte Edouard Dillon was mortally wounded at the head of his regiment at Laffeldt. 35 The Marquis Charles Laure MacMahon was the uncle of Marshal Patrice de MacMahon, Duc de Magenta and

> As the inspector of the Royal Regiment of Lorraine, in which the Marquis served, I saw him two years in succession. His zeal was not limited to the care of his own company and he stood as the best example for all the regiment ... he has made trips at his own expense to acquire military instruction. As he was taken in 1782 by M. le Marquis de Lafayette as his aide-de-camp, I saw him at Rochefort when I was about to embark aboard *l'Aigle* [...]; he has been noticed during the naval fight that this vessel had with HMS *Hector* as well as during the sinking of this vessel [*l'Aigle*], when he helped preserving the two million *livres* that we have saved, he showed as much calm as courage, as much intelligence as modesty – his behaviour has been perfect in all circumstances and his financial sacrifices have been increased as a result. These are the reasons why I think he deserves the goodness and interest of M. le Maréchal de Ségur.[36]

At the beginning of the French Revolution, a difficult period when traditional authority was turned on its head, Charles Laure de MacMahon managed to keep order and discipline in his regiment and when he resigned on 19 August 1791 he could express his 'satisfaction to give back his regiment perfectly intact without anyone having deserted his duty or the rule of strict discipline[...] as well as his respect and commitment to the king and his officers'.[37]

One could also examine the cases of Charles Edward Jennings, baron of Kilmaine, or Isidore Lynch or Charles Edouard MacDonald to show that the fame of the Irish Jacobite officers was well-deserved and that the part they played in the American War of Independence should be remembered. However, the Jacobite involvement in this war also took the form of diplomatic actions, which are even less well-known.

JACOBITE DIPLOMATIC INTERVENTION IN THE AMERICAN WAR OF INDEPENDENCE

This diplomatic intervention was due to a Jacobite of English origin, Jean Holker, who was the son of John Holker, an inspector-general of factories who was responsible for the introduction of mechanisation in the French textile industry. Jean Holker, who was born in Rouen in 1745, succeeded his father as an inspector-general of factories in 1777, as his father had been granted the living of his office by ordinance of the Conseil. Thanks to the connections of his father and his own activities, Jean Holker knew a lot of people who were

president of the French Repulic. 36 SHD, Vincennes, 4Y d3479/MC/1, MacMahon, Dossier personnel. 37 Ibid., annotation of 19 July 1783.

well-connected and he travelled widely in France, England, Switzerland and in Austrian Flanders.[38]

At the beginning of the reign of Louis XVI, Vergennes, the foreign affairs minister, wanted to send to America someone who would be capable of gathering information and even to negotiate. This person had to be a businessman engaged in his own business, so as not to draw the attention of the British authorities. He asked Leray de Chaumont, a businessman who had already been involved in business in America during the war, to find him someone. De Chaumont knew Jean Holker and suggested his name to Vergennes. Soon thereafter, the mission of Holker was defined as having to:

> Find out the dispositions of the members of Congress and its leaders as well as of the provincial congresses, to try to determine their resources in money, army and naval troops and to find out the commercial possibilities of their harbours; he would have to inquire what treatment the United States would give to a nation ready to sustain their cause; and eventually he will have to inform us as precisely as possible as to the real situation of the English.[39]

On 25 November 1777 Leray de Chaumont was given a note with the instructions for Jean Holker. Holker was to make sure that the Continental Congress would accept no proposition from England, neither about peace nor about a new union – whatever the English government was willing to offer.[40] His mission would be to 'divert the majority of the members of Congress from accepting peace proposals on any basis other than total independence and with the condition that other nations would guarantee this peace, which otherwise would be too easy to break by this continental power (England) that was still master of Canada, Florida and New Scotland'.[41]

At the same moment that John Holker was leaving for America, it was known in France that General Gates had won the battle of Saratoga on 17 October 1777. The misson of the new diplomat was made easier since he was able to express more clearly to the Americans 'the actual proof of the favourable disposition of the house of Bourbon'.[42] Despite his promises, the new *charges de mission* was not taken seriously and Congress sent queries to Paris regarding Holker's legal status as he carried no official letters. Meanwhile, the king of France had signed a double treaty of trade and friendship with the United States on 6 February 1778. On 31 March 1779 the royal administration informed Jean Holker of the development of the relationship between France and America:

38 Archives du ministère des affaires étrangères, personnel, volume 39, 1st series, 'Dossier Holker'. 39 Archives du ministère des affaires étrangères, Etats-Unis, volume 2, 'Correspondance politique'. 40 Henri Doniol, op. cit., ii, p.616. 41 Ibid. 42 André

> When you went to northern America you were charged with affirming to the members of Congress all the affection that the king bore to the United States and the deposition of His Majesty and that you could give them proof of this affection as soon as circumstances allowed. The effects have soon followed the promises that you were carrying, since His Majesty has signed on 6 February last a treaty of trade of friendship with the deputies of the Congress. I thought that I had to inform you, Sir, so that you can consolidate the trust you have been given by the Americans.[43]

This first, almost secret, mission has left few official records and it was only in the following year that Holker was officially nominated by Girard de Raynal, the minister plenipotentiary to the United States, as the French consul in Philadelphia, with an annual income of 10,000 *livres*. Eventually in December 1779 Holker was made the consul general for the states of Pennsylvania, Delaware, New Jersey and New York with his residence in Philadelphia.[44] In fact, his functions were broader than the diplomatic field because he was the 'agent general de la marine royale de France' and also very soon he was involved in personal business ventures. It becomes difficult to distinguish his professional and personal occupations, which can mainly traced due to the papers he left in the United States and which are now kept in the Library of Congress.

As consul general for France, Holker was in charge of dealing with the financial organisation of the military operations of the French expeditionary force. A bill of payment of 1780 for the first months of that year came to a sum of over 1.87 million French *livres* of which 64% was spent on the hiring of transport wagons.[45] Even if a tight system of controlling expenses had been organized, the functions of John Holker were those of a banker, a money-changer, a state account, paymaster-general and *controller de gestion* in naval, military and civilian capacities.[46] Soon enough he developed in America his own businesses, both on his own account and on behalf of some of his Jacobite associates, including men of Irish origin such as Antoine Garvey, an important china manufacturer in Rouen, and the Comte de Clonard, with whom he appears to have had business connections.[47] Even if some later records cast doubt on the level of his success in business in America, we know that John Holker maintained business connections in the United States and

Lasseray, op. cit., pp 56–7. **43** Ibid. **44** Archives du ministère des affaires étrangères, personnel, volume 39, 1st series, 'Dossier Holker'. **45** Library of Congress, Jean Holker papers, AN 415 Mi to 415 Mi 20. The Library of Congress holds 41 volumes of Holker's papers, dating from 1777 to 1805. These archives are also available on microfilm at the departmental archives of Gironde. **46** Departmental Archives, Gironde, Jean Holker Papers. **47** Ibid.

that, even if he spent two long periods in France in 1796 and 1800, that he died in Springsburg, in the Commonwealth of Virginia, on 13 April 1822. Such was the end of the largely unknown but original mission of a Jacobite and former inspector-general of factories who, in his own way, had taken part in the American War of Independence.

'Semper et Ubique Fidelis'

GEORGE MARTINEZ

INTRODUCTION: THE IRISH BRIGADE IN THE SERVICE OF FRANCE[1]

'Always and everywhere faithful' such was the new motto that ornamented the flags of the three Irish regiments that were raised in the 'Army of the Princes' army in Koblenz in 1792.[2] These regiments represented the remnants of the famous Irish Brigade of the eighteenth century royal army, which included six regiments of infantry and the cavalry regiment of Fitzjames cavalry regiment.[3] The impact of the high casualties suffered by these regiments in the War of the Austrian Succession (1740–8) and the Seven Years War (1756–63), combined with a drop in Irish recruitment and the financial restrictions, led to the end of the Scottish regiment, which were merged into the Irish regiments.[4] The *ordonnance* of 26 April 1775 reduced the six Irish regiments into three regiments of two battalions, Berwick, Dillon and Walsh.[5] During the period of peace that prevailed after the Seven Years War, the Irish regiments suffered the same fate as French regiments and had to endure the '*tour de France*' of the garrison towns, which were mainly in the western coast, and they also served in the few colonies left to France after the treaty of Paris.

In spite of the Irish names that they still bore as regimental titles, these regiments were no longer mostly composed of Irishmen. When it came to the rank and file, most of the men were from Liege, the areas that we know now as Belgium and Benelux, and the Northern countries. The others volunteers, which included Germans, Italians, Corsicans and Swiss were sent into specific units. Thanks to the research carried out by Eoghan Ó hAnnracháin on the Regiment of Clare at the battle of Fontenoy, it can be shown that during this period the regiment contained no more than 39.84% Irish-born soldiers.[6] The regiment also included 10% born in France, 23.36% Belgium, 13.76%

1 See J.C. O'Callaghan, *History of the Irish brigades in the service of France* (Dublin, 1854) and E. Fieffé, *Histoire des troupes étrangères au service de La France* (Paris, 1854). See also J. Sarsfield, *Histoire de la brigade irlandaise* (1974) and David Murphy, *The Irish brigades, 1685–2005* (Dublin, 2007). 2 Service Historique de la Défense (SHD), Vincennes, Guerre/ Xu 1–22 : organization of the *armée des Princes*, registers of officers and men, nominations, 1792–185. See also Pinasseau Papers, SHD, Guerre 1K45. 3 Berwick, Bulkeley, Clare, Dillon, Lally and Rooth. See David Murphy, op. cit., chapter 3: Irish regiments in French service', pp 10–40. 4 The regiments were the Royal Ecossais, Ogilvy and Albany. 5 SHD, Guerre Ya 388: Berwick; Ya 389: Dillon; Ya 390: Lally; Ya 393: Walsh. 6 Eoghan Ó hAnnracháin, 'Casualties in the ranks of the Clare regiment at Fontenoy', *Journal of the Cork Historical and Archeological Society* (1994), 96–110.

French, 11.68% English, 5.92% German, 3.52% Scottish and 1.92% from other countries.[7] These figures serve as a good illustration of the composition of the Irish regiments at the height-point of the Irish Brigade's service.

It can also be shown that the officers were born in Ireland or were from Irish origins.[8] After more than one century, the first wave of Irish emigrants had founded their family lines in France, while the link with Ireland remained strong and many families tried to leave one of their sons in Ireland to preserve, as much as possible, the familial estate. During the same period, these Irish families sent their younger sons into the French king's army.

The Irish regiments were regarded as foreign regiments but enjoyed a specific status among the foreign regiments in Bourbon service. Their regimental colours recalled their Irish origins and their uniforms were different from the distinctive French *habit rouge garance*.[9] They were paid the highest wages of all the foreign troops. In 1704, Louis XIV stated that 'he looked at the Irish Catholics that had come to his Kingdom with the late king of England as his own native subjects, without having to receive naturalisation letters'.[10] This principle was confirmed in 1736 by a sentence of the *Bureau des domains*, which rejected the complaints of the farmers of the *Domaine* who protested that the rights of the Jacobite exiles were not based upon actual law. Louis XV confirmed the same rights of French citizens to the Irish by a letter written in Lille on 25 of March 1741. Both the Parlement of Paris and the Conseil confirmed that 'since 1690 the Irish serving France have never been regarded as strangers and have had no other country but France'.[11] A certain lack of legal ambiguity remained and the Irish still had to ask for naturalisation letters to be able to reside in France without being submitted to the *droit d'aubaine* under a preferential status.[12] Actually, statements that proved ten years service were leniently granted, allowing for naturalisation under the *ordonnance* of 1715.[13] If the Irish were anxious when letters of naturalization for British people were cancelled at the start of the Seven Years War, they were soon reassured as their rights were confirmed by royal authority.

7 Three Americans, two from Luxembourg, one Welsh, one Dane, three Dutch, one Polish, one Portuguese and one Spanish. 8 This remained the case until the last years of the Irish Brigade, whed the recruitment of the rank and file became more and more disparate (P. Joannon, *Histoire de l'Irlande et des Irlandais*, p. 156). 9 See *Les étrangers dans l'armée française*, *Les Carnets de la Sabretache* (Paris, 2006). 10 Louis XIV to James II. 11 This lack of stability was also due to the necessity of having letters of naturalization renewed each time a new king came to the throne, which involved some financial aspects: new taxes were to be paid then, as well as each time the said taxes were increased. 12 The *droit d'aubaine* was the royal right which made the king of France the legal heir of all the strangers dying in France and under which he inherited their belongings in France. See Nathalie Genet–Rouffiac, *Le Grand Exil*, pp 278–80. 13 By his *ordonnance* of 1715, Louis XV gave automatic naturalization to all the foreigners – officers as well as rank and file – who could prove more than ten years in the service of France.

THE FRENCH REVOLUTION

On the eve of the Revolution, the three Irish regiments were based in France itself rather than in further flung garrisons. In inspection reports they were regularly commended for the quality of their appearance as well as their military qualities. The Regiment of Dillon was settled at Berghes, in north of France, then in Lille from April 1790. In January 1790, the second battalion of the regiment was sent to Saint-Domingue (San Domingo), from where it never returned. The Regiment of Berwick was stationed at the camp of Saint-Omer, before moving to Berghes and then Landau, on the eastern boarder. The Regiment of Walsh was sent to Brest, and embarked on 20 July 1788 for Ile de France. It would later provide a detachment for an expedition in Cochinchine (Indochina). When it returned to France it was based at Lorient and was given the duty of repressing the first royalist riot, which tried to take Vannes.

During 1791–2 there was increasing turmoil in France and within the army itself. As power had passed from the king to the new national assembly, measures were taken to create a new army. The revolutionary leaders were certain that the existing army was royalist and that it represented a potential threat for the newly-born government. Actually, it appears that the army, in its large majority, did not act to repress popular riots.

On 22 October 1790, a decree ordered that the tricolour ribbon be placed on all the colours of all the regiments and on 1 of January 1791, the former names of the regiments were replaced with a new numerical system. The Regiment of Dillon became the 87th Infantry Regiment, the Regiment of Berwick the 88th Infantry Regiment and the Regiment of Walsh became the 92nd Infantry Regiment. On 22 April 1792, a new regulations ordered the destruction of the ancient colours of the former Bourbon regiments and these were burned on the Place de Grève in Paris, during an official ceremony on 13 August 1793. After April 1792, the regiments began to be dissembled and to be amalgamated with volunteers' battalions, one of the few immediately positive measures.[14] General Arthur Dillon tried unsuccessfully to plead for the survival of the Irish regiments during a famous speech between the members of the Assemblée Nationale, a speech which indirectly led to his undoing.[15] Eventually all of these discriminatory measures created a deep discontent among the officer class and led to a migration of officers.

Regardless of common belief, this emigration was not a specifically aristocratic movement.[16] All of the social levels of the population of France

14 According to Corvisier, *L'armée française au XVIIIe siècle* (1954), the main legacy of royalty was a very strong military force which became the basis of the armies of the Revolution and the Empire. 15 See the entry for Arthur Dillon (1750–94) in Marie-Nicolas Bouillet and Alexis Chassang (eds), *Dictionnaire universel d'histoire et de géographie* (Paris, 1878). 16 See Ernest Daudet, *Histoire de l'émigration pendant la Révolution française* (Paris, 1904) and Henri Forneron, *Histoire générale des émigrés pendant la révolution*

were involved, including more than 20,000 farmers of Alsace and Lorraine who fled to the lands of Bade or the Palatinate. Among this phenomenon of mass migration, which touched up to 150,000 people, it has been suggested in some sources that the army lost 6,000 out of 10,000 officers while the navy lost three quarters of its officers. Many regiments based along the northern and eastern boarders and in the Pyrenees or in Italy simply melted away as many of the rank and file choose to follow their officers and warrant officers into exile. The regiments composed of foreigners seem to have been even more attached to the idea of military honour and tradition and many of them joined the 'Army of the Princes'.

As early as 16 July 1789, the Duc d'Artois was advised, or maybe even ordered, by his brother the king to leave France. Travelling through Switzerland, he reached Turin with his wife and his two sons and met his father-in-law, the king of Sardinia. The princes of Condé were already there. Being informed that a large number of emigrating officers and gentlemen had gathered on the northern and eastern boarders, he travelled to Koblenz, where his uncle, Clement Wenceslas, prince of Saxe and elector of Treves, was then living. He arrived on 15 June 1791. While there, the Duc d'Artois tried to recreate the splendours of the court of Versailles. In the process he squandered the financial support coming from the European countries that had ranged themselves against revolutionary against France.[17] This league of nations did not include Britain at this time and obviously the proposal to re-create an Irish Brigade would not have pleased the British government. Due to the duke's outlandish spending, money was soon in short supply and his debts began to accumulate.

Against this backdrop of total confusion and financial mismanagement, preparations began for a new campaign and forage began to be stocked and new equipment contracts signed. On 3 July 1792, the duke of Brunswick arrived in Bingen where he had an interview with the princes, the Duc de Condé and Xavier de Saxe. In a later conference held at Mayence, two decisions were taken that were to have ominous consequences. The first one was the well-known 'manifest of the duke of Brunswick', actually written by the marquis of Limon but seriously modified, the terms of which would galvanize the energies of the newly-born republic. The second decision, and certainly the most ill-conceived one, was the decision to split the forces of the émigrés into three units which were to be joined to the allied armies on the boarders in the north, in Lorraine and in Alsace. The overall political aim of the allies was obvious. For some France could still was the natural enemy, regardless of what its political regime was. But the aim of re-establishing the monarchy in France united the allied powers in their sense of purpose.

française (Paris, 1889). 17 In June 1792 these funds included 400,000 golden *livres* from the king of Prussia, 1,000,000 *livres* in letters of credit from Spain and 700,000 *livres* from Russia.

In 1791, the Regiment of Berwick was in garrison in Landau. The officers that were still on the rolls establised contacts with the Bourbon princes, in spite of the reluctance of the German authorities which wanted to avoid providing a 'casus belli' for France and were trying to discourage desertions among the regiment's soldiers. When the regiment was sent to Wissemburg, 200 soldiers left the ranks, put pieces of white cloth in their hats and began shouting '*Vive le roi et les princes*', without any interference from their officers. The regiment then went to Ettenheim where it was welcomed by the Prince de Condé himself, who had been warned of the regiment's arrival by Colonel O'Mahony, colonel of the regiment.

On the northern boarder, thirty-six officers of the 1st Battalion of the Regiment of Dillon reached Tournai. These included one lieutenant-colonel, one adjutant-major, one adjutant, six captains, seventeen lieutenants, ten sous-lieutenants, eleven sergeants and eight caporals. On 25 October 1791, all the officers from the Regiment of Dillon were sent to Koblenz where the regiment was re-created following a local recruitment of rank and file. In May 1792, it was reviewed by the Vicomte La Tour du Pin La Charce and it then included one colonel proprietor, the Comte de Dillon, one lieutenant-colonel (Greenlaw), one major (Coghlan), one adjutant-major, two adjutants, one tambour major, two battalions of eight companies of fusiliers with one company of grenadiers in each battalion.

In the case of the Regiment of Walsh, a part of the first battalion come from Longwy to join the Condé army on 22 May 1792, after a march of thirty-seven days from Vannes in Brittany. It was then transported by river to Koblenz. At its first review it was noted that the regiment included its colonel-proprietor, the Comte de Walsh Serrant, the colonel commandant, the Vicomte de Walsh Serrant, two majors (MacCarthy and the Chevalier de Walsh), the lieutenant-colonel (Galway), the adjutant major, the quarter-master/treasurer, the chaplain, 8 captains, 8 lieutenants, 6 sous-lieutenants, 13 under-officers and 44 rank and file spread between two companies of fusiliers and one of grenadiers.

Despite the financial support provided by the headquarters of the army of the princes, the number of men in the three regiments never reached the establishment stipulated by the *ordonnance* of 1788. At the same time, equipment and weapons were of poor quality.

The re-formed Irish Brigade took part in the campaigns in Argonne and in Champagne, fighting alongside the Austrians and Prussians and was present with the army before the battle of Valmy (1792). It would not, however, be involved in this battle. As was the case with the rest of the army of the princes, it was kept in the area of Suippe with no supplies, no forage or tents. It was only used once at the village of Tanlay. In reprisals against the local inhabitants, who were hostile due to the pillage carried out by Prussian troops, five villages were burnt, in addition to the village of Voncq, which was totally destroyed. In the aftermath of the battle, the entire army of the princes found itself in an unenviable position with its luggage and supplies plundered by fleeing Prussian troops while also enduring dreadful weather conditions, remembered later by soldiers as a time of 'mud up to the knees'. The retreat did not stop until 25 October 1792 when the army reached the walls of Liège. On 6 November, General Dumouriez won the battle of Jemmapes and this victory opened the doors of Belgium to the French. At the same time, General Custine threatened Koblenz. This double threat put an end to financial help from Austria and Prussia. Austria had previously promised to integrate the *émigrés* into its own army but now went back on this promise. The only remaining option was to disband the army of the princes and this occurred 21 November 1792. Thousands of men, women and children were cast from the Bourbon service for a second time and found themselves on the road to exile, in a state of utter destitution.

THE 2ND BATTALION OF THE 87TH INFANTRY REGIMENT AND THE IRISH REGIMENT OF DILLON

On 7 January 1792, the 2nd Battalion of the 87th Infantry Regiment, under the command of Lieutenant-Colonel Richard O'Shee, embarked at Le Havre to Saint-Domingue (San Domingo). It was composed of nine regular companies, totalling 629 men, all ranks. On its arrival, the regiment was used against the rebellion led by Toussaint Louverture. One year later, due to casualties, desertions and diseases, it numbered less than 200 men. It was by then positioned on the mole Saint-Nicholas, known as the 'Gibraltar of the Antilles'.

When Lieutenant-Colonel Whitecloke landed in command of a British force on 19 July 1792 it was at the request of the French royalist leaders of the island. Among his officers was a former artillery officer of the Bourbon army from Neux who was now served as an émigré officer in the British army. This émigré officer established contact with Capitaine O'Farrell, the commander of the 2nd Battalion of the 87th of Infantry Regiment. The latter agreed to hand over the his garrison, on the condition that it was to revert to the possession of the king of France.

The 2nd Battalion of the 87th Infantry then re-adopted its old title and was referred to as the 2nd Battalion of the Regiment of Dillon. Capitaine O'Farell was granted a commission in the British service. The regiment then included 8 captains, 16 lieutenants and sous-lieutenants, 4 men of lower rank in the headquarters and 150 men. Fifty three men and a few officers refused to enter British service and were sent back to France as part of a prisoner exchange.

In 1794 and 1795, the regiment took in a series of engagements, repelling the attempts of Republicans and Spanish forces, then allied to the 10,000 men of Toussaint-Louverture, to re-take control of the island. In August 1795, the battalion numbered no more than 56 men. Despite remaining loyal to the Bourbon cause, it was not integrated into the 3rd Regiment of the Irish Brigade (the Regiment of Dillon of Pitt's Irish Brigade), which was later stationed on the island. This was reportedly due to 'its permanent good behaviour'. But by 1796, only twenty men remained. It was disbanded, the eleven surviving officers being spread between the other units.[18] Chevalier Charles Dillon was presented with the regimental colour, with its traditional design of quarters in red and black, the inscription *In hoc signo vinces* and superimposed crown and Celtic harp designs.[19]

THE IRISH CATHOLIC BRIGADE IN THE BRITISH ARMY – 'PITT'S IRISH BRIGADE'

After the disbandment of the 900 men that remained in the Irish Brigade of the army of the princes, many officers and men joined the Damas Legion in the service of Holland while others joined the British army. In March 1794, the Comte Daniel Charles O'Connell and the Chevalier Henry Dillon suggested to the British government that they could raise Catholic or mixed regiments in Ireland for the Bourbon cause and in this way they might attract Irish Catholics recruits. William Pitt, then prime minister, obtained permission from George III to begin raising Irish Catholic troops, despite protests from Ireland's Protestant population. These troops were to serve in regiments commanded by exiled Irish officers who had formerly served in the Irish regiments of the Bourbon army. The duke of Portland acted as intermediary in this scheme and he invited the Duc de Fitzjames to enter British service, taking the remnants of the Irish Brigade of France. In November 1794, it was decided to raise two more regiments and ultimately six regiments constituted were gathered in the so-called 'Irish Catholic Brigade' or 'Pitt's Irish Brigade'.[20]

Recruitment was limited to counties in Ireland with a predominantly Catholic population and the term of enlistment was unlimited. Recruits could leave their regiment but only to transfer to another regiment of the Irish Brigade. The establishment of each regiment was settled at one colonel, one

18 These included the Chasseurs du Duc d'York, Chasseurs de Clarence, Régiment de la Reine. **19** His descendant, the Vicomte Dillon, handed presented this colour to the National Museum of Dublin on 29 December 1949. **20** These were the regiments of Fitzjames, (Comte de) Walsh-Serrant, Dillon, O'Connell (Vicomte de) Walsh-Serrant and Conway. See David Murphy, *The Irish brigades*, op. cit, pp 31–3.

lieutenant-colonel, one major, 7 captains and captain lieutenant, 10 lieutenants, 10 ensigns, one warrant officer, one quartermaster, one surgeon and his assistant, 32 sergeants, 30 corporals, 20 drum players, 2 fife players and 420 men.

Under the pressure from Irish Protestants, the British government assigned these regiments to permanent service in garrisons of Britain's overseas possessions. In April 1796 of the six regiments, four were based in the Antilles where they opposed French Republicans, their Spanish allies and the troops of Toussaint-Louverture. According to some sources, the troops still present in Ireland in 1798 at the brigade depot took part in the repression of the rebellion of that year in Ireland.

After several assignments in the islands of the Antilles, the regiments of Pitt's Irish Brigade were re-called to England and disbanded between November 1797 and 1799. The men from the disbanded regiments were either dismissed or integrated into other regiments.

THE INFANTRY REGIMENT OF EDWARD DILLON – THE LAST FRANCO-IRISH UNIT IN THE BRITISH SERVICE

At the end of 1794, the British minister for war, Lord Windham, reported to the secretary of state, Lord Dundas, that a large number of French émigrés had left France for Northern Italy and a suggestion was put forward to recruit from this exiled French population in order to raise a regiment to protect Corsica. The command of the newly-raised force was given to Edward Dillon, formerly an officer in both the Bourbon service and Pitt's Irish Brigade. In Bourbon service, he had previously served in the Regiment of Dillon and also the Regiment of Provence. The new regiment that Dillon was to command was raised in Livorno on 1 January 1795. The officers were of French origin and included former officers of the Irish Brigade in the French service. The rank and file came from northern Italy, Nice and Corsica. In August 1795, the regiment landed at Bastia where it was reinforced with the remnants of five 'White Ribbon' regiments that had been recruited among French émigrés.[21]

The Regiment of Dillon was engaged against Corsican insurgents until the British army evacuated the island. During its service there were numerous desertions and a part of the regiment was made prisoner by the French Republicans. The remnants of the regiment was sent to the island of Elba, and it then numbered 500, all ranks. Dillon himself then dismissed around 100 more men as being unsuitable and it was then planned to send the Regiment of Dillon to Portugal.

21 These were the regiments of Vioménil, Autichamps, Béthizy, Broglie and Laval-Montmorency.

On 21 June 1797, the regiment arrived in Portugal and was inspected by General Stuart, who declared 'never had he seen a regiment less worthy of the name of a British regiment than that of Dillon and Roll (a Swiss regiment)'. Reorganized, it was submitted to a new inspection by the general who then found it 'in excellent order'.

In Portugal, the regiment was never involved in any major battles and the total of its combat casualties was 24 men. It was based in Minorca in 1799 and left Gibraltar for Egypt, where it arrived on 8 March 1801. It took part in the campaign against French troops under General Menou and acquitted itself well. It was twice mentioned in dispatches during its campaign in Egypt. The regiment was presented with a regimental colour, ornamented with a badge shaped as a sphinx and the battle honour 'Aboukir'.

The Regiment of Dillon was then sent on garrison duties in Malta from 1802 to 1808. At the end of their terms of enlistment, the rank and file were progressively replaced by men of Mediterranean origins. These included Italians, Spanish, Albanians, Greeks and Turks. From 1808 to 1812, it was involved in a lacklustre campaign in Sicily and Calabria. In June 1812, the regiment was sent to Spain, where it is confined to minor engagements and tasks. This Regiment of Dillon was finally disbanded on 24 December 1814 in Gibraltar. It had been the last unit to include officers from the former Irish Brigade of the French army.

Following the restoration of the monarchy in France, the Irish regiments were not re-formed in the French army, despite the efforts of the surviving officers of the Irish Brigade. Louis XVIII opposed this idea for political reasons. A new and different direction was taken with the creation of the Foreign Legion or *Legion étrangère* on 9 March 1813. This new force became the heir to all the foreign regiments that had formerly been in the service of France.

CONCLUSION

In light of new research on the Irish in France, it is now necessary to question some of the long-held assumptions that have been held about that exiled population. It can be shown that after the Revolution that there was another phase of Irish emigration as the Franco-Irish officers and their families fled the new government and sought to re-establish the Bourbon monarchy. The men of this new diaspora would see service across Europe and also in the Antilles, where they served in émigré regiments and the regiments of Pitt's Irish Brigade.

By 1789, all of the original Jacobite exiles had founded new family lines in France and they had become real natives, totally integrated in French society.

This was even if they had maintained some connections with Ireland – connections that, on occasion, have persevered to modern times. However, not all of the Irish officers or officers of Irish origins fled from France at the Revolution. A section of the Irish military community remained. When it came to officers coming from Ireland or of Irish origins, these men were divided in two different social categories. The descendants of the aristocratic families of Ireland saw it as being natural to rally to the cause the French monarchy and the majority of them stayed faithful to the king of France. They took this loyalty to the point of entering the service of the king of England who, through seen as the oppressor of the Irish people, was now the enemy of Republican France. The Dillon family was an example of this inner conflict of loyalties – while many of the family served in the exiled French regiments or even in the British army – two of its members died as officers in the Republican army.

Many others did not come from such aristocratic backgrounds and realized that their fortune could only be won with their own swords. They did not leave France and many of them served the Republic and then the Empire. At least ten of them became general officers. At the Restoration, fourteen *lieutenants généraux* of Irish origins could be found in French army. Perhaps one explanation is that these officers had become sons of France as well as Ireland and were part of the main intellectual and philosophical movements of their times.

There are interesting subtexts to this phase of the Franco-Irish military connection and many of these would benefit from further research. There were, for instance, also Masonic influences at work in these processes. In the old Irish brigade, each regiment had his own lodge and it is at Saint-Germain-en-Laye, in the regiments of James II, that the Scottish Masonic rite came to prominence. During the many moves of these regiments during the eighteenth century, officers who were masons could temporarily join local lodges. The few archives available on the matter are explicit on this point, especially during the stay of the Regiment of Dillon in Carcassonne in 1764. During this year, eighteen officers, including Theobald Dillon, come 'knocking as masons' to the door of a respectable lodge known as 'Saint Jean de la parfaite vérité' in Carcassonne. One officer, named Aubrian (O'Brien) of the Regiment of Berwick, and who had been received as a mason in Dublin, was admitted to this lodge during a private visit.[22] It is often claimed that the forces behind the French Revolution were informed by Freemasonry. It would be interesting to investigate if any of the Irish officers who remained to serve the Republic had been influenced by the humanist ideals that were current in Masonic lodges in the decades before the Revolution.

22 M.P. Costeplane, 'Les Irlandais à Carcassonne', lecture given at the Académie des arts et sciences de Carcassonne, 2003.

As well as the difficulties experienced due to being exiled from their native country, these officers were often driven to a benevolent neutrality in the face of new ideas and the disruption of their times. Others followed their convictions through, which can only inspire admiration and respect. All of them, whatever their choices, showed the qualities that they had always been known for and maintained the fame of the Hibernian warriors.

From a wider political point of view, perhaps the most important development during this period was the creation of the Pitt's Irish Brigade. As Britain needed thousands of new soldiers to wage war against Revolutionary France, the penal laws that had prohibited Catholics serving in the army and the navy were relaxed. This was a significant reversal of penal legislation and the regiments of Pitt's Irish Brigade were raised during this period. As a result, they were part of a wider move towards the re-enfranchisement of Catholics and this movement would gather pace as the nineteenth century progressed. Due to their military service, Britain had to reward Catholics. It is interesting that these soldiers included returned émigrés whose families had fled to France at the end of the seventeenth century to escape the new political dispensation imposed by the victorious Williamite government.

Total war? Revolutionary France and United Irish strategy in the 1790s

HUGH GOUGH

The relationship between France and Ireland was radically changed by the revolutionary and Napoleonic wars which, in Chateaubriand's words, 'equalled many centuries' in their impact on Europe. There had been long bouts of warfare in Europe previously, but nothing on the same scale and intensity and nothing with the same explosive ideological content. For the revolutionary wars were not just traditional dynastic conflicts for territory – although elements of that were certainly involved – but a prolonged struggle for the domination of Europe fuelled by French expansionism and revolutionary ideology. The resulting challenge to the European state system forced countries to adopt new alliances and eventually, in the Congress of Vienna in 1815, devise new strategies for managing international diplomacy. The military and diplomatic aspects of the conflict have fascinated historians for generations, and the Irish aspects deeply embedded into the nationalist narrative. Two recent books have also revisited its significance in the development of the theory and practice of 'total war', which reached its peak in the twentieth century.[1] This article attempts to revisit all three aspects, examining the development of an Irish strategy in revolutionary France, the nature of French intentions towards Ireland and the relationship of Irish events to the theme of 'total war'.

In the best of all possible worlds France would never have needed an Irish strategy. In May 1790 the National Assembly issued a universal declaration of peace, abandoning offensive war and stating France's intention of living at peace with all its neighbours. Yet the revolution was based on an Enlightenment vision of universal nature and reason which, because of France's size, power and cultural prestige was bound to attract interest from intellectuals and opposition groups in most European states. It also appealed to political refugees who had fled to Paris after abortive uprisings in Geneva, the United Provinces, Liège and the Austrian Netherlands in the 1780s. Several leading figures in the Girondin war party that emerged in the Legislative Assembly during the winter of 1791–2, and who saw war as the solution to the revolution's domestic difficulties, were either members or

1 Jean-Yves Guiomar, *L'Invention de la guerre totale: XVIIIe–XXe siècle* (Paris, 2004); David Bell, *The first total war: Napoleon's coup and the birth of modern warfare as we know it* (London, 2007).

friends of such groups and Girondin rhetoric reflected their influence in its call for a war of conquest and liberation that would spread revolution to the rest of Europe. For a short time after the outbreak of war, during the winter of 1792–3 after the historic victory at Valmy, it became official policy to spread revolution in the wake of military conquest. A decree of 19 November 1792 promised French assistance to all foreign patriots and a second in December promised destruction of the feudal order wherever French armies went.[2]

That policy met resistance following the French occupation of Belgium and the German Rhineland in the winter of 1792 as local populations in both areas resented French military behaviour, economic stripping and religious destruction. The decrees were quickly withdrawn in April 1793 and instead, during the terror, the aim of liberation gave way to scepticism towards the optimism of foreign revolutionaries and xenophobia against the republic's major opponents, Austria and Britain. In January 1794 Robespierre renounced the idea of international revolution in favour of assisting individual uprisings as and when they occurred.[3] Instead, from the summer of 1794 onwards, as French armies took the offensive again, the idea of helping foreign revolutionaries was replaced by a determination to expand French territory to its so-called 'natural' geographic frontiers of the Rhine, the Alps and the Pyrenees. Belgium was annexed in the autumn of 1795, the Rhineland occupied in the same year and over the next three years satellite republics created beyond the natural frontiers, in the Dutch republic, Northern Italy, Switzerland and Rome. The expansion was controversial, as political opinion within France was split between conservatives who wanted to restrict it in the interest of negotiating a peace and left republicans who favoured the growth of the 'great nation'. But a military coup in September 1797, the 'fructidor' coup, resolved the argument in favour of expansion.[4]

For much of this time France had no real strategy towards Ireland. The outbreak of the revolution had revived Irish reform agitation that had begun during the American War of Independence, and led to increased contacts with France. It also attracted support from many of the Irish living in Paris, but by no means all because many army officers, merchants and priests rejected it because of its social and religious policies. Some supporters gathered in White's Hotel on 18 November to celebrate French victories and went with British reformers on 28 November 1792 to the bar of the National Convention to welcome the invasion of Belgium and assure deputies of the massive

2 T.C.W. Blanning, *The origins of the French Revolutionary wars* (London & New York, 1986) chapters 3–4. 3 F.-A. Aulard, *La Société des Jacobins. Recueil de documents pour l'histoire du club de Jacobins* 6 vols (Paris, 1889–96), vi, 633–5; Norman Hampson, *The perfidy of Albion: French perceptions of England during the French Revolution* (Basingstoke, 1998) pp 141–2. 4 See Marc Belissa, *Repenser l'ordre européen (1795–1802). De la société des rois aux droits des nations* (Paris, 2006) pp 65–84.

support for revolution in England, Scotland and Ireland.[5] In early January 1793 some Irish joined English and Scottish exiles to form a Société des Droits de l'Homme which appears to have petered out quite rapidly.[6] The Minister of Foreign Affairs during the winter of 1792–3, Pierre-Henri Lebrun believed that Ireland's catholicism would inhibit its potential for revolution potential, but an agent, Eleazer Oswald, was sent to Ireland in the following March to sound out the situation and intermittent contacts were kept up over the next two years.[7]

Sympathy only moved towards strategy when Anglo-French relations took a turn for the worse with the outbreak of war in January 1793. Britain's central role in creating the First Coalition and supporting the civil war against the republic in the Vendée revived deep rooted anglophobia in France, making the British government – and then the English people as a whole – rivals to Austria as the focus of collective French hatred. In a draft version of a new Declaration of the Rights of Man in April 1793, Robespierre included the clause: 'Those who make war against a people to prevent to progress of freedom and to destroy the rights of man must be attacked by everyone, not as normal enemies but as assassins and brigands.' On 7 August 1793 the Convention declared William Pitt, the British prime minister, the 'enemy of the human race' and ordered the arrest of anyone in France from the British Isles. On 9 October the importation of produce from Britain, Scotland and Ireland was banned and on 29 May 1794 French armies were ordered to kill any captured English or Hanoverian prisoners on sight.[8] Speaking for the Committee of Public Safety, Barère told his fellow deputies: 'Between all nations and societies there is a kind of natural law known as the "rights of people". But it is unknown to the supervised savages of Great Britain, making them a people foreign to Europe, foreign to humanity. They must disappear.'[9] The order to murder prisoners of war was carried out in only small numbers and the decree was repealed before the end of the year, but anglophobia intensified with the role played by the British in the émigré landing at Quiberon Bay in the summer of 1795.

Anglophobia and tactical necessity led to an Irish strategy. By the spring of 1796 the disintegration of the First Coalition left Austria and Britain as the only powers aligned against France. The Directory's main military thrust lay to the south, with a land offensive against Austria through southern Germany, but it stalled and instead Bonaparte's victories in northern Italy forced Austria

5 *Convention Nationale. Adresse des Anglais, des Ecossais et des irlandais résidents et domiciliés à Paris à la Convention Nationale* (Paris, 1792); Marianne Elliott, *Partners in revolution: the United Irishmen and France* (New Haven & London, 1982), pp 54–5. 6 *Réimpression de l'Ancien Moniteur* xv, 7 January 1793, p. 58. 7 P. Coquelle, Les projets de descente en Angleterre', *Revue d'Histoire Diplomatique,* 5 (1901), 618–25. 8 *Moniteur* xviii, no. 21, 20 vendémiaire II p. 86; Hampson, *The perfidy of Albion,* pp 141–2. 9 Sophie Wahnich, *L'impossible citoyen: L'étranger dans le discours de la Révolution française* (Paris, 1997)

to an armistice, then a peace, during the summer and autumn of 1797. Meanwhile Hoche's success in bringing the counter-revolution in the Vendée under control freed up troops from the Armée des Côtes de l'Océan for a possible invasion of Britain. Invasion plans dating back to the mid-seventeenth century lay gathering dust in the archives of the defence ministry and several new projects had been submitted in 1793 when the war with Britain first started.[10] These included one from Hoche, submitted in October just before he was promoted to general, which proposed appointing a courageous leader – probably himself – to take charge of a hundred battalions of infantry, supported by cavalry and artillery, to 'fairly ravage the country and soon force the tyrant coalition to sue for peace'. No skill and cunning were needed, just 'iron, fire and patriotism' in a plan that involved crossing the channel by lining up merchant ships: 'Cover the sea with the ships of the merchant marine. Let them be armed for war; let them form a bridge from the coasts of France to arrogant Albion.'[11] Hoche's plan was ignored, perhaps wisely, and a decision of the Committee of Public Safety on 1 vendémiaire II (22 September 1793) to assemble an invasion army of 100,000 men was later abandoned because of lack of manpower was. Even a minor attack on the Channel Islands had to be abandoned through lack of resources.[12]

In the spring of 1796 manpower and those resources were available and Hoche worked with one of the five Directors, Lazare Carnot, to replicate a Vendée-style guerrilla war in England by planning to land former convicts and captured rebels – chouans – from the Vendée and Brittany, into Britain. Carnot also envisaged a landing in Ireland and, after Wolfe Tone's arrival from the United States of America in February 1796, the Irish expedition was taken seriously. Tone convinced Carnot that a French led landing would enable an independent Irish republic to be established that would end Irish recruitment to the British army and force Pitt to sue for peace. In the words of the Directory's instructions to Hoche outlining the mission on 1 messidor IV: 'To detach Ireland from England will reduce England to the position of a second rate power and eliminate most of its maritime supremacy. We should not underestimate all the advantages that Irish independence would bring to France.' Carnot himself stated: 'I see in the success of this operation the fall of the most dangerous and most irreconcilable of our enemies.'

The invasion plan initially had three stages. A fleet en route to India would land a demi-brigade of 5,000 men in Galway Bay in August and take control of Connaught (excluding Leitrim) and part of Co. Clare. A second expedition would bring six thousand more men into Galway Bay, including a number of

pp 255–6. **10** Edouard Guillon, *La France et l'Irlande pendant la Révolution* (Paris, 1888) pp 53ff. **11** Donald R. Come, 'French threats to British shores, 1793–1798', *Military Affairs*, 16:4 (winter, 1952), 75–6; Guillon, *La France et l'Irlande*, pp 72–5. **12** Come, 'French threats', 76–7.

English speaking officers, and a 'free corps' of 5,000 criminals, former rebels and undisciplined soldiers. A third wave from Dunkirk would follow. The composition of the free corps included in the second wave was to be 'such as to rid France of many dangerous individuals and there is a lot to be said for including former chouans who are prepared to co-operate'. The idea for it appears to have come from an officer in the Armée des Côtes de l'Océan, General La Barolière, drawing on his experience of guerrilla warfare in the Vendée, and from General Jean Joseph Amable Humbert.[13] Hoche originally planned to use them in Ireland but Tone persuaded Carnot against the idea because their behaviour could alienate support for France and instead it was decided to send them to Britain to cause diversionary chaos there.[14] Hoche quickly recruited 1600 troops with poor discipline records in Brest – 'the worst subjects we can get rid of from France' – 60 of whom were subsequently arrested for criminality before the end of the year. Another group of 1200 – the 'Legion of the Free' – was assembled in Saint-Malo under Humbert and a third group made up of 5,000 prisoners of war and deserters was assembled in Dunkirk under General Quantin.[15] Quantin's unit was to sail to Newcastle in north-eastern England while Humbert's was to land in Cornwall, destroy bridges over the Tamar and detonate an insurrection among tin miners, before burning the city of Bristol and joining up with the third group which was to land in south Wales.[16] Quantin's force finally left Dunkirk on 10 November in small boats built from a prototype developed by a Flemish sailor, Joseph Muskeyn, for the Swedish navy for use in the Baltic. However it only got three leagues out to sea before being becalmed and blown back onto the beaches of Dunkirk. Muskeyn's own boat, the *Colombe*, sank with the loss of two lives, the soldiers were sent back to their units and the deserters and prisoners force marched to Brest.[17] Humbert's expedition to Cornwall was called off and his troops incorporated into the Bantry expedition, while the troops assembled in Brest went on the ill fated Fishguard expedition in February 1797 under the command of William Tate. It landed in south Wales but was rounded up outside the Royal Oak pub in Fishguard and sent back to France in exchange for British prisoners.[18]

From the start the Bantry expedition was dogged by delay. The Italian campaign always remained the first priority and Hoche complained bitterly

13 Edouard Desbrière, *Projets et tentatives de débarquement aux Iles britanniques* 4 vols (Paris, 1900–2), i, 61–6; Raymond Guyot, *Le Directoire et la paix de l'Europe, des traités de Bâle à la deuxième coalition (1795–1799)* (Paris, 1911), p. 161; Guillon, *La France et l'Irlande*, pp 84–5, 88–91. 14 Antonin Debidour, *Recueil des Actes du Directoire exécutif*, 4 vols (Paris, 1910–17) ii, 660–2, 688. 15 For Humbert, see Henry Poulet, *Un soldat lorrain méconnu: Le général Humbert (1767–1823)* (Nancy, 1928) chapt. II passim. 16 Debidour, *Recueil des Actes*, iii, 173, 177, 453, 524, 625 & 699; Guyot, *Directoire* p. 161. 17 Desbrière, *Projets et tentatives*, ii, 80–2. 18 E.H. Stuart Jones, *The last invasion of Britain* (Cardiff, 1950).

when detachments from his army were sent to reinforce Bonaparte in Italy. But the navy was the real problem. Vice-Admiral Villaret-Joyeuse, commander of the Brest fleet, disapproved of the Irish expedition and would have preferred to sail directly for India to take on British power there. He also faced real problems with an acute shortage of ship building materials and experienced sailors caused by the social upheaval of the revolution. The Indian part of the expedition was finally abandoned in October and the reluctant Villaret-Joyeuse replaced by Morard de Galles in November. The invasion plan was then simplified to a two stage landing, with the first wave to be led by Hoche and the second by general Houdeville, but the preparations remained slow and Hoche showed little sympathy for the problems that the navy had. The fleet that finally left Brest on 16 December carried 14,450 troops and was made up of 17 ships of the line, 13 frigates, 6 corvettes and 20 transport ships.[19] Another 15,000 men and a cavalry unit were to follow. Yet the ships were overcrowded, the sailors inexperienced and Hoche found himself separated from the main body of the fleet during an ill thought-out manoeuvre shortly after leaving Brest. When the main fleet arrived off the south west coast of Ireland it encountered rough seas and heavy gales and after several days attempting to tack its way into the bay in storm force conditions, turned back and headed for home. By the time Hoche arrived off Bantry it was too late and he followed the main fleet back to Brest.[20]

The Bantry failure proved pivotal because the navy lacked the capacity and will to mount a second attempt and Hoche was transferred to the army of the Sambre et Meuse where he died prematurely in September 1797. A second invasion planned in the summer of 1797 by the Batavian Republic collapsed when the Dutch fleet was defeated by the British navy at Camperdown. Significantly too, faced with strong criticism of the Bantry fiasco in Paris, the Directory decided that no further expedition would be sent to Ireland until after an insurrection had broken out.[21] Unfortunately that policy change was never clearly communicated to the United Irish leadership in Ireland, which postponed its plans for a rising in the expectation of a second invasion that never came. Instead of Ireland, in late 1797 the Directory returned to the idea of a direct invasion of England after the peace of Campoformio. Bonaparte was appointed commander of an Army of England to be assembled on the Channel coast, but he quickly realized that the navy could not provide adequate cover and switched his attention instead to an attack on British power in the Levant by the invasion of Egypt. He set sail from Toulon in May with 38,000 soldiers, just as the United Irishmen launched their insurrection

19 Steven T. Ross, *Quest for victory: French military strategy 1792–1799* (London, 1973), chapter 5. **20** Hugh Gough, 'Anatomy of a failure', in John A. Murphy (ed.), *The French are in the Bay* (Cork & Dublin, 1997), pp 9–24. **21** F.-A. Aulard, *Paris pendant la réaction thermidorienne* 5 vols (Paris, 1898–1902), iii, 682, 698; iv, 28.

in Dublin. The Directors responded as best they could to the insurrection by scrambling together a four pronged invasion force under General Rey at Dunkirk, General Humbert at Rochefort, General Hardy at Brest and Dutch ships out of Texel. Yet, as in 1796, preparations were slow and Humbert compounded the problems by setting sail first, without co-ordinating his colleagues in Brest and Dunkirk. By the time he arrived in Killala Bay on 22 August the rebellion had already been crushed and his defeat at Ballinamuck on 8 September ended a brief and futile campaign. Rey's expedition briefly landed Napper Tandy and a group of fellow exiles in Donegal in late September, while the ship carrying Tone from Brest was captured off Lough Swilly in mid-October.[22]

The United Irishmen continued to lobby for a third invasion after the 1798 failure but internal feuds undermined their authority and they constantly overestimated their bargaining power. French officers also sent in new invasion plans, most of which were totally unrealistic. An American, Robert Fulton, proposed using small submarines equipped with three men and two candles to attack British naval vessels. Humbert urged the simultaneous landing of tens of thousands of men in Ireland, Scotland and the south coast of England, apparently oblivious of the manpower and logistical problems.[23] Yet in government circles the optimism of 1796 had long since vanished and when Bonaparte took power at the end of 1799 he was only interested in using the threat of invasion as a bargaining counter to get peace with Britain. At one stage he even proposed an exchange of French émigrés in England for political exiles – including the United Irishmen – in Paris.[24] In 1803, after the breakdown of the peace of Amiens, invasion plans were briefly revived as a diversionary tactic for an invasion of Britain, and 12,000 troops assembled in Brest for a diversionary expedition under General Augereau. Bonaparte even set up an Irish Legion, under the command of Arthur O'Connor, with Irish officers in command. However, the British blockaded Brest and the plans were put on hold. The legion based at Morlaix survived but was racked by conflict between supporters of Arthur O'Connor and Robert Emmet and had no troops for the officers to command anyway. In the summer of 1806, when invasion planes were finally abandoned, it was incorporated into the main Napoleonic army and took part in the repression of the revolt in Madrid against Joseph Bonaparte in the spring of 1808.[25]

22 Elliott, *Partners in revolution,* pp 214ff.; Harman Murtagh, 'General Humbert's futile campaign', in Thomas Bartlett, David Dickson, Dáire Keogh & Kevin Whelan (eds), *1798: a bicentenary perspective* (Dublin, 2003), pp 174–87. **23** Elliott, *Partners in revolution,* pp 243ff; Desbrières, *Projets et tentatives* ii, 260 & 282–5. **24** Ibid. ii, 274. **25** Lt.-col. Pierre Carlès, 'Le corps irlandais au service de la France sous le Consulat et l'Empire', *Revue historique des armées,* 2 (1976), 25–54; John P. Gallagher, *Napoleon's Irish Legion* (Carbondale, IL 1993), pp 21–52.

French strategy towards Ireland clearly had two strands: invasion to initiate an insurrection as in 1796, and invasion to assist an existing insurrection as in 1798. Neither strand succeeded but an evaluation of the strategy as a whole requires consideration of what would have happened if they had. There are clear dangers with counter-factual history but there is enough information to show what French intentions were, even if their eventual implementation remains speculative. French plans, and the actions of French armies in other areas of conquest, suggest that although the Directory wanted its commanders and troops to behave with sensitivity in Ireland, a successful invasion would have ended in the creation of some kind of Irish puppet state serving French interests. In Tone's early discussions in the spring of 1796 Clarke, the minister of war, was vague about the structure of any future Irish republic, believing that public opinion would prefer a restored Jacobite monarchy.[26] But the secret instructions issued by the Directors to Hoche on 19 July 1796 (1 thermidor IV) were quite specific, noting that although the Irish appeared ready for revolution, it would only serve French interests if it resulted in a French style republic. Hoche was therefore ordered to mould public opinion in Ireland, control events on the ground and prevent local adventurers – '*aucun intrigant*' – from assuming power. If a mass rising broke out when troops landed, he was to channel it in the right direction and take command of any Irish army that was formed. He was to hold power and control all government officials while French troops remained in the country and until a general European peace had been concluded. Members of the Catholic Committee could be recognized as authentic representatives of the Irish people until free elections were held to a National Convention, to draw up a constitution. But the Convention was to be small, with two or three deputies from each county, and it to be dissolved if the deputies turned out to be favourable towards England and hostile to French interests. Any constitution would have to be a republic, modelled on the Directory, and if people demanded a strong leader they could have one as long as he was Catholic and anti-English. Since the Irish would be getting a republic with French help, they would be expected to play their part in the war against England and agree favourable trade terms with France. On the religious question, the Directory recognized that the Irish were not ready to abandon Catholicism and advised Hoche to allow the three main denominations – Catholic, Anglican and Presbyterian – to remain in existence, but with no state support. The Catholic Church should be stripped of influence because its principles were opposed to the 'healthy ideas of philosophy and morality and to the progress of science'. Catholics should not be allowed to convert to Presbyterianism or Anglicanism, as this would bring them too close to the English, but should be encouraged to adopt deism instead.

26 Elliott, *Partners in revolution*, p. 85.

Eighteen months later, in 1798 Hardy was instructed to maintain strict troop discipline and to respect both religion and property. But he was to set up a provisional government with the United Irish leaders and appoint civil or military commissioners in towns and rural areas. He was to encourage hatred of the English and do his best to encourage pro-French sentiments among a population that was 'perhaps not sufficiently enlightened to know all the advantages of Liberty and Equality'. In the meantime he was to avoid relying on any Irish who were just politically ambitious, condemn cowards who failed to support the republican cause and encourage a respect for French authority.[27] Clearly French success on either occasion would have resulted in an Irish republic, but embedded in these instructions are the trip wires over religion, political control and the needs of French occupation that were already leading to conflict in Belgium, the Rhineland and northern Italy. There is little doubt, therefore, that an Irish republic would have been a French puppet state with many of the inevitable disagreements and conflicts that French occupation brought elsewhere.

TOTAL WAR

To what extent do plans for the invasion of Ireland tie in with the development during the revolutionary and Napoleonic wars of the tactics of 'total war'? Like all simple phrases, the concept of total war is more ambiguous than it looks. No war has ever been utterly total in the sense of ignoring all civilized restraints but its nature has changed radically since the late eighteenth century in a way that underpins much of the narrative of modern warfare. Briefly stated the theory of total war argues that the nature of warfare changed from limited warfare in the seventeenth and eighteenth centuries, based on aristocratic rules of conduct and the tactics of manoeuvre and surrender, to total war since 1789 involving massive civilian and military mobilisation for the utter destruction of enemy forces and society.[28] Two recent books, by Jean-Yves Guiomar and David Bell have explored its relevance for the revolutionary period from different angles: Guiomar by examining its impact on Franco-German relations from the late eighteenth to the early twentieth centuries and Bell by looking at the background to war and the Vendée and Napoleonic wars, with sideways glances to the Balkans in the 1990s and the US invasion of Iraq.[29] Both argue that the revolutionary wars mark a clear

27 Archives Nationales BB^4 122 pièces 251–5. 28 Roger Chickering, 'Total war: the use and abuse of a concept', in Manfred F. Boemeke, Roger Chickering & Stig Forster (eds), *Anticipating total war: the German and French experiences 1871–1914* (Cambridge & London, 1999), pp 13–28; J.F.C. Fuller, *The conduct of war 1789–1961: a study of the impact of the French, Industrial, and Russian revolutions on war and its conduct* (London, 1961). 29 Guiomar, *L'Invention*; Bell, *First total war.*

break in the nature of warfare by being ideological conflicts that mobilized entire civilian populations and ignored the cultural restraints of previous wars. Less a conflict between dynasties than a struggle between ideologies and peoples, this new style of war required large scale mobilization, demonization of the enemy and the pursuit of total victory at all costs. Politicians for the first time spoke of war as a process of transformation, a war to end all war: 'this war will be the last war' claimed the French politician and general Dumouriez in the summer of 1792. The result was a ferocity that marked the guerrilla struggles in the Vendée, Napoleonic Spain and the Austrian Tyrol and saw mass civilian massacres. One fifth of all the major battles fought in Europe between 1490 and 1815 occurred between 1789 and 1815, and the death toll across Europe for the period 1792–1815 may well have topped five million.

Not everyone agrees with Guiomar and Bell that the change towards total war was that abrupt and all embracing, but most agree that an attitudinal shift did take place that reflected the cultural and political shifts of the revolution. If so, how did it affect Franco-Irish military relations? Clearly Ireland did not experience total war. The physical and material damage of the 1790s was relatively light, and Ireland suffered none of the mass casualties of continental Europe. Humbert's invasion was short-lived and the government repression of agitation and rebellion in 1797 and 1798 left only around 20,000 dead, with the post-rebellion repression adding a further 500 rebels executed.[30] Those deaths were significant in a population of little more than four million, but probably no more than in the 1641 rebellion and very few in comparison with the casualty figures for contemporary wars in continental Europe. In the Vendée and Napoleonic Spain the scale of repression was far harsher. Over 200,000 died in the counter-revolutionary insurrection in the Vendée in western France alone and as many as 50,000 were killed in the siege of Saragossa during Napoleon's war in Spain in the winter of 1808–9.

Yet although Ireland escaped the casualty figures of total war several aspects of French and Irish strategy reflected its assumptions and values. Firstly, the plans which Tone supported, to send guerrilla units into Britain in 1796, accepted the abandonment of normal moral constraints on the waging of war and the deliberate involvement of civilians in warfare in the way that had already happened in the Vendée and other areas of French conquest. Its purpose was to terrify civilians, destroy property and break morale in order to force a total surrender. In the British case it was also underpinned by a lethal cocktail of xenophobia and anglophobia that had developed since 1793. As

30 Michael Durey, 'Marquess Cornwallis and the fate of Irish rebel prisoners in the aftermath of the 1798 rebellion', in Jim Smyth, *Revolution, counter-revolution & union: Ireland in the 1790s* (Cambridge, 2000), pp 128–45; Thomas Bartlett, 'Clemency and compensation: the treatment of defeated rebels and suffering loyalists after the 1798

early as the autumn of that year, in a comment on Britain's role in the Vendée, Hoche had queried what rules of war could be applied in a struggle with 'barbarians who are fighting us with poison, assassination and fire?' In a letter to Carnot in the summer of 1796 on the projected invasion of Wales, he admitted that: 'the war I propose to wage on our rivals is terrible [...]. Every one of (the recruits) knows the effect of rape, pillage and murder in our own country. What will it be like in a foreign country?'[31] Tone shared both his anglophobia and his readiness to use guerrilla tactics. He translated Humbert's instructions for the planned attack on England in the autumn of 1796 that included the plan to burn Bristol: 'The truth is I hate the very name of England; I hated her before my exile; I hate her since, and I will hate her always!'[32] In August 1797 he even reluctantly agreed to the use of guerrilla warfare in Ireland, against his better nature, admitting that 'there is something in the proposed expedition proposed more analogous to my disposition and habits of thinking, which is a confession on my part more honest and wise, for I feel very sensibly that there is no common sense in it'.[33]

Secondly, Tone shared with revolutionaries in France an apocalyptic view of the revolutionary wars. Dumouriez, foreign minister when war was declared in the spring of 1792, called the war against Austria a 'war to end all wars' because the revolution heralded a new age of harmony and progress. That rhetoric, so typical of Girondin euphoria, later permeated the mood behind the Irish expeditions. Carnot told Hoche in the summer of 1796 that victory in Ireland would see 'the fall of the most irreconcilable and dangerous of our enemies' and ensure 'tranquillity for France for centuries'.[34] Tone shared that view when attached to the Texel expedition in the summer of 1797: 'There never was, and there will never be such an expedition as ours if it succeeds.' Its object was neither dynastic nor territorial but rather to 'change the destiny of Europe, to emancipate one, perhaps three nations; to open the sea to the commerce of the world; to found a new empire, to demolish an ancient one, to subvert a tyranny of six hundred years'.[35]

Thirdly, Tone shared the French view that France was the mother country of revolution despite the clear manipulation and exploitation that French expansion brought in the 1790s, and linked Irish independence to the strength of the new French empire. Ireland would gain independence but also play its part in a new European order dominated by French republican values. As the experience of the Batavian, Swiss and north Italian republics showed, this meant close control and continual intervention from Paris in the politics of client states. Tone welcomed this, oblivious of its implications for Ireland, supporting the French coup in the Batavian Republic in 1797 because the

rebellion', ibid., pp 105–19. 31 Guillon, *France et l'Irlande,* pp 91, 102–3. 32 Marianne Elliot, *Wolfe Tone: prophet of Irish independence* (New Haven & London, 1989) p. 316. 33 Elliott, *Tone*, pp 349–50. 34 Debidour, *Recueil des actes,* ii, 688–9. 35 Elliott, *Tone,* p. 347.

Dutch had failed to solve their own problems and welcoming the invasion of Rome in February 1798 as equal in importance to the French Revolution itself. He viewed the blatantly puppet regime of the resulting Roman Republic as the rebirth of ancient Rome. His anticlericalism led him to regard the papacy as the embodiment of evil and the removal of Pius VI as the work of a 'special Providence guiding the affairs of Europe [...] and turning everything to the great end of the emancipation of mankind from the yoke of religious and political superstition, under which they have for so long groaned.' It seems reasonable to assume that, despite his concerns about French interference in 1796, Tone would have accepted similar French intervention in the domestic politics of an Irish republic on the lines dictated to Hoche by the Directory, if and when the inevitable conflicts with France arose.[36]

Fourthly, Tone was also dazzled by the military ethos of revolutionary France. He was enrolled into the French army for the Bantry expedition, enjoyed Hoche's company in Brest while waiting for departure and was delighted with his military uniform and army life. In late November 1796, he wrote to his wife Matilda that, if Bantry succeeded, he would want to be commissioned in the Irish army.[37] For him, as for many in the 1790s, the new military ethos of the revolution, with its linking of martial courage to the civic virtues of political republicanism, was seductive and attractive. Hoche whom he admired was, like Bonaparte, one of a new breed of political generals who had risen from the ranks through their talents and their commitment to the republican values of the revolution. For both Hoche and Bonaparte, politics and war were interlinked, and while Hoche was involved in planning a coup d'état in Paris shortly before his death, so Bonaparte went the whole hog and came to power that way in 1799.

CONCLUSION

Tone and the United Irishmen embraced the ideology of war that the revolution introduced into Europe, and in doing so linked their aspirations to those of the French. That link has been reinforced by historians and political activists over the last two hundred years fitting Tone's ideas, actions and death into a republican tradition designed to transcend social and religious boundaries in the interests of an independent Irish state. Yet Irish and French republicanism were always an imperfect match because the radicalism and secularism built into the French model had limited application or appeal in an Ireland where political discourse was more closely linked to the Anglo-American world and religion seen as an essential element of social and national

36 Ibid., p. 364. 37 Elliot, *Tone*, pp 313 & 321.

identity. The experience of French occupation in regions as far apart as Belgium and the Roman Republic in the late 1790s shows how secularism and anticlericalism generated a popular hostility and resistance that would have been difficult to avoid in Ireland if the invasion attempts had been successful.

In other ways the revolution weakened Franco-Irish links. It abolished the Irish army regiments and closed down the Irish College. The college was re-established after 1815 but the regiments never were and throughout the ninteenth century recruits from Ireland joined the British army in much larger numbers than they had ever joined the Irish regiments in France. Paris remained a place for Irish political exiles but as Britain and France never again went to war after 1815, the opportunities for Irish republicanism to exploit Anglo-French conflict vanished, just as emigration patterns changed the flow of Irish exiles from the European continent and towards Britain and North America rather than the European continent. As the links with France faded, those with Britain were tightened. Harman Murtagh noted of 1798 that: 'The whole episode showed that retaining control of Ireland was a far greater strategic necessity for Britain, especially in wartime, than "liberating" it was for the French'.[38] Certainly one of the main effects of the revolutionary decade in Ireland was the ending of parliamentary autonomy and the passing of the Act of Union in 1801. In the aftermath of rebellion, Pitt decided that the security of the two islands required territorial consolidation of the kind that France was in the process of implementing in continental Europe. Just as revolutionary France pushed the pope out of Avignon, gobbled up the Austrian Netherlands, absorbed a hundred small German states, devoured the kingdom of Savoy, and crunched numerous Italian republics into a handful of client states, so too the Act of Union rationalized the map of the two large islands off north west Europe. Territorial consolidation, as Edward Cooke, under-secretary in Dublin Castle, noted, was the flavour of the day: 'France well knows the principles and force of incorporations. Every state which she unites to herself she makes part of her empire, one and indivisible [...], as we wish to check the ambition of that desperate and unprincipled power, and if that end can only be effected by maintaining and augmenting the power of the British empire, we should be favourable to the principle of union, which must increase and consolidate its resources.'[39] War rationalized the political map of Europe and that rationalisation affected not only France but also the countries that France came into contact with. Ireland was no exception.

38 Murtagh, 'Humbert's futile campaign', p. 187. 39 Quoted in Smyth (ed.), *Revolution, counter-revolution and union*, pp 16–17.

'*Un brave de plus*': Theobald Wolfe Tone, alias Adjutant-general James Smith: French officer and Irish patriot adventurer, 1796–8

SYLVIE KLEINMAN[1]

And now I'm drinking wine in France
The helpless child of circumstance.
To-morrow will be loud with war,
How will I be accounted for?[2]

> Satisfied that [Ireland] was too weak to assert her liberty by her own proper means [...] I sought assistance [...] in France, where [...] I have had the honour to be [...] advanced to a superior rank in the armies of the Republic [...]. I have had the confidence of the French government, the approbation of my generals, and the esteem of my brave comrades.[3]

One crisp day in March, 1796, a French naval officer visited the new Panthéon in the heart of Paris. Aristide Aubert Du Petit Thouars was a former aristocrat who had finally returned from exile to serve his country, and was showing the sites of the capital to one James Smith, an American merchant he had befriended during the transatlantic crossing.[4] The two men were of an age, and though privately Smith was a committed republican who despised unearned privilege, he could not help but like the congenial Frenchman who had known many adventures and reverses of fortune. So he

1 I am grateful to the Irish Research Council for Humanities and Social Sciences as this article is part of my IRCHSS Post-Doctoral project, an illustrated book on Tone's life in Europe, 1796–1798. With thanks to my mentor Professor David Dickson (Trinity College, Dublin) and Bernard Gainot (Institut d'Histoire de la Révolution française) for their ongoing support. 2 Francis Ledwidge, 'Soliloquy', in Dermot Bolger (ed.), *Selected poems of Francis Ledwidge* (Dublin, 1993), p. 74. 3 *Minutes of the court martial held last Saturday, 10th November [1798], at the barracks of Dublin, with the speech made upon that occasion by Theobald Wolfe Tone, Esq.*, in T.W. Moody, R.B. McDowell & C.J. Woods (eds), *The writings of Theobald Wolfe Tone 1763–1798*, iii, *Jan. 1797 to Nov. 1798* (Oxford, 2007), p. 397; afterwards Tone, iii. 4 Amiral Bergasse du Petit Thouars (ed.), *Aristide Aubert Du Petit Thouars, Héros d'Aboukir 1760–1798: Lettres et Documents inédits* (Paris, 1937). See Joseph Clarke, *Commemorating the dead in Revolutionary France revolution and remembrance, 1789–1799* (Cambridge, 2007) for a thorough discussion of the Panthéon.

may not have shared with his tour guide the emotional response to the Panthéon he consigned to his diary that night, moved at this dignified republican tribute from a grateful French nation to its 'Mighty Dead'

> Certainly nothing can be imagined more likely to create a great spirit in a nation than a repository of this kind, sacred to everything that is sublime and illustrious and patriotic.[5]

France and Britain were at war and within weeks, Du Petit Thouars would be reinstated in a weak French navy which sorely needed competent officers, regardless of their royalist origins. Almost a year later, fate would throw the two men together in what, to the Frenchman, seemed the most improbable of settings. Going about his duties in the naval port of Brest in January 1797, as a *chef de division* assigned to compile reports on the ill-fated expedition to Ireland,[6] Du Petit Thouars discovered Smith had not come to France to sell American grain, and wrote in consternation to his sister

> I've been deep in thought after a most extraordinary encounter, and I was oppressed when I got back. I had told you of a Smith I had crossed the Atlantic with [...]. Well I just met him again on the ships returned from the Irish expedition, sporting the uniform of a chef de brigade. He told me he was Irish, and that [...] he had attempted to deliver his country [... and that] if he were taken, he would be hanged.[7]

Smith was of course a *nom de guerre*, as he was really Theobald Wolfe Tone, whose daring mission to France remains one of the most cherished episodes of Irish history. Though his noble efforts ended in tragic failure, his tenacious lobbying for a French invasion of Ireland and his role as a French infantry officer in the campaigns from 1796 to 1798 were considerable achievements. Only recently have the insights from Du Petit Thouars been discovered, and they corroborate the clandestine nature of Tone's exile, while reinforcing the sense of romantic adventure. Like Tone he would not survive the year 1798, when Napoleon sailed to Egypt rather than Ireland, having fought bravely but perishing heroically on the ship he commanded at Aboukir. It is possible many fellow officers whose respect Tone had earned did not learn of his true identity until reading in the papers how tragically his life had ended in Dublin in November 1798. Defiantly he had defended his honourable status as a

5 T.W. Moody, et al. (eds), *The writings of Theobald Wolfe Tone*, ii: *America, France and Bantry Bay, Aug. 1795 to Dec. 1796* (Oxford, 2001), p. 120; afterwards, Tone, ii. 6 SHD/Marine: inter alia, CC7/778: Aristide Du Petit-Thouars (1760–98); BB4 Campagnes navales 1796, Registre 103/Expédition d'Irlande. 7 Bergasse Du Petit Thouars (ed.), *Aristide Aubert Du Petit Thouars*, p. 454, author's translation.

French officer, requesting to be shot by a firing squad and not to be forced to mount the scaffold as a traitor. When this was denied, he cut his own throat and died in his prison cell. Overnight, Tone became a martyr of liberty, according to contemporary French phraseology, having sacrificed all that he held dear for his country. Thus his visit to the Pantheon is steeped in fateful symbolism, as Tone himself would later undergo Irish 'pantheonisation'.

Tone's arrest and death are defining moments in the story of Irish freedom, and his life for ever more became associated with that of a French officer. In fact there are no portraits of the adult Tone in civilian dress. He commands a special place in the Irish collective memory, because of Ireland's nationalist heroes he seems so familiar, as his biographer Marianne Elliott has underlined. This is thanks to his detailed diary which combines keen observation with utter candour.[8] Embedded in his journals and correspondence is a vivid portrayal of life as a foreign officer in the army of the French Republic, against the backdrop of Napoleon's rise to power. His military career may have been brief, yet despite its significance at a major point in Irish history, and the wealth of insights he provides for scholars of French history, it has never been the subject of a dedicated study.

This may be due to the particular course of Irish political history, and how Tone's hagiographical image as a separatist was later misappropriated by militant republicans who conveniently removed Tone's virulent anti-English phraseology from its eighteenth-century context. But with its new-found maturity, Irish society can now confidently honour Irish involvement in the armed conflicts of the past, as Minister John O'Donoghue TD appropriately stated in his speech opening the National Museum of Ireland's major military history exhibition, *Soldiers and Chiefs*.[9] It was time to take the politics out of war in Ireland, he affirmed. With the recent publication of the third and final volume of the scholarly edition of Tone's *Writings*,[10] painstakingly footnoted and indexed by a dedicated team of editors, it is timely to examine Tone's military career as a French officer, and redress the imbalance in his historiography.

'IT IS A LIFE I HAVE DESIRED; I WILL THRIVE'[11]: FULFILLING A CHILDHOOD DREAM

Irish military migration to France had, and would continue to be, motivated by socio-economic reasons. Tone had left his peaceful exile in America to negotiate as an unofficial United Irish envoy with the French Directory, and in the army of the new French republic his patriotic and ideological

8 Marianne Elliott, *Wolfe Tone, prophet of Irish independence* (New Haven and London, 1989). 9 Speech by the Minister for Arts, Sports and Tourism, *Irish Times*, 5 Oct. 2006. 10 As note 3.

commitment to Ireland could be fulfilled. But his savings dwindled fast and enlistment also became a material necessity. However, his writings reveal that his new life also held the promise of romantic adventure. From the day he received his first commission (18 July 1796), Tone's combined diary and correspondence were written from the perspective of a soldier somewhat lost in France, and facing an uncertain future. To 'divert the spleen which was devouring' him during the interminable wait for his posting to Brest, he also wrote a brief autobiography, which reveals that ever since boyhood he had entertained an 'untameable desire' to become a soldier. He had regularly 'mitched' (skipped class) to attend parades and reviews of the Dublin garrison (some at the same barracks where his life and his brother Matthew's would tragically end), and was enthralled by the splendid raiment of the troops and the pomp of military displays.[12] Tone turned twelve in 1775, and his adolescence coincided with the American Revolution and the craze for Volunteering. Then, by the age of seventeen, at an age when many had already enlisted, he recalls

> [It] will not be thought incredible that *woman* began to appear lovely in my eyes, and I very wisely thought that a red coat and a cockade with a pair of gold epaulettes would aid me considerably in my approaches to the objects of my adoration [...]. I began to look on classical learning as nonsense [...]; in short I thought an ensign in a marching regiment was the happiest creature living.

This allows us to appreciate how unashamedly he recorded his first day in uniform

> Put on my regimentals for the first time; as pleased as a little boy in his first breeches: foolish enough, but not unpleasant. Walked about Paris to [show] myself. Huzzah! Citizen Wolfe Tone, *chef de brigade* in the service of the Republic![13]

His daily ritual in Paris was to go to the Tuileries gardens to watch the changing of the guard, and his observations perfectly captured the intrepid but somewhat fanciful appearance of the citizen-soldiers of the Republic:

> I am more and more pleased with the French soldiery notwithstanding the slovenliness of their manoeuvres and dress. Everyone wears what he pleases; it is enough if his coat be blue and his hat cocked. They are fond of ornamenting themselves, particularly with flowers [...], a little

11 Tone ii, 206. He quotes one of Falstaff's soldiers in Shakespeare's *Merry Wives of Windsor*, I, iii. 12 Ibid., p. 268, and following. 13 Tone, ii, p. 307.

> bouquet in his hat or his breast [...] most frequently in the barrel of his firelock [...]; they all seem in high health and spirits, young active and fit for service.[14]

But he never lost sight of his ultimate goal, that the French expedition would lead to the establishment of an independent Irish army. From Brest, on 30 November 1796, he wrote with pride and determination to his wife Matilda

> I wear at present a fine embroidered scarlet cape and cuffs on my uniform, and a laced hat, which is only permitted to the general officers, but I shall be happy on the first occasion [...] to change my blue coat for one as green as a leek.[15]

The proceedings of Tone's court martial somewhat mockingly described him as 'splendidly dressed in the French uniform – a blue coat highly ornamented with gold'.[16] The gold referred to the embroidered borders of oak leaf, a highly distinctive feature of French revolutionary uniforms, both civilian and military. The noble tree had long been a symbol of vigour and martial valour, and in portraits of the director for war Lazare Carnot, General Lazare Hoche or the younger Napoleon Bonaparte one sees their lapels and collars embellished with oak leaves.[17] Not only did Tone proudly wear a similar uniform, he came to know all three of these major figures of French history.

Tone's writings meander between giddiness and fatalism, bravura and a detached insouciance which some Irish readers have interpreted as his personal penchant for flippancy. But this is to ignore that these are recurring features in the journals and letters of soldiers of all ranks, in an era when they were increasingly verbalising their reactions to war.[18] Seen in a broader European context, his narrative is far from unique, though its particular resonance for Ireland is evident. But Tone partially relieved his anxiety by embedding quotes on soldiering from Shakespeare and other playwrights, as well as romantic fiction and song, in so doing demonstrating how warfare had permeated cultural life. His motto was 'nil desperandum' and on the French stage he found inspiring the inimitable character of the dragoon Montauciel, who vowed never to desert. Tone thought he captured the gaiety, pride and high spirit so 'characteristic of the *French soldier*.'[19]

Tone frankly admitted to the French he had no military experience, dismissing his brief time in the Belfast Volunteers as no service at all. Though

14 Ibid., p. 208; his comments are not subjective and are corroborated by Jean-Paul Bertaud in *La Révolution armée* (Paris, 1979), iii, p. 281. 15 Ibid., p. 406. 16 Tone, iii, 393. 17 See images in T. Bartlett, K. Dawson & D. Keogh (eds), *Rebellion: a television history* (Dublin, 1998), pp 48, 52 and 62 respectively. 18 See Alan Forrest, *Napoleon's men: the soldiers of the Revolution and the Empire* (London and New York, 2002), inter alia pages 105–31. 19 Tone, ii, 148–9.

this may have been problematic, he was assured the French would appoint him to any rank he wished. Despite its unfortunate history, future practice in Napoleon's Irish Legion sheds lights on the matter. It offered commissions at the rank of captain to Irishmen in France who enjoyed a 'certain standing or reputation' back home, and though 'previous military experience was helpful', it was not essential as the men would be trained.[20] An established name and the ability to muster and lead the radicals and discontents in Ireland were more important, and Tone certainly fulfilled this criteria.

Hoche was appointed commander of the Irish expedition on 20 June 1796, the day of Tone's thirty-third birthday. By then Tone had had numerous strategic discussions with Henri Clarke, the son of an Irish exile who had served in the Irish brigade and was now head of the government's topographical and military bureau. Clarke knew of Tone's increasingly embarrassing financial predicament, and reassured him on 23 June a commission was imminent. Then Tone raised the vexed question of the Irish among the English prisoners of war in France, frequently discussed with Nicholas Madgett. Another Irishman, he was head of the translation bureau at the ministry of external relations and an *ad hoc* advisor on Anglo-Irish affairs. Madgett had offered to locate his brother Matthew among the prisoners and have him released[21], but on 22 March Tone had dismissed as 'damned nonsense' his pilgrimages around the depots to propagate the republican faith among the Irish soldiers and seamen. He surmised it was really a 'job *à l'Irlandoise*', that is, underhand crimping by Madgett to line his own pockets. But the French needed to man the Irish expeditions and were keen to rid themselves of prisoners difficult to exchange. By June however Tone had twigged, and asked Clarke if there remained many Irish around the depots, who could be usefully employed after the landing? Clarke teased him: 'I see you want to form your regiment'.[22] Tone wished: 'to command two or three hundred of them [...] formed into an advanced guard of the army not only as soldiers [...] but as *éclaireurs*', that is, scouts who could also inform the locals. Tone and Clarke corresponded, and evidently conversed, in English, so it revealing that he used the specific French word, only recently introduced into the language.[23] This front-line contact role for bilingual scouts had evidently captured Tone's imagination. Though it never came to pass, we must recall Tone's proposal that his scouts be formed into a specifically '*Irish corps* [his emphasis] of hussars, in green jackets, with green feathers, and a green standard with the Harp, surmounted with the Cap of Liberty', to which Clarke reacted enthusiastically.

20 Thomas Bartlett, 'Last flight of the Wild Geese? Bonaparte's Irish Legion, 1803–1815', in Thomas O'Connor & Mary Ann Lyons (eds), *Irish communities in early modern Europe* (Dublin, 2006), pp 161–2. **21** In fact Matthew had been released in May 1795 and was then in America. See Tone, iii, pp 575–6. **22** Ibid., pp 210ff. **23** First used in the newspaper *L'Ami du peuple* on 29 Dec. 1792 (p.7), it became widespread after 1797 in narratives of Bonaparte's foreign exploits.

CHRONOLOGY OF TONE'S SERVICE: 'I DID NOT CALCULATE FOR A CAMPAIGN ON THE RHINE, THO' I WAS PREPARED FOR ONE ON THE SHANNON.'[24]

Probably because of the brevity of his service, there is no unified record ('état de service') for Tone in Vincennes,[25] but dates in his diary, correspondence, and individual service letters all tally. On 17 July 1796 the minister for war (Pétiet) wrote confirming Tone's commission would be effective from the 19th.[26] The amalgamation of the former royal army with levied infantry units had partially replaced regiments with *demi-brigades*, and the rank of full colonel being renamed 'chef de brigade'[27], to which Tone was initially appointed. From the outset his fate was determined by the patronage of Hoche, the iconic military hero of the French Revolution. After their first meeting (12 July), Hoche stated he would apply for the rank of adjutant-general for him when his regiment was formed, though Tone never did command his own unit. This promotion was made in October, and Tone remained at that rank, between colonel and general of brigade and also an innovation of the Revolutionary army for staffs of larger units.[28]

The new French Constitution of the Year III (1795) was one of the first books Tone had bought while lounging about the bookstalls of the Palais-Royal in Paris. Its article 287 stipulated that foreigners could not enlist, unless they had served in one or more campaigns to set up the Republic or been naturalized.[29] But Tone had not been in France long enough (seven years) to apply for citizenship,[30] and if he was granted honorary citizenship this has yet to be corroborated. Even in 1797 Tone alluded to the constitutional impediment, which the authorities had circumvented by setting up foreign legions, such as the Légion franche which Humbert headed on the Bantry campaign.[31] His enlistment and rank were great honours bestowed on him by the French republic, but he remained uncertain of how and if they would protect him 'in case the fortune of war [...] should throw [him] into the hands of the enemy', especially on Irish soil.[32] As a lawyer, he knew he had

24 Tone to Matilda, Paris, 11 February 1797, reprinted in Tone, iii, p.22. 25 SHD/Tone's file, 17y[d] 14 Généraux prétendus; in contrast to, e.g., Henry O'Keane (1763–1817) or Bernard MacSheehy (1774–1807), alphabetical files in Sous-série 2Y[e] (1791–1847). 26 See Tone, ii, 240. 27 John A. Lynn, *The bayonets of the republic: motivation and tactics in the army of revolutionary France, 1791–1794* (Boulder and Oxford, 1996), p. 74. 28 André Corvisier, trans. by Chris Turner, *Dictionary of military history* (Oxford, 1994), p. 356; William J. Macneven, *Of the nature and functions of an army staff* (New York, 1812), p. 6. 29 See www.conseil-constitutionnel.fr/textes/constitution/c1795.htm, accessed 26 May 2008. 30 SHD/Sous-série 2Y[e]/John Sullivan (1767?–1801?): Madgett's nephew, also a translator and much valued as such by Tone, he had been naturalized, as had Henry O'Keane; they served as captains under Humbert in Mayo (1798) and so after Ballinamuck were exchanged and returned to France. 31 Alain Corvisier & Jean Delmas et al. (eds), *Histoire militaire de la France* (Paris, 1992), ii, 247. 32 Tone, ii, 142.

committed high treason, having gone to France while open war raged with Britain, and there enlisted despite being 'a natural born subject of [our] Lord the King'.[33] His initial rank appears in the bilingual *Cartel* for prisoner-of-war exchanges, in force at the time of his arrest, as equivalent to a brigadier-general, and worth fifteen men.[34]

Tone frequently relieved tension with one invaluable quotation, especially during the agonising wait to report at Brest. It had nothing of the profound or the sublime, being but a verse from a drinking song, but he thought it wore like steel: 'Tis but in vain, for soldiers to complain.'[35] On 17 September, he left Paris to join Hoche at army headquarters at Rennes. He mused that Mr. Pitt probably knew where he was, and possibly for internal reasons Hoche told him he would have to remain for some time Mr Smith, an American. Though maintaining the *incognito* increased his reliance on Hoche, a few trusted associates learned of his real identity and motivations, such as the congenial Henri Sheé (1739–1820). A maternal uncle of Clarke, born in France to an Irish father and wise with years of military experience, he became a useful mentor and trusted family friend. When Tone arrived in Brest (29 October) he learned of the arrests of Thomas Russell and other United Irishmen for high treason. Their fate would persistently haunt him for the next two years. Tone embarked on the *Indomptable* on 2 December, and for the record he was temporarily in command of the troops, being the highest in rank. After the failure of Bantry, Tone returned to Paris on 12 January 1797. In a rather perplexing move, having previously referred to them as sad blackguards, he confidentially offered his services to Colonel William Tate, the American head of the Légion noire sanctioned by Hoche to lead his desperadoes to raid England and burn Bristol. But Tone's family had arrived safely in Hamburg from America, and wisely he remained in Paris. Hoche, now commander of the Armée de Sambre et Meuse, offered Tone a position among his staff as plans for another Irish expedition were simply suspended. On 9 March 1797, Pétiet ratified Hoche's nomination and Tone's appointment.[36]

When Tone arrived at Hoche's headquarters in Cologne (7 April) he found many *connaissances expéditionnaires* – former comrades in arms – from the Bantry expedition. The Army of Sambre et Meuse was famed for its *esprit de corps* and a certain Waudré expressed satisfaction that Tone was now 'of' them: 'Eh bien, c'est un brave homme de plus' (Well, we have one more brave man among us).[37] Though the Rhine campaign was about to begin, Tone took

33 *Minutes of the court martial*, Tone, iii, 386. 34 SHD/Marine/FF1/33/V 1, agreed 13 Sept., 1798. Matthew Tone had also gone to France after the war had started, in 1794, thus committing treason, but as a captain in Humbert's force, was theoretically worth six men according to the Cartel. He was executed on 29 Sept. 1798. 35 Tone, ii, 309. He is quoting from 'How stands the glass around' in *The Buck's bottle companion* (London, 1775), pp 211–12. 36 Reprinted in Tone, iii, 31ff and 26. 37 Ibid., p. 49.

leave and joined his family at Groningen on 7 May, mindful not to advance beyond the Dutch border into Hanoverian territory. He had left Cologne on 20 April and had travelled from Nimeguen to Utrecht, and Amsterdam to the Hague. As a French officer he was privileged to travel through parts of the Continent closed to the leisurely tourist due to the war, and his diary (as in Paris in early 1796) emulated the travel literature of the age. Like many a soldier, his service had transformed him into an accidental tourist.[38]

Tone returned to headquarters in Friedberg on 1 June, then Hoche dispatched him to The Hague on the 27th, to meet the Batavian Committee of Foreign Affairs. Under new plans for an Irish campaign Hoche would command a force from Brest and Tone would sail with the Dutch fleet from the Texel. Shee had settled Tone's family near his own just outside Paris, and so his mind was at rest. On 8 July he to take up his post on board the aptly-named *Vryheid* (Freedom) of 74 guns under the Dutch commander-in-chief, Hermann Wilhelm Daendels, another instrumental protector of both Tone's career, and later his family. Hoche had asked that Tone be employed at his rank and assured his Dutch counterpart 'he will be most useful to you, for it is from him that I have partly drawn the intelligence which has served me to date to organize the expedition.'[39] Tone's rank, with the honours of a staff officer, was confirmed on 15 July 1797 by Paris.[40] Mention in the diary of having new regimentals made up prompts us to wonder if he now wore a Dutch uniform.

But success once again escaped him. Though news from Ireland was increasingly alarming, by the end of July Tone knew it was unlikely they would sail. Political tensions were mounting in France, and on 6 August Hoche informed the Minister of War he would not go to Brest: 'No longer will I go play Don Quixote on the seas to the great pleasure of a few men who would prefer to see me at the bottom.'[41] On 3 September, the eve of the Directory's *coup d'état* (18 fructidor), Daendels despatched Tone to inform Hoche of a revised Dutch plan. But by the time he reached Wetzlar, the general's health had deteriorated to the point of no return, and he died on 19 September, aged 29. It is said Tone was among the few privileged at Hoche's bedside, and he was given a lock of Hoche's hair, and asked to contribute 'honourable memories cherished by every friend of liberty' to a forthcoming biography.[42] So reliant were officers on the patronage of their generals that

38 See my 'The accidental tourist: Theobald Wolfe Tone's secret mission to Paris, 1796', in Jane Conroy (ed.), *Cross-cultural travel: papers from the Royal Irish Academy Symposium on Literature and Travel, NUI Galway, Nov. 2002* (New York, 2003), pp 121–30, and 'Rough guide to Revolutionary Paris', *History Ireland*, 16:2 (Mar.-Apr. 2008), 34–9. 39 Minutes of the Dutch committee for Foreign affairs, 5 July 1797, cited in Tone, iii, 99 n. 1, author's translation. 40 SHD/Tone: *Ampliation*, Département de la Guerre 15 July 1797. 41 Cited in Marie-Louise Jacotey, *Le Général Hoche L'ange botté dans la tourmente révolutionnaire* (Langres, 1994), p. 227, author's translation. 42 Chérin to 'Smith', 22 Oct.

Tone's connection to the Army of Sambre et Meuse was severed overnight. Daendels wrote references for Tone, instrumental when he returned to Paris, *c.*30 September 1797. The image of the United Irishmen in Paris was severely undermined by Napper Tandy's posturing and overt hostility to Tone, and the French would not sail to Ireland until a rising took place. But Tone's hopes were raised again as the twists and turns of French political life were, paradoxically, favourable to him. Hoche's demise left Napoleon Bonaparte as the leading and unrivalled French general, and he shrewdly exploited the government's reliance on the war machinery to keep it afloat. To Talleyrand, now minister of foreign affairs, he confessed his ambitions: 'France should profit by the opportunity to seize England *bodily* – in Ireland, Canada, India.'[43]

And so a new episode in Tone's extraordinary personal history opened, as he was to have dealings with both these towering figures in European history, as would his family in the early years of the nineteenth century. As Tone's son William would later observe: 'whatever character may be assigned to him [that is, Talleyrand] in history, we certainly owe gratitude for the lively and disinterested part which he always took in our fate'.[44] On 15 October Talleyrand confirmed how well adjutant general Smith had served Hoche and trusting Tone's character reference for Irish refugees, forwarded it to the Minister for Police (Sotin).[45] So when the Directory decreed the formation of a new Army of England (26 October 1797), with Bonaparte as its commander, Tone had re-established his credentials at the highest level. If his memory was honoured posthumously by the support several Frenchmen and exiled Irishmen accorded his widow, one must also look to how various people demonstrated *during* his lifetime that he had earned their respect. Hoche's motto had been *Res non verba*, and after his death General Hédouville, chief of staff on the Bantry expedition, actively assisted Tone, personally introducing him to the Director La Révellière-Lépeaux and Talleyrand (who both conveniently spoke English), the minister for war (Schérer), and finally Général Berthier, Bonaparte's chief of staff in the Army of Italy. Berthier promised to speak to his commander, and on 12 November, 1797, Tone penned a brief 'job application' to the brilliant soldier who had been so jealous of his former protector's successes. Attached to the late General Hoche for the last eighteen months of his life, and in charge of his foreign correspondence, 'James Smith' was now offering his services in the newly formed Army of England, hoping General Bonaparte may find him useful.[46]

1797?, reprinted in Tone, iii, 172, author's translation. 43 Cited in Robert Asprey, *The rise and fall of Napoleon Bonaparte*, i (London, 2000), p. 242, my emphasis. 44 'The Tone family after 1798', in Thomas Bartlett (ed.), *Life of Theobald Wolfe Tone memoirs, journals and political writings, compiled and arranged by William T.W. Tone, 1826* (Dublin, 1998), p. 890; afterwards Bartlett (ed.), *Life*. 45 Reprinted in Tone, iii, 160. 46 Tone Papers,

On 20 November Tone was favourably received by the dashing Desaix, at twenty-nine a military hero after his victory with the Army of the Rhine at Kehl, and interim commander of the Army of England. Then, on 21 December 1797, Desaix introduced Tone and the United Irish envoy Lewines to Bonaparte, returned triumphant from Italy. His praetorian army was fast becoming a political powerbase, carried by the propaganda it had so cleverly developed. Bonaparte's plans for a strike at England were intense, and Tone's hopes were raised to unprecedented heights. On 12 January 1798, the minister for war (Schérer) confirmed to 'James Smith' his appointment to the rank of adjutant general.[47] The next day Tone had his last meeting with Bonaparte, and mentioned he was listed among the senior officers of the Army of England, but regretted he could only be most serviceable to his commander 'on the other side of the water', having neither the requisite military knowledge nor experience. Bonaparte interrupted him: '*Mais vous êtes brave*' (But you are brave). Tone replied that this would emerge 'when the occasion presented itself'. 'Well, that is enough,' Bonaparte replied.[48]

On 26 March Tone was shocked to learn that the entire Leinster committee of the United Irishmen had been arrested at Oliver Bond's. Every man he knew and esteemed in Dublin was now in custody, and his mind grew more savage by the hour: 'Measures appear to me now justified by necessity which six months ago I would have regarded with horror.'[49] This explains his mounting frustration after reporting to headquarters in Rouen (5 April), and finally he faced the mounting rumours, that major preparations were under way in the ports of the Mediterranean. Bonaparte had conceded that the French were not yet masters of the sea, and could not attempt such a difficult and hazardous operation as a landing on English soil, but Tone was totally baffled by the change in Bonaparte's destination, presumably India. In Rouen he had met the congenial General Kilmaine, the Dublin-born son of an Irish Jacobite exile, now interim commander of the Army of England. Kilmaine was very supportive and obtained a captaincy for Tone's brother Matthew, in Paris since the previous autumn. By 26 May it was clear there was no more a chance of a strike at England than there was one on the moon, so Tone wrote confidentially to Kilmaine, stating he was offering to serve the French in India, where he could achieve his goal

> My first object, undoubtedly, is to assist in the emancipation of my own country; if that cannot be attained my next is to assist in the humiliation of her tyrant […] in whatever quarter of the globe.[50]

TCD/MS 2050, f.18 [author's translation]; reprinted in Tone, iii, 173. **47** SHD/Tone: Lettre de service, Schérer to 'Smith', 12 Jan. 1798. **48** Tone, iii, 191. **49** Ibid., 220–1. **50** Ibid., 263.

But two days later he was named adjutant to Général Béthencourt, commander at Le Havre, as the English were shelling the northern ports. At least there Tone was relatively active, touring the posts and batteries, walking the ramparts with fellow officers, then compiling reports for the chief of staff (Rivaud) at Rouen, for which he was afterwards praised. He came face to face with his avowed enemy on 8 June, as an English attack seemed imminent and all the batteries were manned, but it was just mischief. The rising broke out in Ireland on 24–25 May and the Directory launched an invasion plan on 25 June. Tone was recalled to Paris and arrived around 23 June, shortly after his thirty-fifth birthday. His diary ends on 30 June 1798, after which we are reliant on correspondence and official archives. On 15 July he was ordered to report at Brest to serve under General Jean Hardy, arriving on 1 August and embarking on the flagship *Hoche* of 74 guns. They finally sailed out of Brest on 16 September, and Tone knew the British were fully alerted of his presence on board. On 12 October the *Hoche* was engaged in a fierce and furious sea battle of almost ten hours with Admiral Warren's squadron off the Donegal coast. His fellow officers tried to convince him to escape, anxious at what his fate would be in the hands of the enemy. But Tone replied: 'My friends, do you wish me to return to France and there announce that I saw the French fighting for the emancipation of my country, while I shamefully fled?'[51] He had ended his diary entry for 1 July, 1796 on a humorous note with his characteristics 'Huzzah!'s, then 'I hope to see a battle before I die.'[52] On board the *Hoche*, Tone had commanded guns and fought bravely. It was his first, and only genuine battle.

On 3 November, Tone set foot on Irish soil at Buncrana, Co. Donegal. Much to the delight of the British arresting officers, he was 'styled in the role d'équipage 'Adjutant-Général Theobald Wolfe Tone, (called Smith), Co. Kildare, Ireland'.[53] On 17 July, Tone had written to Schérer asking for his service letters to be re-issued in his real name, as he wished to appear among his compatriots legitimately.[54] Efforts were made and moving testimonials hastily penned on the French side to save his life, which may have been possible had there been more time. To the French, Tone became a martyr of liberty, but he had not seen sufficient combat to warrant a pension for his widow, though Matilda did get one in 1799 as part of an overall pension reform.[55] The state also paid for his son William's education at the prestigious cavalry school in Saint Germain-en-Laye, and he later served in the

51 SHD/Tone: *Précis* by Matilda Witherington Tone, n.d., author's translation; William takes this up (in English) in Bartlett (ed.), *Life*, p. 871. 52 Tone, ii, 220; source of this quotation not identified. 53 Report in the *Londonderry Journal*, 6 Nov. 1798, see Tone, iii, 357, and following, p. 362. 54 'Smith' to Schérer [Minister for War], Paris, 17 July 1798, author's translation; reprinted in Tone, iii, 313. 55 SHD/Tone, and J.P. Bertaud, *La Révolution armée les soldats-citoyens et la Révolution française* (Paris 1979) pp 294–5.

Napoleonic wars.[56] One of the great ironies of Irish republican culture and ritual is that it was originally Lucien Bonaparte who called for an annual commemoration by 'the independent people of Ireland' of Tone's sacrifice.[57]

COMPETENT OFFICER AND FUNCTIONAL BILINGUAL: 'I AM THE OFFICER "CHARGED WITH THE GENERAL'S FOREIGN CORRESPONDENCE." THAT HAS A LOFTY SOUND'![58]

Because it is a recurrent indicator discussed in studies of the Irish regiments in France under the ancien régime, a brief physical description is justified. Tone was slender, and certainly no description of his build conjures up the 'ruddy cheeks and strapping thighs' of Irish recruits into the royal regiments.[59] But the exacting records of the internal police at Le Havre confirm that he measured '5 pieds 4 pouces', i.e. exactly five foot eight inches or just under one metre seventy-three.[60] This was the minimum height a British recruiting officer could accept in peacetime, but the average Frenchman was shorter. The taller and sturdier men from the northeastern regions of France (for example, Lorraine, where Humbert hailed from) were more numerous in the National Guards and the cavalry, for which Tone was just over the threshold.

Tone had joined a fencing club in Paris, but despite combining fact and anecdote in his diary says nothing whatsoever about having received any other training (that is, of firearms) following his enlistment. Clarke had suggested a possible cavalry position (where the pay was highest) and given Tone's extensive travels around Ireland for his political activities we must presume his horsemanship was adequate. He had been devouring military books all his life and while 'lounging about the bookstalls' of the Palais-Royal picked up many 'dog cheap', including a cavalry regulation which he duly studied. Apart from his evident logistical role in the Irish expeditions, it may at first appear Tone had only been given nominal positions, particularly in the Sambre et Meuse army. When told in Cologne he was in charge of arming, equipping and clothing the troops, he joked that he knew 'no more than my boot what I shall have to do, save look after some 80,000 troops'.

56 See Tone, iii, 537–9 for further reading on William. 57 See Bartlett (ed.), *Life*, Appendices A, B & C, pp 886–990 for the Tones post 1798, and an English version of the original *Motion d'ordre faite par Lucien Bonaparte pour la veuve et les enfants de Théobald Wolfe-Tone* (1799), reprinted in *Life*, pp 895–6. Lucien's assistance was genuine, but it should be noted he was overseeing reforms and this *Motion* was only one of many; see www.bnf.fr. 58 Tone, iii, 26. 59 Colm Ó Conaill, 'Ruddy cheeks and strapping thighs': an analysis of the ordinary soldiers in the ranks of the Irish regiments of eighteenth-century France', *Irish Sword* 14:98 (2005) 411–25; www.tcd.ie/CISS/mmfrance.php for the complete data base. 60 Archives municipales du Havre, [register of internal passports], *PR I² 35, 9 Feb. 1796.

An adjudant general was an adjunct to a brigade or divisional general, and usually charged with administrative and liaison duties in preparing campaigns. The French had identified Tone as an ideal staff officer quite capable of assuming quarter duties, and rightly so, as his intellectual abilities largely compensated for any other weakness. He wrote with a clear hand and had advanced literacy and numeracy skills, high on the wish list of recruiting officers, and especially in France where illiteracy and regional dialects were still common. Unlike some of his French comrades, and indeed his mentor General Hoche, he was college educated and had practised as a barrister. Secretarial experience gained with various committees, and organisational skills, were put to good use and Tone was fastidious about keeping copies of his correspondence, even numbering his private letters.

Among his privileges were naming two of his own adjutants, and we note that he commented on preferring candidates who spoke English, wishing Matthew could be one of them. So for some time he had reluctantly appointed the fully bilingual but infuriatingly antagonizing Bernard MacSheehy in this position. We are easily amused by Tone's indignant outpourings against the unfortunate MacSheehy, repeatedly dismissed as an arrogant blockhead, but it is none the less a prophetic portrayal of the failed commander of the future Irish Legion. However Tone may have envied MacSheehy's ease in the French language and so this leads us to the vexed question of how he himself coped with the language barrier. One of the most enduring facets of Tone's legend as a doomed romantic hero is that he arrived on his clandestine mission in revolutionary France not speaking the language of his hosts. From February 1796, his journal would be peppered with self-deprecations demonstrating his experience of what two centuries later would be called culture shock. Barely making himself understood, his French was but a 'detestable, execrable jargon', he lamented, and generations of readers have taken his subjective evidence at face value. But who would not warm to a hero who admitted his own faults? One need only look to that most historic occasion, his first meeting with Napoleon Bonaparte on 21 December 1797, and how Tone recorded the end of their interview

> Buonaparte [...] then asked me where I had learned to speak French? To which I replied that I had learned *the little* that I knew *since my arrival in France*, about twenty months ago.[61]

It is not surprising that confusion has surrounded this vexed question.[62] Yet in an earlier, equally historic encounter, one gets a different feel of how he managed. During his first meeting with Carnot, Director for War, in the

61 Tone, iii, 185, author's emphasis. **62** On the question of Tone's proficiency in French see T. Bartlett, *Theobald Wolfe Tone* (Dundalk, 1997), preface, and Marianne Elliott, *Wolfe*

sumptuous setting of the Luxembourg Palace, he mischievously juggled with status and power relations, exploiting the language issue

> I *began* the discourse by saying, in horrible French, that I had been informed he spoke English. [Carnot] *answered*, 'A little, Sir, but I perceive *you speak French*, and if you please we will converse in *that* language.'[63]

Historians have generally overlooked how the language gap was bridged in international relations, too easily assuming key players either struggled and relied on often untrustworthy translators, just got on with communicating as best they could, or were bilingual and therefore did not have to rely on others. But scrutiny of the communications between the United Irishmen and France makes clear that some individuals were 'chosen' by history and became mediators and *ad hoc* translators and interpreters at key moments.[64] The careers of various Irishmen having served France during the 1796 and 1798 campaigns demonstrate that apart from their ideological motivations and military prowess, their command of English (and possibly Irish) rendered them highly useful in the French conflict with Britain, at a time when there was no formal training for military translators and interpreters.[65] In fact despite his histrionics, Tone's narrative is a major source of insights on intercultural communicative behaviour, and allows us to reconstruct the types of settings when translation was a factor during political negotiations and military campaigns. In terms of his own competence, he left behind a substantial corpus of diverse material in French to dismantle the myth he had constructed, though his initial anxieties were justified given how much was at stake.

His written French is of a remarkably high standard despite the odd grammatical error or awkward turn of phrase. When discussing a future role with Madgett, Tone had stated he 'could be of use' in the general's staff, 'speaking a little French, to interpret between him and the natives' after the landing.[66] So even his own weak bilingualism was an asset, and when Clarke had stated, 'It would be absolutely necessary the general *en chef* should speak English,' Tone (having identified a role for himself) replied, 'It would undoubtedly be convenient, but not absolutely necessary.'[67] We have one insightful anecdote, when, on 4 November 1796, Hoche needed his assistance to interrogate the 'American' captain of a Liverpool vessel taken by the

Tone: prophet of Irish independence (New Haven, 1989), p. 287. See also Sylvie Kleinman, 'Pardon my French: the linguistic trials and tribulations of Theobald Wolfe Tone', in E. Maher and G. Neville (eds), *France-Ireland: anatomy of a relationship* (Frankfurt, 2004), pp 295–310. **63** Tone, ii, 75, author's emphasis. **64** Discussed in my PhD thesis, 'Translation, the French language and the United Irishmen, 1792–1804' (DCU, 2005), unpublished. **65** Inter alia, the instrumental role played by Bernard MacSheehy, Henry O'Keane, John Sullivan, Bartholomew Teeling and Mathew Tone, discussed in my thesis, as note 63. **66** Tone, ii, 142. **67** Ibid., 205.

French at Brest and assess his intelligence that the revolution had taken place in Ireland. Tone may have translated the questions formulated by Hoche and the Admiral Villaret de Joyeuse, or indeed Hoche may have allowed him to lead the questioning, being a former lawyer. Tone concluded that the informer was a liar, because he had prevaricated so much, but especially because 'he was a Scotchman [*sic*], with a broad accent'. Though the Frenchmen probably understood some English, this was a nuance they may not have picked up on, and doubtless there were many other such incidents.

Shifting from a foreign language into one's mother tongue is always easier, and this Tone did when Hoche asked him to translate his orders for Tate to undertake his commando raid into England.[68] But Tone's role also overlapped with that of what we now call a communications officer and/or propagandist. An essential question in preparing the landing was the composition *and* translation of various proclamations and addresses, informing the Irish of French intentions, calming their fears of dominational or religious intolerance, or exhorting the militia or Irish in the British navy to join their ranks. In the carton of manuscripts relating to the Bantry expedition in the archives at Vincennes there is an original in Tone's hand of the *Proclamation to the people of Ireland* from the French commander in chief which was to have been distributed immediately after the landing.[69]

But when the fleet was dispersed he and Chérin on the *Indomptable* were forced to write up a new, shorter version to be printed, because the originals were with Hoche on the *Fraternité*. Tone specifies that they worked 'writing and translating proclamations', also keeping a record in French of what Tone had rewritten, from memory, in English.[70] Then he assisted General Grouchy with drawing up a new commander's proclamation in French, which was to have been translated into English. Two years later, the same exercise would be repeated at Brest with General Jean Hardy. Copies of Tone's English translation were found among his papers, but possibly to ensure there was a copy on record, Tone wrote one out in French in his distinctive hand, and it too is filed with the other 1798 expedition material at Vincennes.[71] We note how he emphasized the key phrase '*L'Irlande sera libre pour toujours!*'; it is underlined and in larger letters in the French, and in capitals in the English version:[72] 'IRELAND SHALL BE FREE FOREVER!' The preservation of such invaluable testimonies of Tone's contribution to Irish history, in France, is a moving testimony to the fraternal relations between the two countries, and to the vital role of archivists. In short, Tone had become a fully functional and balanced bilingual, though ironically when he was stationed in Le Havre in 1798, his reports to Rivaud were in French, and in French only.

68 Ibid., 397. 69 SHD, 11B[1], première expédition d'Irlande (1796), transcribed in Tone, ii, 196–8. 70 Tone, ii, 423–4, and following, p. 426. 71 SHD/11B[2] et [3]: 2e expédition d'Irlande (1798). 72 Tone, iii, 317.

There was one particular task Hoche singled Tone out for, for which rhetorical command of French patriotic jargon, but in *English*, was vital. On 13 November Tone duly followed his general's orders and went among the prisoners of war at Pontanezen (near Brest). He was to offer them their liberty if they agreed to serve aboard the French fleet (no destination mentioned), this being 1796 when no international conventions banned this form of political 'harassment'. Sixty were taken in by Tone's proselytizing, of which fifty were Irish. He didn't deny his methods

> I made them drink heartily before they left the prison, and they were mustered and sent aboard the same evening. I never saw the national character stronger marked than in the careless gaiety of these poor fellows. Half naked and half starved as I found them, the moment they saw the wine before them, all their cares were forgotten [...]. They all said they hoped I was going with them, wherever it was.[73]

Having discovered a flair for leadership and team building, Tone noted a few days later he had been hard recruiting, and had picked up another twenty stout hands, for which the republic owed him five louis. Derision of Madgett's 'scheme' was a thing of the past, Tone having understood the necessities of the 'job *à l'irlandoise*'.[74] Later, in July 1798, fights had broken out in various depots between English prisoners and those 'attached to the patriotic party' (that is, the Irish) and these were triggered by the news from Ireland. Tone was asked by the minister for marine to go and sort out the situation, helped if necessary by Charles MacDonagh, another Irish officer formerly of Berwick's regiment.[75] It is not clear if he ever did go, and one wishes we knew more about these unfortunate Irishmen caught up in the political turmoil of the 1790s, so far from home.

A CHRONICLER OF FRENCH HISTORY: 'IT IS A DROLL THING THAT I SHOULD BECOME ACQUAINTED WITH BUONAPARTE'

Tone was gifted at capturing the moment on paper, and his frank impressions of people and places give a certain quality to his narrative which distinguishes it from other accounts of the revolutionary decade.[76] This makes Tone's a vital historical record , as stressed by Thomas Bartlett, for whom 'his account of the Bantry Bay expedition is a minor classic of military reportage'. The ability to convey to the reader a sense of immediacy was a literary skill characteristic of two highly influential genres of eighteenth-century writing,

73 Tone, ii, 373–4. 74 See note 23 above. 75 Tone, iii, 310–1. 76 See the insightful editorial introduction in Bartlett (ed.), *Life*, p. xlvi.

romantic fiction and travel writing. But stylistically they also left their mark on officer's writings, apart from the increasing coverage of campaigns in the press. Tone even acknowledged it: 'my journals [...] are written at the moment and represent things exactly as they strike me'.[77] His written legacy in this sense was no different than other innumerable officer's narratives of the Revolutionary and early Napoleonic period, though until now it has not been acknowledged. But these are also gripping human stories, and a recent and welcome development in military history is the study of soldiers' writings. Their diaries and letters convey reactions to events around them over which they have no control, and even if these are not always tales of heroism, they have a considerable appeal and complete the 'guns and trumpets' histories. Though one could not say that as a staff officer Tone's narrative was history from below, it was written from a peripheral perspective, that of a foreigner always mindful of not transgressing protocol and appearances. Furthermore, his writings perfectly express 'that sense of adventure which soldiers have always shared [as] travellers faced with new experiences and alien cultures'.[78]

Though we naturally have a different perspective looking back over two centuries at the events Tone lived through, broadened by the insights of specialist historians of the period, it appears he was conscious of having become a chronicler of French history:

> I am now a little used to see great men, and great statesmen, and great generals [...]. Yet, after all, it is a droll thing that I should become acquainted with Buonaparte [...], the greatest man in Europe.

Tone's gift as a military diarist emerged during those historic passages relating the Bantry expedition. The following excerpt has been chosen to illustrate the human element of warfare that Tone brings out in his evocative descriptions of French army life. Even during the wait to set sail for Bantry the troops kept their spirits high, and were 'as gay as if they were going to a ball.'[79] Tone was a keen classical musician and recognized the importance of camp entertainment and maintaining troop morale, having even written out several traditional Irish songs and ballads for his French brothers in arms.[80] Efforts to dispel gloom and melancholy among the men were common, but it was important to improve poor relations between the army and the (famously weak and badly equipped) navy. Posterity can be grateful for this most evocative vignette, while the fleet was still anchored at Brest. The band of another regiment had come on board his ship, and Tone went down in the great cabin and where the officers messed to hear the music

77 Tone, iii, 174. 78 Alan Forrest, *Napoleon's men*, p. xi. 79 Tone, ii, 415, and following. 80 This unique cultural document has never been located.

> The cabin was ceiled with the firelocks intended for the expedition; the candlesticks were bayonets, stuck in the tables; the officers were in their jackets and bonnets de police, some playing cards, other singing to the music, others conversing, and all in the highest spirits [...]. At length Watrin and his band went off, and, as it was a beautiful moonlight night, the effect of the music on the water, diminishing as they receded from our vessel, was delicious.

Throughout Tone's narrative there is the constant reminder of his privileged existence as a staff officer. He often dined in state, and rarely spoke of the rank and file, so we find no depictions here of the legendary brotherhood of the cauldron of the jacobin era of the French Revolution.[81] But we frequently read of the gaiety and carelessness of the officer's mess, even in Hoche's presence: 'All very slovenly and soldierlike; but nobody minds a dirty plate or thing of that kind here. "*On est à la guerre, comme à la guerre*", as the French say.' Though he only once evokes the famous *ordinaire*,[82] or mess unit, we constantly sense his loyalty to it, as when they are bawling for him to come and join them.When there was a 'furious penury of beds', he gave a good example with Hoche's aide de camp Privat, and 'lay rough on a mattress on the floor. Lay awake half, the night laughing and making execrable puns [...]. I like this life of all things'.[83] If the puns were in French, his competence was obviously up to it.

Tone's depiction of the militarization of the public imagination is corroborated (if not cited) in the works of the foremost French historian of this period, Jean-Paul Bertaud, namely in his examination of how adept Napoleon had become at exploiting martial valour. Since the dawn of time this had been a potent and binding social value, but in France reached new heights when it became institutionalized during the Revolutionary period.[84] When Tone arrived in France in 1796, demonstrations of public support for the army necessarily spilled over into cultural life, beyond reviews and official festivals. His diary contains many evocative descriptions of republican secular ceremonies, or *fêtes*, which marked the anniversary of major events of the Revolution and perfectly capture the transitional mood of this period between the fall of Robespierre and the rise of Napoleon. Citizenry and the duties of soldiering went hand in hand as stipulated by the Constitution of 1795, and though men could only vote if they paid a certain level of tax, this was waived if they had served in one or more campaigns to establish the Republic.

81 See 'Les Compagnons de la marmite', in Jean-Paul Bertaud's seminal *La Vie quotidienne des soldats de la Révolution, 1789–1799* (Paris, 1985), pp 104–48. 82 Its importance in terms of the men's commitment to a primary group is discussed in Lynn, *Bayonets*, pp 163–82. 83 Tone, ii, 359. 84 Inter alia, J.-P. Bertaud, *Quand les enfants parlaient de gloire L'armée au cœur de la France de Napoléon* (2006), p. 249.

Enlistment in the reserve national guards was now a duty incumbent on French citizens, the individual having obligations towards the state.[85] Thus soldiers were given an honoured part in the public festivals which promoted republican virtue but also countered the re-emerging royalist propaganda, and Tone was enthralled with the military dimension to public events. This included a role in theatre performances, which also ended with the *Marseillaise*. Initially his perspective was that of an ideological civilian, only recently arrived in his new country of asylum, as illustrated by the festival of youth he attended on 30 March 1796.[86] The scene must be imagined as he witnessed it, republican iconography having replaced Christian symbolism inside the secularized church of St Roch. Sixteen-year old youths were presented to the municipality to receive their arms, and the 21-years-olds to be enrolled as citizens in order to vote:

> The church was decorated with the national colours and a statue of liberty with an altar blazing before her. The procession [...] consisted of the Etat-major of the sections composing the district, of the National Guards under arms, of the officers &c., and finally of the young men.

Following speeches on the citizen's duty and the honour of bearing arms in the defence of France, guns and sabres were distributed and the young men congratulated by their loved ones. Tone did not 'wonder at the miracles which the French army has wrought in the contest for their liberties'. But the poor Irish were a brave people, and could also show the same enthusiasm, if they had the opportunity. By September he had become a participant, attending the feast of 1 vendémiaire in Rennes to celebrate the establishment of the republic. After the review he met Hoche, who asked him had he heard the cannonade? Anticipating his landing in Ireland, his general elaborated: 'you will soon hear enough of that.'[87] Exactly a year later, his protector was no more, and Tone must have had a heavy heart when he attended another vendémiaire *fête* only three days after Hoche's death. But the episode is typical of the paradoxes which punctuate Tone's life in Europe. He found himself in Bonn to take part in a ceremony to proclaim the *République cis-Rhénane*, i.e. the part of occupied Germany between the Meuse and the Rhine annexed by France. Tone was in the procession, and Shee's six-year old son even helped him to plant a tree of liberty. They then dined in state with the Municipality, and 'drank sundry and loyal toasts'. But Tone must have wondered if he would ever plant a Tree of Liberty on Irish soil, with his own children at his side.

85 Alan Forrest, 'Citizenship and military service', in Renée Waldinger, Philip Dawson & Isser Woloch (eds), *The French Revolution and the meaning of citizenship* (1993), pp 156–7. 86 Tone, ii, 136. 87 Ibid., 317.

Tone's impulse to write and record, his powers of capturing the seemingly mundane, and his mere presence at certain historical junctures transformed him into one of those rare accidental witnesses of history so valued by the great French historian Marc Bloch. Because of the density of detail and constant shift in topic matter, the specialist seeking insights on a specific dimension must patiently sift through Tone's writings, but expect disappointment. Yes, Tone alludes to unpaid, unkempt and probably hungry soldiers, but only in passing. More frequently, he notes arrears in his own pay, and constantly cites the recurrent financial crisis of the Directory as a material impediment to the Irish expeditions, the 'military chest [being] heinously unfurnished'.[88] Incapable of provisioning its armies, the regime was none the less reliant on ongoing campaigns, as dues were levied on occupied territories. Parallel to the respect Hoche had earned in the Rhineland, he was exacting monies, and we note Tone's subtle comment after attending Easter Mass in Cologne cathedral: 'I fancy they have concealed their plate and ornaments for fear of us, and they are much in the right of it.'[89]

Timelags between an appointment and the issuing of the precious *brevet* (commission) by the ministry of war were, as discussed by Bertaud, a recurring deficiency in army life at this time, so Tone's case was far from unique. It had become the practice to let generals do as they deemed fit, then just confirm their decisions. The red tape was frustrating for officers and helped contribute to their increasing political detachment from civilian bureaucrats they deemed incompetent, corrupt, and lacking in patriotic virtue. This largely explains why by mid-decade generals were establishing their own clientèle of loyal staff.[90] Effective management was to be revealed as one of the great secrets of Napoleon Bonaparte's success, and Tone, placed as he was in such privileged proximity to Hoche, was at the very heart of historic developments in France. The legendary *esprit de corps* which had developed through the decade in the French army certainly encouraged an independent governance removed from politicians and the legislature, though they also shaped foreign policy and all this can be continuously be read through the lines in Tone's narrative.

But on a more human level the vignettes he left us of his personal contact with other officers whose paths he crossed, famous and not so famous, are most revealing. A random illustration of this is his brief acquaintance with Desaix, whose glorious death at Marengo (1800) at thirty-two was exploited by Napoleon to enhance his own personality cult. For a fleeting moment Desaix had energized Irish expectations, and we must savour the scene, when Tone and other Paris-based United Irishmen gave a grand dinner in his honour, probably to herald his role in the new Army of England. The setting

88 Tone, iii, 49. 89 Ibid., 52. 90 Jean-Paul Bertaud, 'Recrutement et avancement des officiers', *Annales historiques de la révolution française*, 210:14 (1972), 532.

was unique, as it was the superb Méot's which had bedazzled Tone only a few days after his arrival, in February 1796, as it exemplified night life in the new revolutionized capital:

> Dined at a tavern in a room covered with gilding and looking glasses down to the floor [...]. It was the hotel of the Chancellor to the Duke of Orleans. There went much misery of the people to the painting and ornamenting of it and now it is open to anyone to dine for 3*s*'.[91]

Could he ever have imagined he would return, in so honoured a position, and in such grand company?: 'Our dinner was superb, and everything went off very well. We had the Fort of Kehl represented in the dessert in compliment to Desaix &c., &c.'[92]

It is not only through the entries in Tone's private journal that one concludes his attachment to Hoche was special. Though it appears he never did contribute to the memoir, it is noteworthy he had been asked to by Chérin, who wrote that no one more than Tone had this right to do so, as he had 'known him intimately'.[93] One particular conversation stands out between Hoche and 'Schmitt', which is how he wrote to him,[94] because in it is interwoven elements of both French and Irish history. Tone was alarmed at what he deemed Bonaparte's 'grossly improper and indecent' proclamation to the government of Genoa, of which he read 'the most obnoxious passages' to Hoche.[95] Such a dictatorial stance would have the most ruinous effect if Bonaparte commanded in Ireland, where the Irish understood their rights. Tone could detect 'a great jealousy', but was re-assured by Hoche's retort. This timeless phrase (translated literally from Hoche's French) has remained embedded in Tone's narrative, totally overlooked by French historians

> I understand you, but you may be at ease in that respect; Buonaparte has been my scholar, but he shall never be my master.

In Hoche, Tone had commented, he had one of the greatest military masters in Europe, possibly envisaging a lifetime as a professional soldier if his dream of Irish independence never came to pass. He must have known that the self-taught Hoche had organized internal exams to tighten up the professional skills of his own corps, and had written a memoir on the reorganization of the army for the Directory.[96] This was shortly after his epic crossing of the Rhine at Neuwied, and though he never mentions his mentor's triumph, Tone was

91 Tone, ii, 51. **92** Tone, iii, 178. **93** See note 42. **94** Hoche to 'Schmitt' (*sic*), Wetzlar 19 August 1797, SHD/Tone file; germanizing 'Smith' made it easier for a Frenchman to pronounce. **95** Tone, iii, 98. **96** 1 May 1797, reprinted as *Un Mémoire de Hoche sur la réorganisation de nos armées en l'an V*, Section historique de l'état-major de l'armée (Paris,

at that illustrious spot a few times during that hectic spring of 1797. Then there came the meeting with Bonaparte, also relatively intimate, as it took place in Josephine's much talked about little house in the rue Chantereine, which Tone described as: 'small, but neat, and all the furniture and ornaments in the most classical taste'. It was to this home that Bonaparte returned triumphant from Italy, and the week after Tone's visit, the street would be renamed, not surprisingly, the 'rue des Victoires'. He left us this image of the 28 year old future tyrant

> He is about five feet six inches high, slender and well made, but stoops considerably; he looks ten years older than he is, owing to the great fatigue he underwent in his immortal campaign of Italy. His face is that of a profound thinker, but with no marks of that great enthusiasm and that unceasing activity by which he is so much distinguished; it is rather [...] the countenance of a mathematician than of a general.[97]

Tone did not survive into those tumultuous years of the early nineteenth century, and unlike his compatriots in the Irish Legion, would not be confronted with the dilemma of swearing an oath to France's self-appointed emperor.

AUX GRANDS HOMMES, LA PATRIE RECONNAISSANTE?[98]

Tone's history remains virtually unknown in France but this does not mean he was not remembered by the individual Frenchmen he had served with, or that his ideological mission did not leave a lasting influence. If in the nineteenth century Young Irelanders and Fenians emigrated to France, and there gained military experience but it would not be in a direct armed conflict with Britain. But one anecdote has many touching echoes of Tone's experience and merits retelling.

In the spring of 1861, the (future) Fenian John Devoy ran away to France to become a soldier. It was his only option after a 'hot argument' with his father, concerned that his 19-year-old was too involved in the national movement.[99] He could only join the Foreign Legion, but military training could also be useful in the cause of Ireland. In Paris Devoy wore down his shoes looking for John Mitchel's address, but eventually tracked down John P. Leonard, a Young Irelander lecturing English at the Sorbonne. Leonard escorted him to the ministry of war, and Devoy enlisted. At a barracks near

1910). **97** Tone, iii, 185. **98** The inscription atop the pediment of the Pantheon: To great men, from a grateful nation. **99** John Devoy, *Recollections of an Irish rebel* (New York, 1929), pp 385–8.

Lyon a sergeant-major of infantry took him under his wing. 'What countryman are you?', he asked, and smiled on hearing he was an Irishman. After soup, the sergeant-major sat down to play cards and Devoy's presence triggered a heated argument about Ireland and England over the game. He was astonished that despite having attended the military school at St Cyr, a lieutenant knew nothing whatsoever about Ireland, alluding in French to 'the same government, same queen, same language'. But the sergeant-major thumped his hand on the table and declared: 'L'Irlande, par sa position géographique, par race, par religion, et surtout par la volonté de son peuple, est une nation' ('Ireland, by its geographical position, by race, by religion, and above all by the will of its people, is a nation').

How did Devoy's sergeant-major acquire such a sympathetic perception of Ireland? We could take the liberty of speculating that he was the descendant of one of the veterans of the 1796 or 1798 French expeditions to Bantry, Mayo, or Donegal. Throughout the late 1790s, many Frenchmen and exiled Irishmen had been thrown together by the fortunes of war and alongside the rigours of the campaigns also experienced an enriching intercultural contact. Possibly a grandfather had sailed to Bantry or Lough Swilly, or indeed served under General Lazare Hoche during his last heroic campaign on the Rhine in 1797. If so, he may have befriended the congenial and diffident Adjutant General James Smith, and through discussions at the mess table gained a deeper understanding of the Irish national question. Or indeed an uncle may have been an archivist at the department of war.

In 1796, Tone's friendship with Du Petit Thouars and their common destiny as soldiers was not accidental. The editor of Du Petit Thouars' journals described his ancestor as possessing an unquenchable thirst for adventure, being 'A quintessential man of the last decades of the 18th century, an engaging figure infused with the sensibility, enthusiasm and anxiety of his times, possessing an incessant imagination and a mind broadened by a combination of reading and travel.'[1] Nothing best captures the essence of Theobald Wolfe Tone, whose life and mission have often been misunderstood, because they have not been framed in the cultural context of his times. The influence of Tone's enlistment in the French army on his perception of the individual, the citizen and the state must not be underestimated. Overnight the son of a bankrupt coachmaker attained an honourable standing in society he may never have achieved in Ireland. He could serve on an equal standing with any aristocrat, rise through the ranks on talent and merit, and not have his career predetermined by the privilege of birth. The sense of belonging to a fundamental institution, linked with the rights and duties of a citizen, spilled over into his writings and his aspirations for Ireland. This correlation between

1 As note 4, p. 163.

Irish freedom and France has been tarnished by the grim legacy of Napoleon's despotism.

Many more insightful dimensions merit discussion, such as Tone's record of the surviving networks of Irish military families, many of his acquaintances bridging the gap – and honourably surviving – the transition between the old régime and the new dispensation. Possibly the most appealing dimension is, as Marianne Elliott underlined, that Tone the revolutionary was never far from Tone the aesthete. Thanks to his French *voyage de guerre*, he came face to face with his adventurous impulses and creative leanings, and wrote one of the most engaging and unusual travel narratives of the golden age of travel writing. Jean-Jacques Rousseau may have once noted that as travellers, soldiers were not the best observers, but this could not be said of Tone. When strolling the banks of the Seine at Rouen, attending a service at Cologne cathedral, enjoying the debates at the Batavian Convention at the Hague, or drinking hock at an outdoor *guingette*[2] on the Rhine, he must be pictured in his French uniform.

This article has attempted to portray a brave soldier's legacy and sacrifice as would any historian of the French Revolution, unhindered by the self-imposed constraints of old-school Irish historical discourse. Multidisciplinary perspectives are adding a new dynamic to what were once sensitive, if not taboo themes, and first and foremost of these is commemoration of the war dead of Ireland. Official tributes to great people who sacrificed their lives for Ireland are, at long last, freeing themselves of associations with legendary divisions, and paramilitarism, but can one truly say that Ireland has demonstrated its gratitude to Theobald Wolfe Tone? He may have provided inspiration to republican nationalists, but the passages relating with fatalism and humour his lonely exile in Europe have moved a very broad readership. Generation after generation have felt 'the poignant vulnerability in every line' because they know how the tale ended.[3] The poet Francis Ledwidge, who died at Messines in 1917, may have remembered reading of Tone's rambles through France and Flanders, and faced with his own inevitable destiny found solace in his inimitable fusion of moral optimism and light-hearted wit, which he claimed defined the French soldier. Possibly Samuel Beckett, when as an interpreter during the Second World War had recalled Tone's own portrayal of the absurdities of exile, and how he came to render himself useful as a translator. Above all it is the humanity of the man faced with an uncertain future that is most infectious. Tone was a conservative egalitarian, and we may speculate he would have been shocked at the thought of a woman 'combattant' dispelling melancholy by citing *his* diary, when writing her own,

2 Outdoor café with dancing. 3 Declan Kiberd, 'Republican self-fashioning: the journals of Wolfe Tone', in *Irish Classics* (London, 2000), p. 221.

in Kilmainham prison in 1916. On 7 May, Madeleine ffrench-Mullen noted her indignation at being fobbed of with a 'tea' which consisted of stirabout and dry bread. Though she does not name her source, we recognize Tone's grip on the national imagination, transported from France to Dublin, as she consoled herself:

> 'However, "'Tis in vain for soldiers to complain", as a good Irishman has already said.'[4]

4 Memoir of Madeleine ffrench-Mullen, Kilmainham and Mountjoy jails, 5–20 May 1916/Allen Library/201/File B, May 7. I am grateful to Darragh O'Donoghue for providing me with a copy of his transcript.

The Irish Legion of Napoleon, 1803–15[1]

NICHOLAS DUNNE-LYNCH

The Irish Legion had a brief life, from 1803 to 1815, during which it underwent many changes of name and organization. Starting as 'La Légion irlandaise', the corps was known successively as the 'Irish Battalion', the 'Irish Regiment', the '3rd Foreign Regiment (Irish)' and, finally, the '7th Foreign Regiment'.[2] Its short life was, however, longer than that of other foreign regiments of the epoch, all of which were formed later and disbanded earlier than *'le 3ème Régiment étranger (irlandais)'*.[3]

Many readers will be familiar with *The Memoirs of Miles Byrne*,[4] which are coloured by the author's Irish republican nationalism, since Byrne was a zealous United Irishman and a leading rebel. This approach also characterizes *Napoleon's Irish Legion* by John G. Gallaher,[5] who shares neither Byrne's background nor his experience, but draws heavily upon him in mood and matter. Other studies, though informative and thorough, present overviews.[6]

CONTEXT

The idea of a new Irish corps in the French service existed in the minds of many long before it came into being. Since the demise of the old Irish Brigade after the French Revolution and, because of royalist sympathies, the defection to the British of the greater part of its cadres, the idea had much currency. Its first manifestation came in the form of a so-called 'Irish Brigade' that sailed with Hoche in 1796. Although it contained some native Irishmen and descendants of Irishmen, this force seems to have been Irish in name only.[7] A proposal for a new force was rejected by Napoleon in 1800, and any hopes that

1 Author's note: This article is based on research to date. Numerical data particularly is the latest available, but research continues. 2 Sometimes more than one of these titles were used simultaneously. 3 The sister regiments of the 3eme Régiment étranger (irlandais) were the 1er Régiment étranger (la Tour d'Auvergne), the 2eme Régiment étranger (d'Isenbourg, and the 4eme Régiment étranger (le Prusse). All had been disbanded by 1815. Many other foreign units had longer service. 4 Miles Byrne, *The memoirs of Miles Byrne*, 2 vols (Dublin, 1907). 5 John G. Gallaher, *Napoleon's Irish Legion* (Carbondale, IL, 1993). 6 See Guy C. Dempsey, *Napoleon's mercenaries: foreign units in the French army under the Consulate and Empire, 1799–1814* (London, 2002), and Lt.-Col. Pierre Carles, President du Centre d'Histoire de Montpelier, 'Le Corps irlandais au service de la France,' *Revue Historique des Armées* (1976/2), 25–54. 7 Carles, p. 26. Also Service Historique de la Défense, Vincennes (SHD), Carton Xh17.

he might change his mind faded with the peace of Amiens, the breakdown of which, on 22 May 1803, opened the way for the formation of the Irish Legion.[8]

The presence in France of exiled United Irishmen and fugitive rebels of 1798 and 1803 was a major impetus for creating the new Irish force. The vision of Irish liberty, shattered in Ireland, was alive among these exiles, who longed to try again with the help of growing French military power. At the very least, such a corps would give employment to political refugees, and be a source of irritation to London and Dublin, but the prospects of military success under the Consulate seemed far better than they had been under the Directory.

The prime movers were Arthur O'Connor and Thomas Addis Emmet, both recently arrived in Paris, each claiming to be sole emissary of the Directory of the United Irishmen in Dublin. The contest was academic, since that body no longer existed in any form that either would have recognized. A mutual hostility between these two men, however, long preceded their arrival in Paris and marred the early days of the Legion. The conflict even predated their incarceration at Fort George as state prisoners (1798–1802). While Emmet represented more cautious elements, O'Connor, a confederate of Lord Edward Fitzgerald, was more militant.[9]

Commissioned to examine the feasibility of a new Irish corps, former Irish Brigade officer Alexandre Dalton,[10] aide de camp to the Minister for War Berthier,[11] charged O'Connor and Emmet separately with sounding out Irish expatriates, since he realized that there could be no cooperation between them. In Dalton's absence, Irish-born General Oliver Harty, a former captain in Berwick's regiment,[12] asked Captain James MacGuire, a veteran of the Hardy expedition, to explore the issue.[13] Based on the response to the various enquiries Bonaparte decreed the formation of the Légion irlandaise on 31 August 1803. That the First Consul acted for any reason other than the necessity of the moment is unlikely, and the belief that he respected the Irish seems to be a myth.[14]

8 25 March 1802–22 May 1803. **9** See Jane Hayter Hames, *Arthur O'Connor, United Irishman* (Cork, 2001). In attributing, as a possible irritation to Emmet, the 'privileged treatment' that O'Connor received at Fort George in that 'Mrs O'Connor and their children' were allowed to remain with him, Gallaher seems to be confusing Arthur with his brother, Roger, also a State Prisoner: see Gallaher (pp 17–29) Arthur was unmarried at that time. **10** Alexandre Dalton (D'Alton-Shee), 1776–1859; Lt (Berwick's Reg, later 88th Line), 1791; Adj. Cdt. 1803; Brig. Gen. 1809; Maj. Gen. 1815; Baron 1810. **11** Louis Alexandre Berthier (1753–1815), Napoleon's chief of staff, duke of Wagram, 1st duke of Valengin, 1st sovereign prince of Neuchâtel, Marshal of France. **12** Brig. Gen. Oliver Harty, 1746–1823, Baron Harty de Pierrebourg. **13** Harty to MacGuire, 8 July 1803, SHD XH16c. **14** Carles, p. 27, and Edouard Desbriere: *1795–1803: Projets et tentatives de débarquement aux Iles britanniques* (Paris, 1900) iii, p. 593.

EARLY YEARS

The intention was that, when the Legion landed in Ireland with a French invasion, the officers would raise, train and lead volunteers. We will look at these officers later. To begin with, the Legion was based in the small Finisterre town of Morlaix and attached to the force intended for an invasion of the British Isles, under General, later Marshal, Augereau.[15] During its early years, the Legion moved to other Finisterre towns, such as Carhaix and Lesneven. Augereau had begun his military career in the old Irish Brigade and was keen on the Legion's success, though his chief of staff, General Donzelot,[16] directed matters. Newly raised to the rank of lieutenant general and now on the general staff,[17] Arthur O'Connor supervised the Legion, while Harty was more directly involved as brigade commander. There were, perhaps, too many midwives for a trouble-free delivery. Later, the interest of these men was superseded by that of Henri Clarke, duke of Feltre and minister for war,[18] of Irish descent and another former old Irish Brigade officer, whose patronage would damage the unit, and who was destined to become its undertaker.

In the quiet atmosphere of the small Breton towns, the Irish unit lacked an immediate role and the officers had few troops to supervise. Since it was intended to be the framework of a force raised in Ireland, the Legion was not actively recruiting. All this combined to bring about the problems of the early days, which included duelling, provoked by political, national and personal differences, both among themselves and with local antagonists. Though such behaviour probably arose out of the deep frustration of exiled and dispossessed men, it damaged French interest. A new outbreak of war with Austria and the British naval victory at Trafalgar in 1805, made a landing in the British Isles less feasible, and almost killed off the corps. The Legion languished until 1806 with about seventy-five other ranks passing through.

Since recruits were initially limited to the Irish or those of Irish descent, and few of either came forward, the intake was very low, the first volunteers coming as deserters from the 18th Foot, later the Royal Irish Regiment, based in the Channel Islands, or from naval prisoners of war. The British Army was

15 Charles Pierre François Augereau, 1757–1816; marshal of France, 1804; duke of Castiglione. 16 Brig. Gen. François Xavier Donzelot, 1764–1843. 17 O'Connor's commission as *général de division* (lieutenant general) is dated 24 Feb. 1804. The rank of lieutenant general existed before the Decree of 21 February 1793 by which it was replaced by *général de division*. Under Napoleon, the higher rank of *général de corps d'armée* became equated with lieutenant general, as it is today. The rank of lieutenant general returned with the Restoration. See O'Connor's naturalization dossier at Archives Nationales de France (ANF) 5635B2 BB/11/149/2. 18 Henri-Jacques-Guillaume Clarke, 1765–1818; Count Hunebourg, 1807; duke of Feltre, 1809; marshal of France, 1816; minister for war, 1807–14, and in 1815.

not active in Europe, so the only Irish prisoners of war who volunteered came from the Royal Navy.

Country of birth	*Number*	*%*
Ireland	35	46
France	28	37
Not given	13	17
Total	76	100

Figure 1: National origins of Irish Legion volunteers up to October 1806[19]

Any suggestion that the Irish flocked to the Legion would be an exaggeration. Those who joined at Morlaix as officers in 1803 and 1804, added to recruits between 1803 and 1806, with Irish places of birth, amounted to less than a hundred. In addition, the archives contain letters from about fifty Irishmen and others seeking admission to the corps, either as officers or in the ranks. Most of these were declined, or did not accept the rank offered.[20]

OPERATIONS[21]

In 1806, the Legion was deployed in Germany and, at last, recruiting actively. The first surge came at Mainz with a contingent of Prussian prisoners of war, among which were many Irish rebels of 1798 and 1803 who had been inducted into the Prussian army as an alternative to transportation,[22] though it is widely held they were sold into slavery in the salt mines.[23]

19 SHD 23YC207 (1803–1806). **20** SHD Xh16b, 16c and 16d. Chief among these was James Joseph MacDonnell, a rebel leader at Castlebar in 1798, who held the brevet of general from Humbert and later turned down the rank of captain in the Irish Legion. **21** Space allows only a very sketchy summary of operations. See Byrne, Gallaher, Carles and others for greater detail. **22** Byrne estimates that 1,500 prisoners from the Prussian army joined, but does not specify the number of Irish. One officer, Charles Mullany, born Co. Donegal in 1777, who had served under Humbert, may have joined with this group. See Byrne, ii, 26–7. Mullany's enlistment date is given as 27 November 1806. See his naturalization dossier, ANF BB/11/136/1 1505B4. A gap in the registers of troops at SHD seems to exist for late 1806. **23** Letters exchanged in 1798 and 1799 between Viscount Castlereagh, chief secretary for Ireland, and Captain Schouler of the Prussian Army discuss military service only, Schouler insisting on able-bodied rebels and no common criminals. See National Archives of Ireland (NAI), Rebellion Papers 620/18a/2/1–11, etc. Rebels were to serve ten years in the Prussian Army, but were liable be sent to the salt mines for breaches of discipline. See *Freeman's Journal*, 1 June 1799. The king of Prussia gave them leave to return home. See *Waterford Mirror*, 6 April 1806, seven months before

Now a functioning military unit, the Legion was deployed in coastal defence along the estuary of the river Scheldt, in malaria-infested marshlands. Coincidentally, on the English side of the Channel, in a similar role, stood a brigade commanded by Major-General Sir Arthur Wellesley,[24] including the Connaught Rangers,[25] destined to become his shock troops in the coming war in the Iberian Peninsula.

In November 1807, the Legion was strong enough to allow the transfer to Spain of a provisional second battalion of 800 men under Louis Lacy, a Spanish captain of Irish descent.[26] This contingent was outside Madrid when 'The Second of May' uprising broke out, and was deployed to restore order. While demoralized by the sudden disappearance of their commander, Captain Lacy, the provisional second battalion was reinforced in late 1808 by 600 men under Captain Jeremiah FitzHenry,[27] who was soon raised to lieutenant colonel and appointed commander of the contingent in Spain, now the 2nd battalion. The officers and men were yet to receive a serious blow to their morale – they discovered that Lacy had defected back to his former compatriots, the Spanish.[28] Far worse was to come.

The former Irish rebels would have relished an encounter with Sir John Moore, whose army was in retreat across the north of Spain in the midwinter of 1808–9. Though Moore had participated in the suppression of the Irish Rebellion in 1798, the Irish respected him because of his humane treatment of civilians and captured rebels.[29] The battalion was, however, diverted to Burgos, where assignments included garrison duty, construction of defences, escorting prisoners and fighting irregulars, a task the Irish executed very well, though very much against their inclinations.

The newly formed 3rd battalion of 800 men under Lieutenant Colonel J.F. Mahony arrived in Spain 1810.[30] Intended to support the 1st battalion, which

they joined the Irish Legion in November 1806. **24** First duke of Wellington, b. Dublin, 1769. **25** The 88th Foot, about which Wellington would later declare, 'Whenever anything very gallant or very desperate is to be done, there is no corps in the army I would sooner employ than your old friends, the Connaught Rangers.' See Sir James MacGrigor, *The autobiography and services of Sir James McGrigor* (London 1861), p. 259. **26** Louis Lacy was born in Andalusia in 1776. His father, Patrick, was colonel of the Irish Ultonia Regiment in the Spanish service. Lacy resigned his own commission in the Ultonia after an incident, and went to France. **27** Gallaher seems to be confusing Jeremiah with northern rebel John FitzHenry, who was born in Co. Derry in 1768 and died at Landernau in 1805. See Gallaher, p. 119. For comparative details, see Nominal Roll, 20 Floréal, Yr 12 (8 May 1804), SHD Xh14. Byrne refers to John FitzHenry as John MacHenry. See Byrne, ii, p. 300. **28** Lacy later became captain general of Catalonia, but was executed in 1816 for alleged republican activity, the king posthumously creating him duke of Ultonia. **29** Byrne, ii, 51. **30** More frequently referred to as 'Mahony' than by his correct name, 'O'Mahony,' which is given in his Legion of Honour file, as 'le Comte O'Mahony, Jean François.' ANF L201803. Up to 1814, his usual signature reads 'J.F. Mahony,' See Mahony to the King, Antwerp, 23 April 1814, SHD Xh16b. The Declaration of Loyalty to the King dated 1 January 1815 is signed 'Le Chevalier de Mahony,' See SHD Xh16a.

had been fighting a British sea-borne attack at Walcheren, the 3rd battalion was diverted to Spain when Flushing surrendered on 15 August,[31] and almost the entire 1st battalion went into captivity,[32] the Legion's first operational disaster. Though a handful of officers and men avoided capture, including William Lawless, who would later command the regiment,[33] the majority remained in captivity until 1814.

Under FitzHenry, the 2nd battalion managed to keep an active force in the field in the Peninsula, mustering between 500 and 550 effectives from April to December 1810 – a respectable number when compared with other foreign units.[34] Though the Irish Legion became 'The Irish Regiment' in May 1809, and Colonel Daniel O'Meara was appointed commander,[35] FitzHenry remained commander in Spain until O'Meara's arrival. However, O'Meara's dismissal for incompetence by Junot,[36] placed FitzHenry again at the top in Spain, but the unit was shocked by a second defection in early 1811, that of FitzHenry himself, this time to Wellington.

By December 1811, the demoralized and reduced battalion had been recalled to France, beginning its withdrawal on Christmas Day. The Irish who had joined at Mainz in 1806, and who had gone to Spain with Lacy in 1808, transferred with their non-Irish comrades to the Prussian regiment. The rest of that Irish contingent had fallen again into British hands at the surrender of Flushing. Thus, the Legion lost its most significant Irish element. Never again would the Irish account for more than about 10% of the rank and file.

Although they had taken part in the assault on Astorga (for which the corps received three awards of the Legion of Honour),[37] in Massena's pursuit of Wellington to the Lines of Torres Vedras, and in the rearguard of Massena's retreat back into Spain, acquitting themselves well in a desperate situation, the Legion had never engaged the British in pitched battle in almost four years in Spain and Portugal. At Fuentes de Onõro, for example, they had to

31 Vlissingen, Zeeland, The Netherlands. 32 The Flushing prisoners were transported to Norwich, England, and to Scotland. 33 Those who also avoided captivity were Capt. William Dowdal (d. of wounds), Capt. Patrick MacCann (d. of wounds), Capt. William Barker, and Lt. Terence O'Reilly. The wounded Lawless and O'Reilly saved the regimental eagle. Lawless' replacement as battalion commander, brevet Lt-Col Joseph Koslowski, b. Poland 1776, arrived back at Antwerp in October 1809 after only two months of captivity. See Koslowski to Minister for War, received 22/11/1809 (SHD Xh16c). Dublin-born Lt. Charles Ryan arrived back in 1812. (Ryan to Minister for War, 27 May 1812, SHD Xh16d). Capt. Arthur MacMahon also escaped. 34 Carles, p. 37. 35 Son of John O'Meara, of the Clare Regiment, Daniel Joseph (b. Dunkirk, 1764) was a twin brother of former Legion officer William. Daniel began his career as a cadet in Dillon's Regiment and retired from the army in 1811. 36 Jean-Androche Junot (1771–1813), duke of Abrantes, commander of the 8th Corps of the Army of Portugal. 37 John Allen (b. Dublin, 1777) captain of the light (voltiger) company, who led the assault, Adj. Maj. James Perry, and a drummer who went on drumming with both legs broken. Byrne, ii, 74. Legion of Honour dossiers: John Allen, ANF L0023016; James Perry L2016285.

stand by while the Connaught Rangers, which had received hundreds of rebel prisoners into its ranks after 1798, led Wellington's Third Division in the final assault.

Nor did the Legion receive a single battle honour for its efforts, while, on the opposite side, the Connaught Rangers topped the Irish regiments with twelve.[38] The greater part of the Legion's losses were due to desertion, and not a single officer fell in combat,[39] while the Rangers lost twenty-seven.[40] One might well wonder whether the Legion had been in the same war.

At the beginning of 1811, the newly formed 4th battalion had been absorbed by the reformed 1st battalion, under Antrim-born Presbyterian, Lieutenant Colonel John Tennent. The returned cadre of the 2nd battalion was now the framework for a new 2nd battalion, while the surplus officers of the old 3rd battalion were also rebuilding. Meanwhile, recruiting for a new 4th battalion was in progress at Landau. Despite attempts to put itself into marching order, the regiment once again languished in 1812, perhaps fortunately so, while the Grande Armée invaded Russia. In early 1813, Napoleon's shattered army rebuilt itself as he embarked on his Saxon campaign. The 1st and 2nd battalions of what was now the 3rd Foreign Regiment, went into action. Ringing hollow was Napoleon's remark that he preferred the unit in coastal defence to save his line troops such arduous duty. In the Irish ranks were the veterans of many fine armies, and they were seen as experienced and dependable among Bonaparte's green levies.

The campaign of 1813 was a disaster. Acquitting themselves favourably in combat at Goldberg, Lowenberg and elsewhere, the Irish battalions suffered serious casualties including the first battalion commander, John Tennent, killed, and the regimental commander, William Lawless, losing a leg, with as many as 400 all ranks killed in action. Soon after, fighting under the wounded Hugh Ware, the Legion was let down by its generals and trapped with Puthod's division against the flooded Bober river, with all bridges cut. In a devastating Russo-Prussian surprise attack and bombardment, as many as 1400 men, the bulk of the two battalions, were cut down, drowned in the retreat across the river, or captured.[41] The debris, mustering about 30 officers and less than 100 men, limped back to Bois-le-Duc. The Legion was to see some action in the defence of Antwerp, but the Bober would remain the most costly engagement.

38 Battalions 1, 2 & 3/27th, The Inniskillings; 2/83rd, later the County of Dublin; 2/87th, the Prince of Wales Own Irish, later the Royal Irish Fusiliers; and 1 & 2/88th, the Connaught Rangers. 39 Capt. Patrick Brangan died of wounds at Bejar, Estramadura, in late 1811, but it is not clear how he sustained them. See Byrne, ii, 294. 40 Between 1808 and 1814, Wellington's Irish battalions suffered, on average per battalion, 600 battle casualties and 10 officers killed. Compiled from C.B. Norman, *Battle honours of the British army* (London, 1911). 41 These figures require further investigation.

Besieged by the British, the unit stayed hemmed in at Antwerp until Bonaparte's abdication in 1814, apart from one sortie under Hugh Ware. Although the officer corps declared loyalty to the Bourbons,[42] the Bonapartists among them hid the eagle they had been ordered to destroy. Now the 7th Foreign Regiment and under Ware, the unit saw no action during 'the Hundred Days' campaign, and the duplicitous action of the Bonapartists certainly contributed to their disbandment in September 1815,[43] with their former patron and minister of war, Henri Clarke, wielding the axe.

THE OFFICER CORPS

The officer corps of the Legion was very diverse, with origins in over eighteen countries. The early officers were overwhelmingly Irish, but this changed over time. Of those who joined during the six months beginning December 1803, Bernard MacSheehy, the first overall commander, had come to France for his education in the 1780s. James Blackwell, the first battalion commander, for the same reason, and William Barker had come in the 1770s. William Lawless and John Tennent, among others, had been members of the United Irish movement and escaped from Ireland before the rebellion of 1798.[44] All of these men had served in the French army before they joined the Legion, some before the Revolution, such as Blackwell and Barker, the latter having served with the old Irish Brigade.

Officers who joined the French army between the rebellion of 1798 and that of 1803 included prominent United Irishmen William Corbet, in 1798, and both John Tennent and William Lawless in 1799.[45] While Lawless was to become the Legion's most famous colonel, and Tennent probably its most

42 Officers to King, 1 Jan 1815. SHD Xh16d. 43 Byrne blames 'Lord Castlereagh and the English influence on the French council' (Byrne, ii, 173) for the unit's disbandment. However, if Castlereagh was to blame, why he had not acted after the first Restoration in 1814? The regiment had done nothing militarily to attract attention during 'the Hundred Days', and it is hard to imagine that the British were in the least concerned, since the Irish contingent was very small. 44 Commentators have remarked upon the anomaly of having a lieutenant colonel or *chef de bataillon* (Blackwell) and an overall commander, an *adjudant-commandant* (MacSheehy). In fact, there is no anomaly. MacSheehy outranked Blackwell, as his rank was was considered to be between colonel and general of brigade, though William Corbet was promoted to *chef de bataillon* in 1813 and *adjudant-commandant* in 1814, well before his colonelcy. It is probable that MacSheehy was intended to command a multi-battalion regiment, while Blackwell would command the 1st battalion. A second battalion was considered, but it not materialize until late 1807. See 'Project d'Organisation en Deux Bataillons', (SHD Xh14), prepared by Adjutant-Majors Alexis Couasnon and Edmund Saint Leger. Though undated, this document originates after 12 September 1804, since Lt. Col. Pettrezzoli had already replaced both MacSheehy and Blackwell. 45 Lawless joined with the brevet of lieutenant colonel on 9 September 1799, Corbet with that of captain on 4 September 1798, and Tennent with the brevet of captain in 1799.

tragic officer, Corbet's tenure was short, but he was destined to rise highest in the French army, retiring a major general.

The names of at least three officers appear in the Fugitives Act of 1798,[46] those of William Lawless, Arthur MacMahon and Valentine Derry, associate of Fr Patrick MacCoigly, executed in England in 1798. In the Banishment Act,[47] the names of at least six appear: Christopher Martin, Bernard MacDermott, Hugh Ware, Patrick MacCann, William MacNeven and Hamden Evans.[48]

The second major category of Irish born officers were those who fought in one or both of the rebellions. Chief of these was Miles Byrne, who served in Wexford in 1798 and in the guerrilla war in Wicklow, before taking part in the rising of 1803. Byrne's account is invaluable, if fanciful at times. Other senior rebels were Austin O'Malley, brevet colonel under Humbert, Jeremiah FitzHenry, a field commander in Wexford, and Hugh Ware, a field commander in north Kildare.[49] O'Malley escaped to France; FitzHenry was amnestied but exiled himself; Ware was compelled to surrender in the face of overwhelming odds. Avoiding the fate on the gallows suffered by prominent Kildare United Irishmen, he was imprisoned and banished for life in 1802. Notable also among this group were John Allen, Alexander Devereux, and Terence O'Reilly.

Neat categorization is impossible, as the early officers had a wide variety of backgrounds, and many fall into more than one category, such as William Barker, who returned to Ireland after the disbanding of the old Irish Brigade, fought in the Wexford Rebellion, and went into exile afterwards.

Most Irish officers came from the upper strata of Irish society. O'Malley, Ware and FitzHenry were landowners, O'Malley being a descendant of Gaelic heroine Grace O'Malley, known as 'Grainuaile'. Austin's cousin, George, would later command the Connaught Rangers.[50] FitzHenry descended from

46 Fugitives Act (38 George III, c.78) calls on rebels to surrender on pain of being attainted of high treason. 47 Banishment Act (38 George III, c.78) pardons named individuals concerned in rebellion, subject to banishment; forbidding return to British dominions or passage to any country at war with Britain. France was not at war with Britain when they exiles took up residence during the peace of Amiens, 1802–3. 48 Arthur O'Connor and Thomas Addis Emmet are also listed. The Hamden Evans listed was also a State Prisoner at Fort George from 1798 to 1802, and was the father of an officer of the same name (b. Dublin, 2 Oct. 1782), who appears for the first time on the nominal roll dated 5 May 1810. However, the senior Evans appears, also a lieutenant, as a signatory on the process verbal of the formation of the Legion on 30 Jan. 1804 (SHD Xh14), but on no later document. The matter remains to be clarified. See Miles Byrne on the Evans family. (Byrne, ii, 178, 324, 336 etc.). 49 While Ware's rebel activities are documented both in the Rebellion Papers and contemporary historiography, FitzHenry's are not, though there was great interest in him after his departure to France. His wife, Mary, was a sister of executed United Irishman and rebel, John Colclough of Ballyteague. However, neither is there significant reference to Miles Byrne who, according to his SHD dossier, commanded 1,500 men, 2,000 in some documents, though he himself does not mention such a figure. 50 Maj.

King Henry I of England, and from Meillor FitzHenry, prominent among the first Normans in Ireland. Others came from the rising middle class, such as James MacGuire, originally a tailor, and John Allen, a partner in a Dublin drapery business. The early officer corps contained several lawyers,[51] doctors (William MacNeven and William Lawless), former officers of the British army (Edward Masterson); the French army (James Perry, Bernard MacSheehy);[52] the Spanish army (Louis Lacy); and the Portuguese army (Robert Lambert). Patrick MacSheehy had served in the Irish yeomanry cavalry during the 1798 Rebellion.[53] There was at least one Presbyterian minister of religion, Arthur MacMahon,[54] and one Anglican clergyman, John Richard Burgh,[55] though no Catholic priests are reported.[56] Several officers, many of them graduates or expelled students of Trinity College, Dublin, had been professors of English in French military academies.[57]

Two early officers had been condemned to prison terms in Botany Bay. The death sentence for treason on Edward Gibbons was commuted to transportation for life, while Michael Sheridan was sentenced to transportation for distributing forged banknotes.[58] Both escaped and made their way to France. Attempts by fellow officers to remove Sheridan on the grounds that he was a convicted criminal did not succeed, probably because, as Dalton notes, he had been a captain of insurgents under Humbert.[59]

Throughout the life of the regiment, the Irish-born officers came mainly from three areas of Ireland: Wexford, Mayo, and the area of the Pale, including the county of Dublin, which produced most, and parts of Kildare and Meath. A few came from Ulster. The origins of the rest cover most of

Gen. George O'Malley (1780–1843) commanded the 44th Foot under Wellington at Waterloo, and became colonel of the Connaught Rangers in 1825. His statue stands in Castlebar, while Austin has no monument. **51** Several had legal training, but Luke Lawless seems to be the only one who practised, and resumed his profession in the USA after his expulsion from France in 1816. **52** Capt. Bernard MacSheehy (b. Paris 1783) a relative of the commander. SHD Xh16c. **53** A cousin of Adj. Com. MacSheehy, Capt. Patrick MacSheehy was born in Co. Kerry in 1770. Byrne, ii, 287. Byrne confirms his membership of the Yeomanry. However, Commander MacSheehy does not mention this and states that, in 1798, his cousin 'fought very actively against the English.' Up the 1798, he had been professor at an academy at Gorey, Co. Wexford. MacSheehy, Nominal Roll of Officers, 27 Floréal, Yr 12, (17 May 1815). SHD Xh 14. Some of the Yeomanry defected to the rebels. **54** Born in Co. Down in 1755, Arthur MacMahon was minister of the parish of Hollywood in that county. His name appears on the earliest nominal rolls of the Irish Legion, and on the latest, though in his 70th year. For example, see Nominal Roll, 1 Sept. 1815 (SHD Xh16.) John Tennent was the son of a Presbyterian minister. **55** B. Dublin 1867. Donzelot, Inspection Report, 19 Vendémiaire, Yr 13. (11 Oct. 1804) SHD Xh14. **56** Valentine Derry is reputed to have been a brother of the Catholic bishop of Derry. **57** Other professions include, land surveyor (Hugh Ware, *mechanicien*, probably an engineer (Joseph Parrot) and shoemaker (John FitzPatrick). **58** Donzelot, Inspection Report, Lesneven, 10 Oct 1804, and Couasnon/St-Leger,'Project d'Organisation en Deux Bataillons', SHD Xh14. Couasnon had left by the end of 1804. **59** Dalton, Supplementary Roll, undated, but probably late 1803, SHD Xh14.

Leinster and Munster, and the officers originating there were Catholic or Anglican, the Connaught men were mainly Catholic, and the northerners, mainly Presbyterian.[60] No reports of religious disharmony appear in archived documents. In fact, religion is hardly mentioned.

Leinster		*Munster*		*Ulster*		*Connaught*	
Dublin	18	Cork	11	Down	4	Mayo	6
Wexford	7	Kerry	5	Antrim	1	Galway	4
Kildare	6	Tipperary	3	Derry	1		
Louth	4	Clare	1	Donegal	1		
Kilkenny	3	Waterford	1				
Carlow	2						
Wicklow	2						
Meath	1						
Totals	43		21		7		10

Figure 2: Counties of origin of the Irish born officers. 80 originate as listed above. The county of origin of some officers has yet to be established. Unrepresented counties are omitted.

To move on to the French born of Irish parentage or descent, and to those with a mixed nationality who espoused Irish nationalism, even if their commitment was less fervent than that of the Irish born, a difference that was to cause some friction. This group included those who simply wished to attach themselves to an Irish military unit, and those who were sent to the Legion because of their Irish origins, however tenuous.

Leading among this group at the formation was old Irish Brigade officer William O'Meara, who left France after the Revolution and served in the Irish Brigade in the British service, returning to France during the peace of Amiens.[61] His twin brother, Daniel, who had remained in France, would become the first regimental commander of the Irish Legion when it became the Irish Regiment in 1809. Another Franco-Irish commander who had served in the British Army was Jean François O'Mahony, who joined in 1809, only to become a very unpopular commander of the regiment in 1813. Others of Franco-Irish parentage include Henry Mandeville, a relative of the Henri Clarke, later minister for war,[62] and Louis Tournier Dupouget, both having

60 Catholics included Bernard MacSheehy, William Lawless, Jeremiah FitzHenry and Austin O'Malley, Presbyterians, Arthur MacMahon, John FitzHenry and John Tennent, and Anglicans, William and Thomas Corbet, John Richard Burgh. James Perry was probably agnostic. 61 William O'Meara later became colonel of the 2nd Foreign Regiment (*d'Isenbourg*), a command he held for some years. 62 Mandeville was serving as

served in the old Irish Brigade. In this group also is Edmund Saint-Leger, the son on an Irishman, who was later followed by two brothers, Patrice and Auguste. Auguste Osmont claimed Irish descent and was admitted, becoming a stalwart officer despite initial rejection by the native Irish. The first quartermaster was also named Bernard MacSheehy, who also experienced similar rejection but, unlike Osmont, soon left the Legion.

Typical in the next group, which was composed of French officers without Irish connections but who were attached to the Legion mainly to supervise training, was Alexis de Couasnon, adjutant-major in 1804. Of the lower aristocracy, Couasnon had served at Versailles as a page to the queen, and as an officer in the King's Artillery. O'Meara, Mahony and Couasnon had all served with the British army between 1794 and 1802. Mahony had fought against Napoleon in Egypt and Couasnon against the French in the Low Countries. His commission is signed by Lord Cornwallis, later lord lieutenant of Ireland.

Most of the officers of non-Irish birth but of Irish origins were born in France, notably the O'Meara twins, William and Daniel; Baron Patrice Magrath and his sons Achille and Louis; the Saint-Leger brothers, Edouard, Patrice and Auguste; William O'Morand and J.F. Mahony. Louis Lacy and Alfred de Wall were born in Spain, and Thomas O'Sullivan in The Netherlands. The fathers of the last three were serving or had served in the armies of those countries after quitting the French army on the dispersal of the old Irish Brigade in the early 1790s.

The leading ten French departments and Irish counties from which Irish Legion officers of Irish birth or descent originated appear in Figure 3. The now defunct *département de la Seine*, which contained the city of Paris, produced an equal number to Dublin. The *département du Nord*, produced eight, behind Cork's eleven, but just ahead of Wexford.

Irish county / French dept	*No.*	*% of total*	*Irish county / French dept*	*No.*	*% of total*
Dublin	18	6.5	Mayo	6	2.2
Seine	18	6.5	Pas de Calais	5	1.8
Cork	11	4.0	Kildare	6	1.8
Nord	8	2.9	Kerry	5	1.8
Wexford	7	2.5	Tipperary	3	1.1

Figure 3: Leading ten counties or departments of origin of Irish or Franco-Irish officers[63]

a second lieutenant in the 111st Line regiment. His father was killed in action with the old Irish Brigade. See SHD Xh14. **63** Figures 3 and 4 are based on a total of 280 officers throughout the life of the Regiment, although the provenance some has yet to be established.

If 41 out of the 46 original officers were Irish or of Irish origin (88%), this percentage declined over the years, and Figure 4 shows the leading eight out of eighteen countries of origin of officers over the Legion's life.

Origin	*No.*	*% of total*	*Officers*	*No*	*% of total*
France	102	36.0	Poland	7	2.5
Ireland	87	31.0	Italy	6	2.0
German States	24	8.5	Scotland	4	1.5
Prussia	10	3.0	England	4	1.5

Figure 4: Leading 10 countries of origin of Irish Legion officers, 1803–15

However, when the wider category of Irish parentage or origin is employed, the table changes and the combined Irish category moves up, but it never reaches 50%. In other words, more than half of the all the officers who served in the Irish Legion had no Irish connection.

The initial surge of Irish officers soon peaked as the pool of exiles was exhausted,[64] and promotions from the ranks, such as that of Patrick MacEgan and Francis Eager, filled some vacancies. Younger men trickled in, such as Dublin-born Samuel Stephens, who made his own way from Ireland in 1808 at the age of 18, starting in the quartermaster's department, before being commissioned and transferring to a company. Anthony Setting, also Dublin-born, seems to have been scarcely fourteen when he enlisted and, having 'passed through all the grades,' was a sergeant major of the light company when he was commissioned at barely 17.

Auguste Saint-Leger and Arthur Barker joined directly from the Irish College in Paris after at least one rejection on the grounds of age.[65] Saint-Leger, who followed in the footsteps of two older brothers, one of whom died at Flushing, managed to be accepted and was already a lieutenant at 18.[66] Born in 1797, Arthur Barker would have had no recollection of the rebellion, but he wanted to follow his late father's example.[67] The regiment thus took on the characteristics or a more mature corps, mainly because it had entered the minds of many Irish as the successor of the old Brigade. However, the shortage of Irishmen soon told.[68]

64 One of the factors that contributed to this shortage was the amnesties of Lord Cornwallis, which reduced the number of refugees. 65 Le Séminaire Collège des Irlandois, Anglois, et Écosses réunis. Today, the Irish Cultural Centre, rue des Irlandais, Paris. 66 However, archive sources disagree on the date of birth of Auguste St. Leger. A memorandum of 12 April 1810, signed by Col. O'Meara and Gen. Solignac, gives 3 Jan. 1791, while a service record of 1 Feb. 1814 signed by Col. O'Mahony gives 4 Jan. 1794. Dates of birth given in archived documents are often inaccurate, with the exception of those signed by the individual. 67 William Barker, b. Co. Wexford, 1759; d. Ghent, 1811. 68 The family groupings are too numerous and too complex to allow a full discussion here.

The vacuum caused by that shortage was filled in the main by Frenchmen, Germans, Prussians and Poles. Unlike the French, many of whom resented being in a foreign regiment since it retarded their promotion prospects, the Prussians had a strong loyalty.[69] By 1814, the dominance of the Irish officers in general was under serious threat, and the transcript below of a nominal roll of 1 March 1814 demonstrates.[70]

Rank	*Total Posts*	*Held by Irish*
Colonel	2	2
	Staff	
Lieutenant colonel	5	4
Adjutant-major	7	3
	Battalions	
Commander (lt.-col. or capt.)	4	4
	Companies	
Captain	24	5
Lieutenant	24	6
Second-lieutenant	24	2
	Grenadiers	
Captain	16	3
Lieutenant	4	1
Second-lieutenant	6	0

Figure 5: Irish representation among the officer corps in 1814

The two colonels listed, William Lawless and J.F. Mahony, were both Irish, the former Irish-born but no longer on active duty, and the latter of Irish parentage and in command. Though four out of the five lieutenant colonels on the staff are Irish, only three out of four adjutants have Irish birth or connections. Further down the staff, among the quartermaster, pay-officer and so on, there are no Irishmen. However, the four battalions have Irish commanders, only two of whom are lieutenant colonels.

Of 24 posts of company commander, 5 are vacant and 5 are held by Irish captains, while 14 are held by French or Germanic officers.[71] Out of 24 posts

69 Having been transferred out in early 1814 as their country was at war with France, a number of Prussian officers, supported by their former Irish comrades, petitioned the ministry of war. They were reinstated three months later. The disbandment of the other foreign regiments – the Irish being spared – brought a wave of officers, which further diluted the Legion's waning Irish character. 70 Nominal roll of officers, 1 March 1814, SHD Xh16. 71 Officers whose origins fell within the present Federal Republic of *Germany* (*Bundesrepublik Deutschland*) and the Republic of Austria (*Republik Österreich*), or those with Germanic names whose provenance lay within the Austro-Hungarian empire.

of lieutenant, only 6 are held by Irishmen and 4 are vacant. Among the second lieutenants, only two officers are of Irish birth and one of Irish origin, the other 22 having mainly Germanic origin. Among the grenadier companies, only 3 Irish captains out of 16 of that rank are present. Only one of four lieutenants is Irish, but none of the six second lieutenants.

The actual breakdown by nationality over the life of the unit, given in Figure 4, above, does not tell the whole story. The Irish dominated the officer corps from start to finish, but, at the end, it was in the senior ranks only. Though it was necessary to bring in some non-Irish officers at the formation and, later, an abundance, even at higher ranks, the unit never lost its Irish aspirations or character. The following table shows that the Irish or those of Irish origin served longer, and, the top Irish officers on the roll on 1 January 1815, had almost eight times the length of service of other senior officers.

Officers Awarded the Legion of Honour[72]		*Officers on roll 1 January 1815*	
Nationality	Months of service in IL	Nationality	Months of service in IL
Irish origins	100	Irish origins	115
Others	51	Others	15

Figure 6: A comparison of length of service between officers of Irish and others

Not a single officer of non-Irish birth or origin, who had joined during its first year, was still present when the corps was disbanded, and the vast majority of officers of non-Irish birth or origin were transient. The core of dedicated Irish nationalists who were there at the beginning, survived, for the most part. Miles Byrne is the best known. Yet, the 40+ officers of Irish origin at the formation of the regiment in 1803–4 had dwindled to less than 10 by September 1815.

THE COMMANDERS

In all its designations, the Legion had six commanders, in three of whom it was fortunate and in three others unfortunate. We will look at these in order.

72 Officers who served with the Irish Legion who won the Legion of Honour, before, during or after their service. At least 35 officers received that decoration. There may have been many more.

From	*To*	*Officer*	*Rank*	*Unit Designation*
Dec. 1803	Sept. 1804	Bernard MacSheehy	adjutant-commandant	*la Légion irlandaise*
Sept 1804	July 1809	Edouard Antoine Petrezzoli	lieutenant colonel	"
April 1809	May 1810	Daniel O'Meara[73]	colonel	*le Régiment irlandais*
May 1810	Feb. 1812	No overall commander in fact. Junot had placed O'Meara on his staff, but Clarke still considered him commander, which the archived documents reflect, until he appointed Lawless.		"
Feb. 1812	Dec. 1813	William Lawless	colonel	*3e Regiment étranger (irlandais)*
Dec. 1813	April 1815	Jean F Mahony	colonel	"
April 1815	Sept. 1815	Hugh Ware	major[74]	*7e Regiment étranger*

Figure 7: Commanders of The Irish Legion in all its designations.

Adjutant-Commandant Bernard MacSheehy[75]

The first commander, Bernard MacSheehy led the corps from December 1803 to September 1804, had been deputy to Wolfe Tone when the latter was

73 Junot dismissed O'Meara in May 1810. Feltre appointed William Lawless on 8 February 1812. 74 Napoleon raised Ware to colonel during 'the Hundred Days', but he reverted to major after the Second Restoration. He again raised in 1831. The anomaly is significant, as the grade of major was normally reserved for a lieutenant colonel on depot duty as distinct from field or overall commander. 75 MacSheehy's 500-word entry in

serving as adjutant-general in the Army of the Sambre. Although Tone seems to have felt that MacSheehy was prone to self-aggrandisement and intrigue, he recommended him to Hoche for a fact-finding mission to Ireland, a task the young MacSheehy seems to have conducted well.[76] [77]

However, Tone's opinion of MacSheehy was very low, and he was not happy with him as deputy.[78] When he sent him to Paris to collect his trunk, Tone wrote to Matilda, 'He is a blockhead, but be civil to him.'[79] When Tone was at last free of MacSheehy, he declared, 'if ever there was a rascal in the world, devoid of all principle, he is one.'[80] Miles Byrne echoes Tone's opinion and blames MacSheehy for retarding his military career.[81] However, other factors may have caused Byrne's failure to advance.[82]

Whatever MacSheehy's administrative and organizational talents, he lacked the qualities necessary to command the Irish Legion,[83] and his inability to manage the internecine nationalist factions marred the early years.

Lieutenant-Colonel Edouard Antoine Petrezzoli

Italian-born Petrezzoli replaced MacSheehy as regimental commander. The greatest asset in the eyes of his commanders was that he was *not* Irish and knew nothing about the political conflict that wracked the corps. He may have been a 'kill it or cure it' remedy for the squabbling Irish. A veteran of the Italian campaigns who had come up from the ranks, a tough light-infantry commander and an able tactician, he had to deal with devastating malaria, intrigue, unwillingness to cooperate and outright disobedience. Always making sure he had the backing of his commanders and the minister for war, he pensioned off incapable officers, dismissed troublemakers, and even had William Lawless, Thomas Markey and other recalcitrant officers sent into what amounted to internal exile *'en mission'* to the maritime prefect at Brest, where they languished a full two years.[84]

Danielle and Bernard Quintin, *Dictionnaire des Colonels de Napoléon* (Paris, 1996) makes no mention of his 10 months as commander of the Irish Legion. **76** Marianne Elliott, *Wolfe Tone: prophet of Irish independence* (New Haven, 1989), p. 320 and endnote 41 to chap. 23. Also C.J. Woods, 'The secret mission to Ireland of Captain Bernard MacSheehy, an Irishman in French service, 1796', *Journal of the Cork Historical and Archaeological Soc.*, 78 (1973), 93–108. **77** Elliott, p. 319, and *Partners in revolution*, pp 334–6. **78** Cited by Elliot, TCD MS 2049/359, see also fos. 131v, 136v and 154 for similar complaints. Dr Elliot notes that 'none of these criticisms appeared in the published Life. Matilda came to know MacSheehy well in Paris': Elliott, *Wolfe Tone,* endnote 36, Chap 24. **79** Elliott, *Wolfe Tone,* p. 347. **80** Elliott, *Wolfe Tone*, p. 370, and endnote 12 to chap. 28. **81** Byrne, ii, 8. The question remains as to how well Byrne knew Mathilda Tone before he met MacSheehy, and whether Mathilda influencd his opinion. **82** Several officers failed to advance, notably Lt. Col. James Blackwell and Capt. James MacGuire. However, others accelerated past Byrne, such as Terence O'Reilly, Edmund St Leger, James Perry, Patrick MacEgan, all of whom were younger. James Perry was a sergeant when Byrne was a lieutenant, but made lt. col. a full 12 years before Byrne. MacEgan was a corporal at 17, in 1804, but had become captain adjutant-major by 1815. **83** Byrne, ii, 6. **84** Alexander

Almost universally despised by the Irish officers, who saw him as a foreigner, and an oppressor because of his strict command, he managed to turn the fragmented unit into an effective light infantry battalion, becoming the Legion's longest serving commander, only ceding command a few months before the surrender of Flushing in 1809, in which he was taken prisoner. His nearly five years of effort were all in vain, as almost the entire unit went into captivity with him. Resuming his army career on his release in 1814, he became a French citizen in 1819.[85]

Colonel Daniel O'Meara

Daniel O'Meara's main qualification to command the Irish Regiment was that his wife was a sister of the duchess of Feltre, wife of the minister for war, the man who appointed him. His second asset was that he was of Irish descent. The fact that he had little experience of field command did not escape the commander of the 8th Corps, General Junot, who dismissed him in May 1810, after less than a year in command and about two months in Spain. Assertions that he was too old are hardly credible as he was the same age as the man he replaced, Pettrezolli.[86] The reason was simply incompetence and a lack of experience. Junot asserted that he was not fit to lead 'a squad of ten men.' Clarke would not learn from this experience. Promoting O'Meara over the head of the very popular and experienced Jeremiah FitzHenry was deeply resented by the battalion in Spain.

Colonel William Lawless

Lawless survived his confrontation with Pettrezzoli and was reinstated to the Legion in 1809, first as a battalion commander at Landau, and then, after Pettrezzoli's transfer to the 43rd regiment, as commander of the 1st battalion at Flushing, which he managed to enter regardless of the siege. Universally popular with his officers and troops, and former professor of surgery at the Royal College of Surgeons in Ireland, Lawless was unfortunate in his superiors. A courageous leader and highly effective administrator, he rebuilt the regiment after its fragmentation in the Peninsula. The campaign of 1813 was an utter disaster. Not only did he lose his brother-in-law, Captain

Devereux and John Reilly were also sent to Brest. Pettrezzoli saw these officers as conspirators who were damaging the unity of the corps. Markey wrote several letters to both ministers of war, Berthier and his successor, Clarke, denouncing Pettrezzoli. The first of these provoked a major inspection by Gen. Donzelot in 1805, which, contrary to Markey's intention, revealed that the Irish officers were abusing their French comrades, which had resulted in a duel between Lt. Patrick O'Kelly and Lt. Denis Thiroux, and the resignation of the latter. Later letters by Markey to Clarke, minister for war from mid-1807, show an great resentment against the Italian, and, playing the Irish nationalist card, may have resulted in the removal of Pettrezzoli in mid-1809. Clarke's reasons for co-opting Markey on to his staff can only be guessed at. **85** Pettrezzoli's naturalization file: ANF BB/11/175, 1140BS. **86** Pettrezzoli and O'Meara were both 45 in 1809.

Hamden Evans,[87] he was also so severely wounded that his army career was cut short, and perhaps his life.[88] He lost his command at Lowenberg and his two senior battalions on the Bober river. Denied the title of baron of the empire conferred by Napoleon, he retired as an honorary major-general.[89]

Colonel Jean François Mahony

Feltre's appointment of French-born Mahony over the head of the acting commander Hugh Ware defied all reason, just as his earlier promotion of O'Meara over FitzHenry had done. Mahony had been dismissed by Junot for incompetence as a battalion commander in Spain, and was reviled by the Legion officers, especially the Bonapartists, mainly because of his British service against France, but also because of his politics, character and behaviour. Mahony was a declared royalist, and was even arrested during the siege of Antwerp in 1814 for collusion with the enemy.

Colonel Hugh Ware

'Brave to a proverb,' according to his obituary in *The Times*, Hugh Ware was 'humane almost to a fault.'[90] Wounded at Lowenberg, he assumed command of the Legion when Lawless fell wounded. Ware commanded at the disaster on the Bober river, where he was again wounded, but the affair was out of his hands. Passed over in favour of J.F. Mahony in December 1813, he became commander in April 1815.

A man of outstanding military talent, Ware had the ideal qualities of a soldier and leader, excelling as commander of the elite units.[91] His divisional commander in Spain, General Solignac, rated him among the finest officers he had ever met.[92] It seems ironic that the only Legion commander to have held a significant rebel field command was to be its last, and that such an active and courageous officer was forced to pass his tenure in frustrating inactivity. His obituary in *The Times* demonstrates how widely he was respected.

RECRUITMENT OF TROOPS

The great shortage of Irish volunteers forced the Legion to recruit among deserters and British prisoners of war of all nationalities, and this is

87 A. Martinen, *Officiers Tués et Blessés pendant les Guerres de l'Empire 1805–1815* (Paris 1889), p. 494, reports the death of Evans, as does Byrne (Byrne, ii, 130.) However, a letter from Evans' wife to the minister for war dated 10 June 1814 suggests he had been taken prisoner, or that she believed that to be the case. SHD Xh16b. **88** Lawless died on Christmas Day, 1824, aged 52. **89** ANF BB/11/99/1 3245 B2. **90** Obituary, *The Times*, London, 27 March 1846. **91** Ware succeeded FitzHenry as captain of the Carabineer (Grenadier) company, of which he had been lieutenant since 1803. **92** Written above Solignac's counter signature to O'Meara's promotion proposal for Ware, Perry etc., dated

mentioned in letters and memoirs of British soldiers. A sergeant in the Gordon Highlanders writes of encounters with recruiters during his captivity, which began in 1809.

It may not have been easy to attract recruits at first, perhaps out of loyalty to their own units and officers, or suspicion of the French. However, after a taste of captivity, and a long march across Spain and France, recruits were easier to win over. A Scots Highland prisoner of war writes, '17th to Gap.[93] Met a large party of British who had volunteered out of the depot into the Irish Brigade of the French service.' Still at Gap, the Irish recruiter's pressure is unrelenting, and the Highlander's irritation shows: 'we were beset by those harpies of the Irish Brigade, Capt Reilly and Sgt-Major Dwyer,[94] offering us brandy and telling us all the evils of a French prison; they got three of our party to join them.'[95]

Unable to find enough recruits among British prisoners, the Legion was forced to try further to the east. In time, this changed the national profile of the troops, so that, by 1813, the Irish character had all but disappeared from the ranks, as the tables below demonstrate.[96]

1st Battalion

Origin	*No. of troops*	*% total*	*Origin*	*No. of troops*	*% total*
Hungarian	99	16.39	Prussian	42	6.95
German	77	12.75	Polish	40	6.62
Irish	65	10.76	Silesian	29	4.80
French	57	9.44	Saxon	21	3.48
Austrian	52	8.61	Westphalian	19	3.15

Figure 8: 3rd Foreign Regiment (Irish), 1st battalion. National origins of troops in 1813, showing the leading ten of twenty-six nationalities (604 men)

The Irish were even less well represented in the second and third battalions, as demonstrated by Figures 9 and 10.

12 April 1810. SHD XH15. **93** Dept. of Hautes-Alpes, eastern France. **94** This would have been Capt. Terence O'Reilly, one of the Legion's most active recruiters, and, probably, Sgt. Anthony Dyer or Doyer, b. Cork, 1775 or 1783, and commissioned in 1812. **95** Daniel Nicol (ed. Mack), *The experiences of a Gordon Highlander during the Napoleonic wars in Egypt, the Peninsula and France* (Glasgow, 1853), p. 217. **96** Figures 8, 9, and 10 are based on returns in SHD Xh16.

2nd Battalion

Origin	*No. of troops*	*% total*
Polish	163	24.36
Austrian	124	18.54
Hungarian	113	16.99
Westphalian	83	12.41
French	52	1.94
Irish	9	1.35

Figure 9: 2nd battalion. A selection of the national origins of troops in Nov. 1812. The Irish are eight out of 14 nationalities (669 men)

3rd Battalion

Origin	*No. of troops*	*% total*
Prussian	262	51.27
Polish	42	8.22
Rhinean	40	7.83
French	19	3.72
Irish	12	2.35

Figure 10: 3rd battalion. A selection of the national origins of troops in Nov. 1812. The Irish are ninth out of 14 nationalities (511 men)

Figure 11 further demonstrates declining Irish recruitment, but this may be partly accounted for by the fact that the military situation had changed drastically. After the First Abdication, the total numbers are too small to enable any conclusions.

Epoch	*Total recruits*	*Number of Irish recruits*	*% total*
Bober disaster to 1st Abdication	200 (s)	4	2
Ist Restoration	100 (s)	3	3
The Hundred Days	78 (t)	6	8
2nd Restoration to Disbandment	26 (t)	1	4

Figure 11: Irish recruitment towards the end of the life of the Irish Legion. ('s' = sample, 't' = total)[97]

DESERTION

Desertion plagued the Irish Legion almost from the very start, and a high proportion of the recruits who joined up to November 1806 deserted.[98] From the first 76 names on the muster rolls, of which 70 were effectives, 12 out of 13 deserters, or 92%, were Irish-born. Ten were apprehended, court-martialled, and sent back to their units, but two deserted again.

97 Compiled from Registers of Troops, SHD 23Yc 205 & 206. 98 Register of Troops SHD 23Yc 207.

Family Name	*Given name*	*Nationality*	*Rank*	*Deserted*
Carrotte	François	French	Drummer	1804
O'Brien	Daniel	Irish	Corporal	1804
Born	Andrew	"	Private	1804
O'Connor	Michael	"	"	1805
Fitzgerald	Thomas	"	"	1805
Goodchild	Thomas	"	Sergeant	1806
Aldwell	John	"	Corporal	1806
Mulauney	Francis	"	Sergeant	1806
Gallagher	John	"	Corporal	1806
Harragan	John	"	"	1806
MacGuicken	Hugh	"	"	1805 & 1806[99]
Malley	John	"	Private	1806x2
Moore	John	"	"	1806

Figure 12: Deserters recorded in the first Register of Troops of the Irish Legion (1803–06).[1]

Desertion grew worse, and the contingents sent into Spain suffered greatly along the way, losing as much as 60%,[2] a persistent figure. Out of an 1809 sample of 100 recruits of all nationalities, 63% deserted.[3] From a cluster of 11 recruits who gave their nationality as Irish, from the depot of deserters on 8 August 1809, 6 deserted (55.5%), 2 by the end of August and 4 by the end of September.[4]

Of a random sample of 60 men recruited from among deserters and prisoners of war between 25 October and 21 November 1813, 37 (62%) are given as Irish-born and 18 (49%) of these deserted, 7 by the end of 1813 and the rest by the end of May 1814. In yet another random sample of all nationalities, 66% deserted.

Some of these deserters, both non-Irish and Irish, joined the Irish Legion simply to return to their own units, or to escape the rigours of captivity, deserting as soon as possible. The outflow from the Legion through desertion was, thus, very high, and came to the attention of the enemy. A British officer wrote in 1810:

99 Probably Hugh Boyd MacGuekin, about whom there is considerable correspondence in SHD Xh16c & d. Rated as 'neither good for officer nor soldier,' he was discharged in December 1806 with permission to immigrate to 'America'. 1 SHD 23Yc 207. 2 Estimated desertion from the contingents going into Spain are: Lacy (1807–8) 60%, FitzHenry (1808) 35%, O'Mahony (1809) 60%, Osmond (1810) 60%. 3 SHD 23YC 208, numbers 601–700. 4 At least one Irishman of the Legion was executed for desertion. Joseph Howard (b. Belfast, 1789), was recruited in Spain on 8 August 1809, probably from

> Deserters still continue to come in [...] 45 arrived here last Monday, some of whom are Irish. They report that they belonged to the Irish Brigade, one regiment of which, composed of 900 English and Irish, entered Spain a few months since, and they had not crossed the Pyrenees six weeks before it was reduced by desertion to 500.[5]

This describes the arrival of the third battalion in Spain under Lieutenant Colonel Mahony. Numbers in the battalion were certainly falling, but such loss en route through desertion was extremely high, and would get worse, as an officer commanding outposts for the British Light Division, observed:

> At the beginning of June 1810, a sergeant who had deserted the enemy's Irish Brigade gave information that the brigade was then in Junot's corps and was commanded by Gen Torny and that the battalions had about 350 men each.[6]

The sergeant's information agrees with a return of the same month giving the total strength in Spain at 735 men and 36 officers.[7] That figure appears to have declined little by February 1811. However, when Byrne declares that the Irish Regiment, with just under 700 men, 'still mustered one of the strongest in the army', he neglects to make clear that the second battalion had very recently absorbed the debris of the third,[8] and the cadre he mentions was the surplus officers and NCOs returning to France.[9] Yet, there is some dispute in the archived sources. In reporting to Napoleon on 11 February 1811, Berthier puts the post-merger strength of all ranks at 505, some 200 fewer than Byrne's figure. Before the merger, desertion had reduced the third battalion to the strength of one company.[10]

prisoners taken at the battle of Talavera (27–28 July 1809). Deserting on 23 March 1810, Howard was caught, found guilty and executed on 28 May. SHD 23Yc 208, #694, p.114. 5 John Aitchison, *An ensign in the Peninsular War: the letters of John Aitchison*, ed. W.F.K. Thompson (London, 1983), p. 93, Letter to his brother William, Vizeu, 6 April 1810. The desertion given is 56%. 6 Sir James Shaw Kennedy, *Diary of Gen. Craufurd outpost operations in 1810*, in Rev. Alexander H. Craufurd, *General Craufurd and his Light Division* (London, 1892), p. 289. The 'Gen Thorny' mentioned is Brig. Gen. Thomieres. 7 SHD Xh15b. 8 In compliance with the Imperial Decree of 28 October 1810, the 1st bn absorbed the 4th, and the 2nd absorbed the 3rd, the latter not until February 1811, as both battalions were in action. 9 Report to the Minister, Bertrier, 11 Feb. 1811, SHD Xh15; Report on 8th Corps of the Army of Portugal, 11 Feb 1811, SHD C7 28; Report 1 Jan. 1811 SHD C7 26. Also Byrne, II, pp 78–9. 10 Berthier to Napoleon, 11 Feb 1811. Many documents mention desertion, e.g. The History of the 3rd Battalion, which states that most POWs enlisted 'to escape' rather than from 'a desire to serve His Majesty the Emperor,' SHD Xh14. This document seems to have been torn from a register of troops, such as 23Yc 208, which covers the early recruitment of the 3rd battalion. An undated and unsigned note in SHD Xh15 gives the strength at Bois-le-Duc on 15 February 1811 as 557 men and 22 officers, with that in Spain in Sept. 1810 at 992 men and 35 officers. However,

Byrne also neglects to mention that total strength had dropped from 1050 rank and file in April 1810, a loss of 33 per cent, mainly through desertion, though losses at Astorga must be taken into account.[11]

THE DIFFICULTIES

The Legion's major difficulties were both built into its make-up and caused by external agents. Better management might have prevented most of these.

Barred from normal recruiting in France or the areas controlled by France, the unit was forced to fall back on deserters or recruits from prisoner of war depots, many of whom deserted again.

The liberation politics of Ireland caused the early disputes, and the inability of the first commander, Bernard MacSheehy, to manage the conflict seriously damaged the unit. The second conflict was the reluctance of many Irish officers to accept a commander or officers that were not Irish born. Italian commander Pettrezzoli suppressed all conflict with extreme measures. However, interference from Minister for War Clarke, son of an Irishman and former officer in the old Irish Brigade, became a kiss of death for the regiment.

This interference of Clarke played a part in provoking the worst crisis in the short life of the Legion, the defection of Jeremiah FitzHenry, commander in Spain, who became the fallen angel in the eyes of his comrades. Junot predicted that the loss of FitzHenry would be the end of the Legion in the Peninsula. Nothing the Legion suffered pierced its very soul as did FitzHenry's defection to Wellington, and for this Clarke's interference, nepotism and vindictiveness were in part at least to blame, as it impossible to attribute his defection to a single cause.[12]

As Junot had predicted, on Christmas Day 1811, the remaining men of the 2nd battalion transferred to the 4th Foreign Regiment (Prussian), and the

the numbers may have been transposed, as a return dated 12 Aug. 1810 gives the strength at Bois-le-Duc at 981 with 26 officers present. See Return signed Lawless, 12 Aug 1810, SHD Xh15. 11 Most of the 80 men killed at Astorga were from the Irish Regiment's Light Company (Voltigers). 12 Other factors probably include FitzHenry's eight-year separation from his wife and family, and the recent death of his father. Born in Co. Wexford in 1774, he is probably the only Irish-born officer of the Legion to die in his birthplace or to be buried in Ireland. His defection will be examined on another occasion. I gratefully acknowledge the research of Lorcan Dunne regarding FitzHenry's career after his defection. William Napier is in error in writing, 'Colonel O'Meara and 80 men of the Irish Brigade were taken by Julian Sanchez; the affair having been, it is said, preconcerted to enable the former to quit the French service': William Napier, *A history of the war in the Peninsula and in the south of France, from the year 1807 to the year 1814* (7 vols, London 1834), iii, 504. However, the substitution of O'Meara's name by either Napier or Wellington, may have been to protect FitzHenry's reputation in Ireland.

officers and NCOs set out on the long winter march to the depot at Bois-le-Duc.[13] Though Byrne puts a brave face on it, there is simply no way to varnish such dishonour.

On 17 September 1815, the last recruit enlisted, an Englishman, John Leoye, aged 20, a blacksmith by trade. On the 29th, the last day in the life of the Legion, now the 7th Foreign Regiment, he transferred with hundreds of his new comrades to the Royal Foreign Legion at Toulon. The register was closed, no doubt with the utmost solemnity, and signed by the Administrative Council in the following order, Staff Sergeant Gugelot (French), Captain Owidsky (Polish), Lieutenant Colonel Braun (German), Captain Miles Byrne and, finally, 'le Major President,' Hugh Ware.[14] The rebel core had held out to the end, but, of the Irish, only Ware and Byrne would stay to obliterate the last vestiges. It must have been with deep chagrin that they watched the burning of the regimental eagle that Ware and his Bonapartist comrades had so reverently concealed on Napoleon's first abdication, the eagle he Ware had saved on the Bober and that Lawless and O'Reilly had rescued from the debacle at Flushing.

THE IRISH LEGION AND OLD IRISH BRIGADE

To compare the Irish Legion with the old Irish Brigade is valid; to equate them would be to ram square pegs into round holes. Legion officers who attempted such an equation probably did more harm than good in reminding the king of the loyalty of the old brigade he had been forced to disperse, contrasted with the disloyalty, as it would have appeared to him, of the Irish Legion.[15]

Though a few Legion officers had served in the old brigade, very few of these were either Irish born or first-generation French, and cannot be described as Wild Geese. The Irish-born officers of the Legion were very different to the Wild Geese, who had fought to support a legitimate monarchist cause and a status quo. They were the exact opposite: rebel, republican, anti-monarchist and, later, Bonapartist and anti-Bourbon. They had little in common with the majority of the old-brigade officers who had defected in 1792 under a royalist 'farewell banner,' proclaiming an eternal and ubiquitous fidelity to the king.[16] These officers had sworn allegiance to the

13 S'Hertengebosch, Noord Brabant, *The Netherlands*. 14 SHD 23Yc206, p. 112. The signatures are immediately below Leoye's recruitment details. 15 For example, The Declaration of Loyalty, dated 1 Jan. 1815 and signed by eighty officers, makes this comparison, organized by royalist Jean François O'Mahony, who may not have been aware that the regimental eagle still existed, in defiance of orders to destroy it. Later attempts to prolong the life of the regiment by evoking the service of the Irish regiments under the ancien régime could only have served to irritate the king. 16 'Semper and ubique fidelis.'

king of France and to him alone; the Legion officers were sworn to the Emperor Napoleon. Not one Irish-born officer of the Legion defected to the royalists, whereas several French-born officers did so during 'the Hundred Days'.[17]

Some former old-brigade defectors of 1792 rose high in the British army, and one of these may have dealt the French the most severe blow of all in the Peninsula. 'The fate of the campaign, and probably of the whole Peninsula, was decided [...] by Colonel Nicholas Trant,' who, with a detachment of Portuguese Militia in March 1810, held the line of the Mondego river against the retreating and desperate Massena.[18]

EPILOGUE

As the sun went down on the Irish Legion, it also set on Irish recruitment to the French army. The trend that had gone on for centuries, surging with the Wild Geese, had dwindled to a trickle. The number of Irishmen who transferred to the newly formed Royal Foreign Legion is hardly worth discussing, and it appears that not a single Irish-born Legion officer was among them. The sum of all the Irishmen who had enlisted throughout the twelve-year life of the Legion had been scarcely enough to fill a single battalion to battle strength.[19]

The British had won the struggle for Ireland's military corpus, if not its very soul. A new wave had begun in 1793, the mass recruitment of the Irish into the British Army. The duke of Wellington is reported to have declared that it was to his Irish Catholic troops that the British owed their military 'pre-eminence'.[20] They contributed greatly to his victories in the Peninsular War and at Waterloo. 'It is not too much to assert,' wrote Home Secretary Sidmouth, 'that the supply of troops derived from Ireland turned the scale on the 18th of June.'[21] The irony is inescapable.

By mid-century, the Irish troops made up more than 40% of the British army, the new symbiosis playing a cardinal role in consolidating the British

17 These were all French-born, through three were of Irish parentage and descent, and one had served in the Irish Brigade. See Appendix. 18 Michael Glover, *The Peninsular War* (London, 1974) p. 144. Descended from a Co. Kerry Wild Geese family of Viking extraction, Trant served as a colonel on Wellington's staff, and, acting as a military agent, he assumed command of several battalions of militia. His contingent acquitting itself very well in several actions, he was raised to major-general in Portugal and, later, in the British Army. 19 Carles, p. 35. 20 Opinion differs as to whether Wellington ever made such a statement. John Cornelius O'Callaghan, *The history of the Irish brigades in the service of France* (Glasgow, 1870) note to pp 615–16, asserts he did, but the balance of the evidence suggests he did not. I gratefully acknowledge the research of Steven H. Smith on this matter. 21 Sidmouth to Whitworth, 24 June 1815, National Archives, Kew, H.O. 100/184 f.204.

Empire, and it was held that the two best things that happened to that army during the ninteenth century were the breach-loading rifle and the Irish foot soldier, who could march all day without tiring and bear extremes of heat and cold like salamanders. Far away from Dunkirk and Belgrade, Irish troops served under a very different banner among the wild hills of India or in the sweltering heat of Africa, a universe apart from the vision of Tone, O'Connor and Fitzgerald, and, indeed, from that of Lawless and his comrades.

The talented men of the Legion who sought to make that vision a reality were scattered across the world and lost forever to Ireland. Fugitives from home, they have been neglected by their own country, and, at first exploited by France, they were all too soon forgotten. The Irish Legion of Napoleon was the inglorious, the bitter swansong of a great movement.

APPENDIX I

List of officers of the Irish Legion (The Irish Regiment, etc.), 1803–15

Ahern, John, *Ireland*
Allen, John, *Ireland*
Aubier, Antoine, *France*
Avenariu, Guillaume, *Germany*
Balthasar, Louis, *France*
Barker, Arthur, *Ireland*
Barker, William, *Ireland*
Behr, Georges Henry, *Germany*
Belaerts, Jacques, *Holland*
Benisch, François, *Bohemia*
Bernhold, Sigismund, *Germany*
Berthomé, François, *Ireland*
Blackwell, James, *Ireland*
Bohnen, Stanislas de, *Sweden*
Bosilio, Maurice, *Italy*
Bourguignon, Jean-Louis, *France*
Brangan, Patrick, *Ireland*
Braun, Antoine, *Germany*
Brelevet, Jean-François, *France*
Brown, Thomas, *Ireland*
Buchwald, Henry, *Austria*
Buhlmann, Jacques, *Germany*
Burghess, John, *Ireland*
Burke, William, *Ireland*
Butler, Alexandre, *France*
Byrne, Miles, *Ireland*
Cabour-Duhay, Edouard, *France*
Campbell, John, *England*
Canillot, Jean-Baptiste, *France*
Canton, Thomas, *Ireland*
Cardaillac, Louis, *France*
Carotte, François, *France*
Cesack, Charles, *Bohemia*
Chatelin, Jean Martin, *France*
Conway, Thomas, *France*
Corbet, Thomas, *Ireland*
Corbet, William, *Ireland*
Cordier, Victor, *France*
Cosgrove, John, *Ireland*
Coüasnon, Alexis de, *France*
Cummins, John, *Ireland*
Dantrass, Edouard, *France*
Debonnaire, Charles, *France*
Decarne, Louis, *France*
Delaheese, Charles, *Prussia*
Delaney, James, *Ireland*
Delaney, Matthew, *Ireland*
Delaplaigne, Adolphe, *France*
Delavieuville, Adolphe, *France*
Delhora, Falian, *Prussia*
Demeyere, Jean-Jacques, *France*
Démon, Jean-Louis, *France*
Demonts, Jean, *Germany*
Derry, Valentine, *Ireland*
DeVerteuil, Stanislas, *France*
Devreux, Alexander, *Ireland*
Dillon, Auguste, *France*
Dowdall, William, *Ireland*
Dowling, Jerome, *Ireland*
Doxal, René, *Switzerland*
Dubourg, Louis, *France*
Dupont, Gilles, *France*

APPENDIX 1 *(contd)*

Dupouget, Louis, *France*
Dyer, Antony, *Ireland*
Eagar, William, *Ireland*
Eckhardt, Chretien, *Germany*
Eliot, Jean, *Lituania*
Engelhard, Philippe, *Germany*
Erbling, Joseph, *Germany*
Erdely, François, *Romania*
Esmonde, Laurent d', *Ireland*
Evans, Hampden, *Ireland*
Farquharson, John, *Scotland*
Ferguson, –, *Scotland*
Ferrary, Dominique, *Spain*
Finney, John, *Ireland*
FitzHenry, Jeremiah, *Ireland*
FitzHenry, John, *Ireland*
Fitz-Patrick, James, *Ireland*
Foyard, Louis, *France*
Gaboulbene, Joseph, *France*
Gaillot, Charles, *France*
Gallagher, Patrick, *Ireland*
Gefort, Jean, *France*
Gerber, Godefroi, *Germany*
Gibbons, Austin, *Ireland*
Gibbons, Edmund, *Ireland*
Gibbons, John, *Ireland*
Gilmer, Joseph, *England*
Giraud, Jacques, *France*
Glashin, Daniel, *Ireland*
Glashin, Jean, *Ireland*
Goetz, Jean-Philippe de, *France*
Gordon, Robert, *Scotland*
Gorrido, Vincent, *Spain*
Gossling, George, *Germany*
Gougis, Charles, *France*
Gourgas, Bernard de, *France*
Gourlay, François, *France*
Gregoire, Victor, *France*
Hamman, Jean, *Prussia*
Heraud, Honoré, *France*
Hertig, Charles, *Prussia*
Hoyne, William, *Ireland*
Hughes, John, *Ireland*
Hupert, Frederic, *Germany*
Igydowitz, François, *Italy*
Jackson, Thomas, *Ireland*
Jeetze, Charles de, *Prussia*
Keller, Godefroi, *Prussia*
Keller, Guillaume, *Prussia*
Kienlin, Louis, *France*
Klembt, Jean, *Poland*
Klopstock, Jean-Henry, *Italy*
Koning, Jean Pierre de, *Belgium*
Koslosky, Joseph, *Poland*
Lablairie, Olivier, *France*
Lacy, Louis, *Spain*
Lalande, François, *France*
Lambert, Robert, *Ireland*
Landy, Michael, *Ireland*
Landy, Richard, *Ireland*
Lawless, Luke, *Ireland*
Lawless, William, *Ireland*
Lefort, Jean, *France*
Lerreuse, Bague, *France*
Levacher, Claude, *France*
Lossell, Patrick, *France*
Lynch, Patrick, *Ireland*
MacCann, Patrick, *Ireland*
MacCarthy, James, *Ireland*
MacDermott, Bernard, *Ireland*
MacEgan, James, *Ireland*
MacGawley, William, *Ireland*
MacGuire, James, *Ireland*
MacMahon, Arthur, *Ireland*
MacNevin, William, *Ireland*
MacSheehy, Bernard (1), *Ireland*
MacSheehy, Bernard (2), *France*
MacSheehy, Patrick, *Ireland*
Magrath, Achille, *France*
Magrath, Louis, *France*
Magrath, Patrick, *France*
Malisieux, François, *France*
Mandeville, Auguste, *France*
Marcelin, Joseph, *France*
Maréchal, François, *France*
Markey, Thomas, *Ireland*
Martin, Christopher, *Ireland*
Masterson, Edward, *Ireland*
Menzer, Joseph, *Germany*
Metz, Charles, *Germany*
Milleville, Bartholme, *France*
Montagu, Maurice, *France*
Montbert, Ragnar de, *France*
Montrand, François, *France*
Morrison, Fecorbert, *Ireland*
Mougenot, Henri, *France*
Mullany , Charles, *Ireland*
Mundt, Charles, *Prussia*
Murray, Paul, *Ireland*
Noel, Félix, *France*

APPENDIX 1 *(contd)*[1]

Nugent, Charles, *France*
O'Brien, Jean, *France*
O'Gorman, Thomas, *West Indies*
O'Kelly, Patrick, *Ireland*
O'Mahony, Jean-François, *France*
O'Malley, Austin, *Ireland*
O'Meara, Daniel, *France*
O'Meara, William, *France*
O'Morand, William, *France*
O'Morand, Edouard, *France*
Onslow, Maurice, *France*
Oppermann, François, *Germany*
O'Quin, Patrice, *France*
O'Reilly, Terence, *Ireland*
Osmont, Augustin, *France*
Osmont, Edouard, *France*
O'Sullivan, Thomas, *Holland*
Owidzky, Patrice d', *Poland*
Parrott, Joseph, *Ireland*
Peeters, Philippe, *Belgium*
Perry, James, *England*
Petrezzoli, Edouard Antoine, *Italy*
Pickert, Leopold, *Denmark*
Plunkett, Christopher, *Ireland*
Poleski, François, *Poland*
Powell, Patrick, *Ireland*
Prevost, Louis, *France*
Ramm, Pierre, *Germany*
Raymond, –, *Switzerland*
Read, Thomas, *Ireland*
Regnier, François, *France*
Reiff, Joseph, *Germany*
Reiffe, Matthieu, *France*
Reilly, John, *Ireland*
Reynolds, Matthew, *Ireland*
Robiquet, Jacques Charles, *France*
Roche, Hercule de, *France*
Ross, Daniel, *Scotland*
Royal, Nicholas, *England*
Ruff, Gottlieb, *Prussia*
Russell, Michael, *Ireland*
Ryan, Charles, *Ireland*
Saint-Leger, Auguste, *France*
Saint-Leger, Edmond, *France*
Saint-Leger, Patrice, *France*
Salomez, Daniel, *France*
Salomez, Jean Henry, *France*
Sanford, François, *Germany*
Schmidt, Johan, *Germany*
Schroeder, Dominique, *Germany*
Schroeder, Jean, *Germany*
Schurmann, Joseph, *Germany*
Segaud, Jean Pierre, *France*
Serisy, Edward, *France*
Setting, Antony, *Ireland*
Sheridan, Michael, *Ireland*
Smith, James, *Ireland*
Souillard, François, *France*
Ste-Colombe, Maurice de, *France*
Stephens, Samuel, *Ireland*
Sturm, Frederic, *Germany*
Swanton, Armand, *France*
Swanton, Robert, *Ireland*
Sweeny, John, *Ireland*
Tennent, John, *Ireland*
Thiroux de St Cyr, Denis, *France*
Thompson, Henry-Jean, *France*
Thuillier, Louis, *France*
Thumerel, Augustin, *France*
Towne, David William, *England*
Tréssan, Louis de, *France*
Tuillier, Jacques, *France*
Tyrrell, Nicholas, *Ireland*
Wagner, Hermann, *Moravia*
Wagter, Godefroi, *Germany*
Wall, Alfred de, *Spain*
Wall, Richard de, *Ireland*
Walsh, John, *Ireland*
Ware, Hugh, *Ireland*
Weichenheim, Charles, *Germany*
Weiss, Michel, *Poland*
Wolff, François, *Germany*
Zapssell, Joseph, *France*
Zelinski, Jean, *Poland*
Zobinsky, Charles, *Prussia*

1 This above is the first comprehensive list of Irish Legion officers to be published. This data is derived from a wide variety of documents at *le Service Historique de la Défense* at Vincennes, and *les Archives Nationales de France*. Some names have been omitted pending further study. The information is provisional, and must not be cited as a reference or reproduced in any form. © Nicholas Dunne-Lynch 2008. Enquiries should be addressed to nicholas.dunnelynch@free.fr

BIBLIOGRAPHY

Full-length works

Miles Byrne, *The Memoirs of Miles Byrne*, 2 vols (Dublin, 1907). All full editions of this work are out of print, but the 1907 edition is now downloadable on Google Books. Demi-Solde Press, San Diego, CA, has produced a facsimile of vol. II, which deals with Byrne's life in France and his military career. http://www.demisoldepress.com/irish.htm

John G. Gallaher, *Napoleon's Irish Legion* (Carbondale, IL, 1993). The only in-depth study of the unit.

Eugene Fieffe, *Histoire des Troupes Etrangères au service de la France* (Paris, 1854) 2 vols.

A. Martinien, Tableaux par Corps et par Batailles des Officiers tués et blessés pendant les guerres de l'Empire 1805–1815 (Paris, 1899).

Danielle & Bernard Quintin, *Dictionnaire des Colonels de Napoléon* (Paris, 1996).

Chapters

Guy C. Dempsey, *Napoleon's mercenaries: foreign units in the French army under the Consulate and Empire, 1799–1814* (London, 2002). Relies heavily on secondary sources.

Articles

John G. Gallaher, 'Irish patriot and Napoleonic soldier – William Lawless', *Irish Sword* 18 (1992), 225–63.

John G. Gallaher, 'William Lawless and the defense of Flushing, 1809', *Irish Sword* 7 (1989), 159–64

E.W. Ryan, 'A projected invasion of Ireland in 1811', *Irish Sword* 1 (1950–1), 136–41.

P. O'Snodaigh, 'The flag of Napoleon's Irish legion', *Irish Sword* 18:72 (Winter 1991), 239.

Thomas Bartlett, 'Last flight of the Wild Geese? Bonaparte's Irish legion, 1803–15', in Thomas O'Connor and Mary Ann Lyons (eds), *Irish communities in early modern Europe* (Dublin, 2006).

Lieut-Col. Pierre Carles, President du Centre d'Histoire de Montpelier, 'Le Corps irlandais au service de la France,' *Revue Historique des Armées*, no. 1976/2, 25–54. Based mainly on archived sources.

Website articles

The Napoleonic Association: Capt. Frank Forde, *Napoleon's Irish Legion*, with translations of the Irish Legion Historical records by Lieut.-Col. Brian Clark, is available on http://www.napoleonicassociation.org/research/articles/Napoleons%20Irish%20Legion.pdf (6 May 2008)

Sympatico: Virginia Shaw Medlen, *Legion Irlandaise (Napoleon's Irish Legion) 1803–1815*, is available on http://www3.sympatico.ca/dis.general/irish.htm. (6 May 2008)

Both website articles provide interesting and comprehensive summaries of the regiments history and operations. Derived mainly from Byrne.

The Irish and the Franco-Prussian war: hopes and disappointments

JANICK JULIENNE

By the late ninteenth century, the links between France and Ireland were old, multi-faceted and deeply set in the memory of the two peoples. This common history, the 'Franco-Irish friendship', was used by different Irish nationalist groups, as well as the French authorities, to put the pressure on England during different periods in history. Yet, after the revolutionary period and the failure of the French landings in Ireland in 1796 and 1798, the closure of the Irish College in Paris and the dissolution of the last Irish regiments in the 1790s, the Franco-Irish relationship may have seemed to be over. But the interest of the French in what was referred to as the 'sister island' did not weaken. On the contrary and by the 1830s, there were abundant publications and the commitment of key French figures like Victor Hugo, Jules Michelet or Flora Tristan, give a strong testimony of the increasing sensibility of French opinion to the 'Irish question'.[1]

But it was not until the 1860s that the links between Ireland and France became stronger due to the influence of the policies of Napoleon III and of the strategy of the Fenians, an Irish nationalist movement, several of which's leaders settled in Paris from time to time. The Franco-Prussian war brought Ireland and France closer; a large number of Irish being eager to offer a material and also military support to the French. This war gave the Franco-Irish friendship a new impulse, in its political, religious and, more than all, military aspects. For Irish nationalists, the Franco-Irish connection still represented a direct threat to England. But the consequences of the Franco-Prussian war spelled an end of these last hopes of French support for nationalist Ireland and would result in a major rift in the ancient Franco-Irish link. In the course of this article, I will examine how the support of the nationalist Irish for the French cause led to the presence of Irish units during the Franco-Prussian War. I will then examine the results of France's defeat and how this affected the Franco-Irish connection.

1 Laurent Colantonio, 'Les usages français de Daniel O'Connell des dernières années de la Restauration à la Deuxième République', in S. Aprile and F. Bensimon (eds), *La France et l'Angleterre au XIXe siècle : échanges, représentations, comparaisons* (Paris, 2006), pp 369–83.

FRANCO-IRISH RELATIONSHIPS DURING THE SECOND EMPIRE

The birth and development of the Fenian movement, which dominated Irish nationalism during the 1860s, had wider links to the international context and more specifically to the tensions between France and England.[2] Napoleon III's international policy had a major influence on Franco-Irish relationships from the 1860s onwards and was largely responsible for the place France held in the strategy of the Irish nationalists in the 1880s.

The personality and personal convictions of Napoleon III had a major influence in French diplomacy during the Second Empire.[3] They led him to support the cause of various nationalist movements. In keeping with the political philosophy of his uncle (Napoleon Bonaparte or Napoleon I), Napoleon III maintained that the governments had to respect the rights of all nationalities to maintain peace.[4] He intended to destroy the diplomatic settlement that was imposed by the Congress of Vienna in 1815, in order to restore French grandeur and to rebuild Europe on a new basis, respecting nationalities and the sovereignty of various peoples. He very early presented himself as the 'champion of nationalities', and this became an abiding feature of his international policies. In 1856, during the Paris Congress, Napoleon III obtained autonomy for the principalities of Moldavia and Walachia (in modern-day Romania), in spite of the hostility of England and Austria.[5] This event gave hope to Irish nationalists.[6]

Furthermore, from around 1860, new tensions began to appear between France and England. From 1857, Napoleon III started a programme of naval re-armament in an effort to restore French naval power and thus the French navy became a challenge to the English fleet.[7] In 1858, Felice Orsini attempted to stir new tensions between the two countries as England then sheltered many French political exiles. The *Moniteur*, the official paper of the government, went so far as far to refer to England as 'the murderers' haunt, who sooner or later will have to be fetched from that island'.[8] It was in such a backdrop of tension between France and England that the Irish revolutionary nationalist movement known as the Fenians was founded. On 17 March 1858, James Stephens received the approval of the Irish in America, who were led

2 R.V. Comerford, 'Anglo-French tension and the origins of Fenianism', in F.S.L. Lyons and R.A.J. Hawkins (eds), *Ireland under the Union: varieties of tension: essays in honour of T.W. Moody* (Oxford, 1980), pp 149–71. 3 P. Bury, 'La carrière diplomatique au temps du Second Empire', *Revue d'histoire diplomatique*, 90 (1976), 295. 4 Pierre Milza, *Napoléon III* (Paris, 2006), p. 88. 5 Ibid., pp 367–71. 6 The Paris Congress ended the Crimean War. 7 Milza, op. cit., p. 376. The French naval re-armament planned for the building of 40 ships of the line, armed with 70 or 90 guns, 20 armoured frigates (*frigates cuirrassées*) of 40 guns, 30 corvettes, 60 light ships and 75 transport ships, which were to able to transport 40,000 men and 12,000 horses. 8 Jean Guiffan, *Histoire de l'Anglophobie en France, de Jeanne d'Arc à la vache folle* (Paris, 2004), p. 125.

by John O'Mahony, to organize a revolutionary movement that later would have links with Europe. Always true to their motto 'England's difficulty is Ireland's opportunity', the Irish nationalists found it natural to turn to France.

This situation explains why, from around 1859, a rumour began to spread that a new French landing on the Irish coast was being planned. In Kilgarvan, Co. Kerry, for example, a member of the Irish Constabulary reported that several of the local population had joined the Fenians and were waiting for a landing of foreign troops.[9] In England the same rumour gave rise to a wave of fear. Many civilians founded volunteer rifle corps and, having armed themselves, began to train and plan to repel a French invasion. A new Franco-English war seemed to be so likely that James Stephens and John Mitchel, the Fenian leaders, who were both wanted in England, decided to go to Paris at the beginning of 1859 to wait for the right moment.[10] While there it would appear that they tried to make contact with the Ministère de l'Intérieur in order to obtain the support of planned insurrection in Ireland.[11]

In Ireland, some decided to take the initiative and created the 'MacMahon Sword Committee', which intended to offer a presentation sword to General Marie Edmé MacMahon, who had Irish origins.[12] On 2 July 1859, the two nationalist daily papers, *The Nation* and *The Irishman*, offered to start collecting funds.[13] Donations soon come flooding in to the organization committee which represented a cross-section of Irish nationalist opinion and included members of the Young Ireland movement, such as T.D. Sullivan and Patrick James Smyth. On 9 September 1860, at the camp of Châlons, General MacMahon received the presentation sword from the Irish representatives with gratitude but also circumspection, being very careful not to proclaim the nationalist sentiment of his guests. William Smith O'Brien very clearly described the implicit aim of the committee – 'giving this sword is meant on purpose as an intimation, meaning that MacMahon would be welcomed here at the head of the French army'.[14] Actually, the whole affair created some sensation, in France as well as in Ireland.[15] Irish nationalists still hoped for a military alliance and looked back to the memory of the help given by previous French governments since the reign of Louis XIV.[16] The actual

9 Comerford, op. cit., pp 153–4. 10 R.V. Comerford, *The Fenians in context: Irish politics and society, 1848–1882* (Dublin, 1985; new ed. 1998), p. 53. 11 Ministère des Affaires étrangères, Paris, *Affaires diverses politiques*, Angleterre, n°35, '1859, Mitchell', correspondance from the Public Security Division to the Foreign Affairs Department. 12 *Mémoires du Maréchal de MacMahon, duc de Magenta, souvenirs d'Algérie*, published by G. de Miribel (Paris, 1932), pp 1 and 14. 13 Comerford, 'Anglo-Irish tension', p. 158. 14 J. Martin and W.S. O'Brien, *Correspondence between John Martin and Smith O'Brien relative to a French Invasion* (Dublin, 1861), p. 5. 15 By example, *La Revue contemporaine*, 20 June 1860, or *La Revue des Deux Mondes*, 15 December 1860 and the pamphlet *Mac Mahon roi d'Irlande* (Paris, 1860). 16 All the articles published on that occasion have been gathered in Martin and O'Brien, *Correspondence.*

political situation of France is not really taken into consideration, only the image of the long Franco-Irish friendship.[17]

Ultimately, it is not impossible that French authorities gave real consideration to the prospect of a new French landing in Ireland. In 1863, the attaché militaire in London, Admiral (Chef d'escadron) d'Andigné, was sent to Ireland to make a military reconnaissance. At the end of his survey, he wrote a report in which he underlined the popularity of the French emperor and the hopes that the nationalist population invested in him.[18] D'Andigné concluded that 'even if I've not been asked this question plainly, I would say that the discontent of a part of the Irish population, however distressed they are, has to be stopped being used in the plans for England, except for the degree of weakening it can cause'. On several points d'Andigné suggested the possibility of a French intervention in Ireland in support of an Irish rebellion, as if such was the question he had been asked to answer. But in his analysis the Irish question is no more than an internal matter for England rather than an issue of French international policies. Despite the expectations of some of the Irish, the idea of a French landing in Ireland was effectively abandoned at this point.

In 1859 and 1860, even if a French intervention seemed unlikely, France was still the focus of Irish aspirations. During the 1860s, Napoleon III occasionally pandered to Irish hopes, alternating phases of support, which were given some significance by Irish nationalists, with phases of pro-English policies. During the period, the Irish question was once more a factor in the deliberations of French politicians and diplomats.

THE IRISH POLICY OF NAPOLEON III: BETWEEN IDEOLOGY AND REALPOLITIK

The good relations that France and England had created since the Crimean War were carefully maintained by both sides in the decades that followed. This diplomatic rapprochement remained a constant figure of the international policies of Napoleon III. His choice was between maintaining an *entente* with England and his ambitions for French supremacy in Europe. Planning to reshape the map of Europe, Napoleon decided to rely on building France as a major power, while at the same time portraying it as being a liberal country. In Napoleon III's mindset, England, as a liberal

17 Comerford, *Fenians*, pp 164–5. 18 'Whether I was talking to a gentleman of the highest quality or to a simple fabric worker, the same questions were coming back again: What is the Emperor doing to help Poland? It's to Paris that each of them is turning for political enlightment': Service historique de la Défense, Vincennes, MR.1435, 'Mémoire sur l'Irlande, rapports et lettres du chef d'escadron D'Andigné', 10 mars–20 juillet 1863.

country, had less to fear from a revolutionary movement of liberal influence. Moreover, the emperor retained an emotional fondness for the country that had sheltered him during his exile.[19] Thus France stood by the side of England during the Crimean War of 1853–6 and was also an ally in the China War of 1858–60. These close bonds were even stronger after the ratification, on 23 January 1860, of a treaty of free exchange between the two countries.[20]

But not everything had been forgotten of the wars and tensions that had punctuated the history 'between two hereditary enemies'. There always remained 'a kind of defiance that can be felt at the smallest disagreement'.[21] During such phases of tension, Ireland became again a trump card that could be used against England. Eager to conciliate political and diplomatic constraints with his personal convictions, Napoleon III oversaw an international policy that knew u-turns and sometimes total contradictions. This held true even when it came to Irish matters. For Irish nationalists, the results were occasionally positive.

While England was still a welcoming place of refuge for French political exiles, France also remained the shelter of the Irish nationalists. As early as 1850, a heterogeneous Irish colony was settled in Paris. This included political exiles of an older tradition, the so-called 'Old Irish' who descended from the Wild Geese. These included Miles Byrne and also also artists, painters, writers, musicians.[22] From the 1860s, this community grew with the arrival of members of the Fenian movement, including James Stephens, who was at the centre of an active group, and also John Patrick Leonard who played a crucial role in Franco-Irish affairs from 1860 until his death in 1889.[23] This Irish colony carried on their activities while in the meantime French authorities made regular but cursory inquiries about them, allowing the Irish in Paris to carry on pretty much as they pleased, without close police surveillance.

When it came to the Irish nationalists, France is not only a haven of peace but also a potential training ground. After the failure of the 1848 rising in Ireland, James Stephens 'made his best [efforts] to get enlisted in French army in order to learn the art of warfare in the hope he could organize insurrection in Ireland later on'. At the same time he advocated reforming an Irish brigade. His demands being refused, he left for America.[24] Around 1858, Stephens came back to settle in Paris in an effort to organize a centre for the

19 Milza, op. cit., p. 384. 20 Ibid., pp 452–4. 21 Jean Guiffan, op. cit., p. 111. 22 These included the four Casey brothers, who were cousins of J. Stephens, and also John Mitchel, Denis Dowling Mulcahy, William O'Donovan and Arthur O'Leary, brother of the Fenian John O'Leary. 23 On Leonard see J. Julienne, 'John Patrick Leonard (1814–89): chargé d'affaires d'un gouvernement irlandais en France', in *Études irlandaises*, 25:2 (Autumn 2000), 49–67. 24 Ministère des Affaires étrangères, Paris, *Correspondance politique des consuls*, Angleterre, vol. 41: letter from the consul Livio, date 31 December 1866. The archives mentioned by Livio could not be found, either in the Affaires étrangères, nor in the Archives nationales.

political and military instruction of Fenian leaders.[25] Some Fenians managed to enlist as individuals in the foreign regiments of the French army and used them as training centres. For example, John Devoy was enlisted on 9 February 1861 into the 2nd Regiment étranger as a 'volunteer' for two years.[26] In spite of the efforts of French authorities to quash the idea of a new Irish brigade, which they knew would be taken as a threat by the English government, the Franco-Irish military tradition still resisted, at least in an unofficial and token form. Fenians also came to France in an effort to get weapons.[27] In this they helped through the unexpected support of a famous French military leader, General Gustave Cluseret.[28] Cluseret later enjoyed the titular title of 'Chief commander of the army of the Irish Republic', which was planned as part of the rising organized by the Fenians in Ireland in March 1867.[29] In the 1860s, the Fenians were enjoying a perfect impunity on French territory, where they were never troubled and seldom sought out by the police.

Furthermore, when the Fenian movement began to grow, the Irish problem took on a larger international significance. The French consulates in Ireland were reminded that the large Irish community in the United States gave its support to the nationalist movement. Thus, France came to expect that that sooner or later the United States would use Ireland as a way to start to pressurize the rest of Europe. So that it is France's interest to entertain Fenian sympathies. During these ten years, Napoleon III offered his protection on several occasions to the Irish nationalists who were wanted by England. On 13 December 1687, after the failure of their attempted insurrection, the Fenians organized the escape of prisoners who were imprisoned in Clerkenwell jail, in London. Among those involved in this prison breakout were James Murphy and Patrick Casey, who both fled to Paris. The English government demanded their extradition, but it was denied by the French on the grounds that James Murphy and Patrick Casey were 'Irish patriots at war against an enemy of their country'. This definition classed Murphy and Casey as political refugees.[30]

25 R.V. Comerford, 'Conspiring brotherhoods and contending elites, 1857–1863' in W.E. Vaughan (ed.), *A new history of Ireland*, v: *Ireland under the Union, I, 1801–1870* (Oxford, 1987), p. 420. **26** Service historique de la Défense, Vincennes, archives de la Guerre, 34 YC 5305, 'Registre matricule de la troupe 2è Régiment étranger, 9 février 1861–27 février 1862', matricule n°7484. His friend, John Moriarty, also served during several campaigns in the French armies (J. Devoy, op. cit., p. 335) Franck McAlevy or the Fenians Edmond O'Donovan and James O'Kelly were also enlisted in France (*Devoy's post bag*, Dublin, 1948, i, 59). **27** Ministère des Affaires étrangères, Paris, *Affaires diverses politiques*, Angleterre, n°35, '1865–66, Mitchel, Stephens, Fenians en France', letter of 5 December 1865 from Livio to Drouyn de Lhuys. **28** G. Cluseret, 'My connection with Fenianism', *Fraser's Magazine*, New Series, 6:31 (1872), 5. **29** Archives Nationales, Paris., Fonds Rouher, 45AP18, 'Affaires diverses politiques', 'Affaire Cluseret, général américain, 1867–68', report of Mr d'Abzac, consul trainee in New York, 17 January 1868. **30** *Devoy's post bag*, ii, 160. It has not been possible to find in the Archives Nationales the file relating

A few weeks earlier, in September 1867, an explosion at Manchester jail has made it possible to break out some imprisoned Fenians, including James O'Kelly who seems to have found refuge in the United States.[31] But in 1868 he received mail in France, these letters having been sent *poste restante* to an address in Le Havre. In January, the English government asked the French authorities to intercept a letter that had been sent to O'Kelly.[32] The letter was intercepted but the French authorities hesitated in giving it to England. The Garde des Sceaux pointed out that 'England offers us no reciprocity. When it comes to political matters, and such is the case, she would dismiss with contempt any of our demands to have letters seized from Mazzini, Ledru-Rollin, Felix Pyat, Victor Hugo, &&'.[33] Moreover, handing over a letter could lead, even indirectly, to the arrest of O'Kelly or his accomplices. France would not unnecessarily endure the reprobation of Irish nationalists and their supporters, who included people in the United States as well as in France. The outcome of this case is not given in the archive file but O'Kelly was not intercepted by the police and was able to proceed with his political activities in England and in the United States.[34] But Napoleon III's policy still was uncertain as he hesitated between his desire to support Irish nationalism and his need not to offend his English ally. In July 1869, he gave orders to seize a French pamphlet denouncing the way Irish political prisoners were treated in English jails, officially to avoid upsetting 'our cordial ally'.[35]

Second Empire France was the focus of many hopes of Irish nationalists and these were often entertained by Napoleon III himself. The strong Franco-Irish links of the 1860s would lead to the involvement of the Irish in the Franco-Prussian War of 1870–1.

THE IRISH IN THE FRANCO-PRUSSIAN WAR: THE REVIVAL AND APOTHEOSIS OF THE FRANCO-IRISH FRIENDSHIP

The commitment of the Irish to the cause of the French during the Franco-Prussian war was a spontaneous development, as if the forces of history were repeating themselves. In the same way the Irish had supported Louis XIV's armies, or Napoleon's, they came to France in 1870 to help the embattled French army. It was mainly in the name of the ancient link between the two countries that sections of the Irish population mobilized to help France.

to this demand of extradition, the BB1 archives being very deficient. But the presence of the Casey brothers, James Stephens' cousins, is testified by several printed testimonies as wall as Stephens' correspondence. 31 D.J. Hickey and J.E. Doherty, *A dictionary of Irish history, 1800–1980* (Dublin, 1987), pp 439–40. 32 Archives Nationales, Paris, BB18 6509, Dossier n°7920 A4, James O'Kelly (1868), letter of 14 January 1868. 33 Ibid, report by the Garde des Sceaux, 16 January 1868 on 'O'Kelly's case, escaped Fenian'. 34 Hickey and Doherty, op. cit., pp 439–40. 35 *The Irishman*, 24 July 1869.

From the beginning of the conflict, the French consulate in Dublin reported that throughout Ireland demonstrations were being organized in support of France.[36] These were often accompanied by an open hostility toward Prussia. In Limerick, by example, in August 1870, 'the Prussian consulate having raised his flag to celebrate a so-called victory, the people threw itself against his house and would have ruined it if the consulate had not followed the advice of the police and had put away those colours, so obnoxious were they to the Irish people'. All during the war, the diplomatic correspondence from Dublin regularly reported demonstrations of the same kind in the towns of Ireland.

Irish people also mobilized to provide material help to France. During the war, the French consulate in Dublin, the consular officials and the French government received a large number of gifts from private citizens. These included money, bandages and other medical supplies. The French consulate archives also records some collective initiatives. On 24 March 1871, the consul sent to the ministry a long dispatch that named the various organisations that had been established in Ireland to help the French.[37] The first one to be mentioned was the Irish mobile ambulance or medical company that toured the French battle fields. The second one was a special committee of the Society of Friends (Quakers) in Ireland. Created with the unique purpose of easing the misery of the civilian victims of the war, this committee collected more than 110,000 francs, which was then given directly to the sections of the French population that were suffering the worst.[38] This philanthropic effort was controlled by delegates sent to France for the purpose. The third Irish initiative was the St François de Sales committee, which was founded by Cardinal Paul Cullen, the archbishop of Dublin. This committee distributed money to wounded soldiers and the affected population using the French bishops and the St Vincent de Paul Society as intermediaries. The fourth committee had been created under the auspices of the lord mayor of Dublin (Patrick Bulfin) as the 'committee for the assistance of the French' and this was based at the Mansion House, the mayor's residence, in Dublin. After slightly more than one month in existence, it had already raised more than 130,000 francs. To these four main associations were added a large number of smaller-scale efforts, for during the whole conflict Irish people kept sending gifts for wounded soldiers as well as for civilian victims of the devastation.[39]

36 Ministère des Affaires étrangères, Paris, *Correspondance politique des consuls,* Angleterre, n° 45, letter of the 22 August 1870. 37 Archives des Affaires étrangères, Nantes, *Consul général de France*, Dublin, n° 7, Diplomatic dispatch, n° 288, 24 March 1871. 38 Such a sum would represent 44,898 days of work from an unskilled worker in the country in 1870 (at 2.45 francs a day) or 100 times the national income by head of the active population in 1875 (Fourastié, op. cit., pp 65 and 43). 39 C. Ronayne, 'The reminiscences of an ambulance surgeon', *Journal of the Cork Historical and Archeological Society*, 24 (1918), 36.

At the same time, first aid was provided by Irish medics to the wounded in the field either in Paris during the siege, or on the battlefields.

THE CARE OF THE WOUNDED IN FRANCE

In Paris, two places of worship are mainly attended by the Irish: the Passionist church, in the 8th Arrondissement, and the Irish College, in the 5th. Both were eager to help French people who were wounded during the siege of Paris. All foreigners were compelled to leave Paris by a law passed on 25 August 1870, but some of the staff of the Passionist church in the city and of the Irish College chose to stay to take care of the wounded soldiers of the French army. They were encouraged to do so by Monsignor Georges Darboy, the archbishop of Paris, who authorized the Catholic clergy to create a hospital facility in each church.[40]

As early as 1870, the priests of the Passionist church organized a temporary hospital, with the help of General MacMahon's daughter, among others. It consisted of six and then eight beds and a French doctor who was assisted by two volunteers. Food and equipment were partly provided by the city, the Beaujon hospital and the Passionist community. During the winter of 1870–1, the Passionist priests shared the same privations as the rest of the population of Paris and had to have recourse to eating dogs, cats, rats or meat of dubious origin.[41] They also regularly visited the wounded outside of their church and later reported their encounters with Irish countrymen in French uniform.[42]

From 15 September 1870 to 20 February 1871, the Irish College in Paris sheltered a military ambulance as well.[43] It operated thanks to the administrator, the Canon Ouin-La-Croix, and a numerous staff, which included around fifty women or religious of the neighbourhood, two doctors (de Rance and Lapeyrère), two medical students (Farge and Brochin) and three chemists (Desnoix, Pelisse-Penner and Lebègue). The medical team was completed by a bursar (Godefroy) and two women who took care of the cooking. The first patients arrived on 17 September 1870. On St Patrick's Day (17 March) 1871 it was decided to disband the ambulance, by this time it had already cared for 300 wounded.[44] The courtyard of the college was also used by the Garde Republicaine to practise drill during the siege.[45]

40 In the ninteenth century, the term 'ambulance' was used to refer to a mobile field hospital or an ad hoc, emergency hospital. There were various ambulances established around Paris during the siege, including some in the various railway stations around the city. Archbishop Darboy was taken as a hostage by the Commune and was later executed by firing squad in May 1871. 41 F. Bamber, 'Note-book of a Passionist during the siege of Paris and the Commune', *The Cross* (June 1912), 54–5. 42 Ibid., p. 58. 43 Archives of the Irish Collège, Paris, n°64J, report on the 'siège de Paris', by Ouin-la-Croix, 20 February 1871, p. 3. 44 Ibid., Report on 'La Commune de Paris 1871', by Ouin-la-Croix, 30 May 1871, pp 2–3. 45 Ibid., p. 6.

If the staff of the ambulance based in the Irish College was mainly composed by French people, its activities were funded from Ireland. As Ouin-La-Croix explained in his address during midnight mass in the chapel on 24 December 1870, 'I simply want to say that I stand here for Ireland, a country that has always been a friend and friendly to France; a country that has also greatly suffered during war and has been obliged to fly from the beloved island since the disastrous battle of the Boyne – the Irish Waterloo, Sadowa and Sedan. A country that has never found on earth a more generous, friendly and steady hospitality than the French one, and is now giving you the shelter of her old College, in a practical and friendly reciprocity'.[46]

As well as the collections from Ireland and the medical help given in the Irish places of worship in Paris, some young Irish surgeons joined the medical units of the French army.[47] Not all of them made their Irish identity obvious and some just wanted to help France as individuals. Such was the case of Major O'Connell, who served in the American ambulance from its creation in August 1870 till its dissolution in March 1871.[48] However, there were others who wanted to create a specifically Irish unit due to political motivations.

Among the different initiatives, one took on a special importance, not only for its activities on the battlefield but also because it was formed by Irish nationalist activists and it had, after the conflict, many political repercussions. This was the Irish ambulance that was created on 7 September 1870 in Dublin with the help of two nationalist deputies, A.M. Sullivan and Patrick-James Smyth.[49] The ambulance enlisted 250 members, who embarked on the French steamer *La Fontaine* on Saturday, 8 October 1870. They landed four days later at Le Havre, on 11 October.[50] They were first based at Evreux but later moved to Rouen, Hécourt and then Pacy.

On 9 November 1870, the ambulance was reorganized and reduced in size, being split into two field ambulances. One of these, with Surgeons Maguire and Cremin, seven students, some nurses and two vehicles, was sent to the Northern army of General Mocquart. The second was sent to the Army of the Loire at Chateaudun, where there had been several engagements and the town itself had been shelled. This section of the Irish unit stayed at Chateaudun for three months and during this period 600 wounded were 'hospitalized', while hundreds more walking wounded were treated. It seems, however, that the Irish work was rather routine and that they grew bored easily. Several reports record absenteeism, insubordination and even some cases of drunkenness while on duty. General Charette complained about the absence of the ambulance at the battles of Patay and Loigny, which both fell within the remit

46 Report on the 'siège de Paris', by Ouin-la-Croix, 20 February 1871, p.13. 47 Ronayne, op. cit., p. 32. 48 Ministère des Affaires étrangères, Paris, *Agents Papers*, 070-Favre, vol. 13, 'Major O'Connell', letter from l'abbé Blanc to Thiers, 9 July 1871. 49 J. Fleetwood, 'An Irish field-ambulance in the Franco-Prussian war', *Irish Sword*, 6:24 (summer 1964), 137–8. 50 Ibid., p. 139.

of its area of operation. Nevertheless, it stayed on duty until the middle of January 1871. The staff was gradually reduced and, on 12 January, the fall of Le Mans spelled the end for the Army of the Loire as well as its Irish ambulance. Most of the Irishmen serving in it went to Le Havre to meet fellow-countrymen who were bound for home.

The second part of the Irish ambulance, which was sent to the Army of the North, under the leadership of a merchant from Le Havre, J.J. O'Scanlon, and including Dr Maguire, chief surgeon, Dr Cremin, seven surgeon assistants and nine volunteers, was stationed in the towns of Bonnières, Vetheuil, Courcelles, Louviers and Elboeuf, before it was stopped by the progression of the Prussian army. Unfortunately, the lack of sources leaves us in the dark concerning this last episode of the existence of the Irish ambulance.

In this ambulance, and on a larger scale during the whole Franco-Prussian conflict, J.P. Leonard played an important role. Of Irish origin, he was close to both Irish nationalists and also leading French personalities, such as Maréchal MacMahon and Bishop Dupanloup. At the beginning of August 1870, Leonard was collecting money and medical equipment from the Irish, both in Ireland and among the Irish community in America.[51] He was then appointed supervisor of the Irish ambulance. On 5 May 1871, he was sent, with Dr Caignet, Maréchal MacMahon's doctor, to visit Saint-Cloud, Suresnes, Sèvres and Puteaux, and to make a report on the condition of the population. On 1 July 1871, Leonard was entrusted by the Suburb Committee with the distribution of aid to war victims at Clichy and Levallois Perret and he continued in this work until the end of August 1871.[52] 'For the commitment of his own person on several battlefields' and for giving aid to the wounded, Leonard was made a Knight of the Legion d'Honneur by a decree of 20 January 1872.[53]

THE REVIVAL OF THE 'IRISH BRIGADE'

The importance of the medical activity of the Irish was not only an expression of generosity from a people known to be very close to the French, but also a kind of reaction to the frustration at being unable to take part more directly in the fight that France was now engaged in. From the beginning of the conflict, the French consulate in Dublin was offered new proposals suggesting the enlistment of Irishmen in the French army. But the French government could

51 Archives de la Compagnie de Saint-Sulpice, Fonds Dupanloup, Letter of Leonard to Mgr Dupanloup, from Luc-sur-Mer, 2d August 1870, and Bibliothèque nationale de France, n.a.f.24695, 'Correspondance de Mgr Dupanloup', vol. xxiv. 52 Archives de la Compagnie de Saint-Sulpice, Fonds Dupanloup, Letter of Leonard to Mgr Dupanloup, from Paris, 16 August 1871. 53 Archives Nationales, Paris, LH1591/68, file of Légion

admit Irish nationals without breaking the neutrality laws.[54] The Irish were invited to enlist individually, on the spot and some Irish did, including some Fenians.[55] While James Stephens did not become involved, some of his companions joined the French ranks; these included Stephens' cousins, the Casey brothers, who were both wounded during the siege of Paris.[56] Frank Byrne, another Fenian who played an important role in the organization in the 1880s, served in the Army of the East under the command of Bourbaki, where he was made a sergeant.[57] Even if it's difficult to estimate the number of these Irishmen, spread among the different units, their importance is mainly due to the publicity that was made of their gesture, and the symbolic way it was used after the war.[58]

But some Irishmen chose to try to recreate an 'Irish Brigade' so as to serve in the French army in a specific Irish unit. The Fenian James O'Kelly, who had already served in the Legion étrangère for several years, went to Paris where he submitted a plan to the French authorities for the creation of an Irish brigade.[59] This plan of O'Kelly was turned down.

Other Irish patriots were more successful on the same demand and the formation of an 'Irish brigade' appeared to be a real possibility. Captain M.W. Kirwan, a former officer of the Irish militia, founded the Irish Company or 'Compagnie Irlandaise', which was closely connected with the Irish ambulance. When the Irish ambulance arrived at Le Havre and was reduced in size, the disbanded Irish were gathered by Kirwan who approached the French authorities and proposed forming a unit made exclusively of Irish soldiers.[60] His initiative was initially coldly received by the French authorities, who let him know that 'by a decree of the 7th September 1870, there has been ordered the creation in Cherbourg of a foreign regiment [organized by one MacAdaras] which, according to the will of the government, is exclusively designed to receive Irish men'.[61] Captain Kirwan was advised to send his men to the French army and that they would have the satisfaction of keeping their national identity and being designated as the 'Irish company'. They were assigned to a regiment of the Legion étrangère and the Irish Company was

d'Honneur of 'John Leonard'. **54** Ministère des Affaires étrangères, Paris, *Affaires diverses politiques,* Angleterre, n°44. **55** Trinity College, MSS 9659d/1–379, 'Letters and papers of James Stephens', letter of Stephens to John Murphy, G. Smith, W. Morov, 2 September 1870. **56** D. Ryan, *The Fenian chief: a biography of James Stephens* (Dublin, 1967), p. 341. **57** Janick Julienne, 'La Question irlandaise en France: perceptions et réactions', PhD dissertation, Paris VII University, 1997, pp 372–85. **58** G.A. Hayes-McCoy, 'The Irish Company in the Franco-Prussian War, 1870–71', *Irish Sword*, 1:4 (1952–3), 279–80. **59** J. Devoy, *Recollections of an Irish rebel: a personal narrative* (New York, 1929), p. 337. There are no French sources that give evidence to his assertion. **60** M.W. Kirwan, *La Compagnie irlandaise: reminiscences of the Franco-German War* (Dublin, 1873), p. 15. **61** NLI, MS 22538, P.J. Smyth Papers, letter from the *sous-préfet* of Le Havre to the *préfecture* of Rouen, s.d..

composed of four officers – Kirwan as the captain, F. M'Alevey and B. Cotter as lieutenants and Dr R. Macken as 1st class surgeon's assistant.[62]

After one month of training in Caen, the Irish left on 16 November 1870 and went to Bourges, where they stayed until 18 or 19 December 1870. Even if they had to endure very harsh conditions, they were initially kept away from any fighting. The Irish soldiers then went to Vierzon, where they stayed until 4 January 1871. At this date, they were sent to Montbeliard and it was in this area of operations that they had their baptism of fire, before retreating with the rest of the French army. Until 27 March 1871, they were stationed in the envoirons of Besançon, long after the armistice was announced on 2 February 1871. The Irish Company was then disbanded and its members trudged back to the coast, where they eventually took ship back to their country.

Thus, the Irish Company of Captain Kirwan enjoyed only a short period of existence and saw only one major action, at Montbeliard. It had been integrated into the foreign regiments of the French army, and ultimately the Irish were not officially distinct from the other units of foreigners. But their decision to use the designation of 'the Irish Company' showed their desire to place themselves within the longer tradition of Franco-Irish military connections – reaching back to the earlier Irish regiments in the Bourbon service. The failure of O'Kelly and Kirwan in their attempts to re-form the famous Irish Brigade was mainly due to the reluctance of the French authorities to sanction one. They were eager to respect the neutrality of England, and the French were undoubtedly aware of the political impact that such a measure would have – the fear that the term 'Irish Brigade' created in England since the eighteenth century was present in all the minds.

In these conditions, the success of MacAdaras was rather surprising. James Dyer MacAdaras was a colourful character, a daring adventurer that managed to recreate an 'Irish Brigade'. Born in Rathmines, Co. Dublin, on 21 June 1838, as soon as he arrived in France he appeared to be an intriguing go-getter.[63] A career soldier, he had taken part in the Crimean War, where he served in the ranks of the British army.[64] During the Indian Mutiny, he had served in the artillery, before resigning in 1866, by then an artillery warrant officer.[65] His activities between this date and 1870 are still obscure, but he

62 5th battalion, 2nd brigade, 1st division of the 15th army: Hayes-McCoy, 'The Irish company', pp 278–9. See also David Murphy, *The Irish Brigades, 1685–2006: a gazetteer of Irish military service past and present* (Dublin, 2007), pp 38–40. 63 Archives Nationales, Paris, BB11 1179, file ni635x71, naturalization of 'MacAdaras, James Dyer', letter of 26 April 1888. 64 Georgetown University Library (Washington, DC), Thomas F. Meehan Papers, Box 1:10, letter of John Mitchel to Th. F. Meehan, 10 July 1872. Archives des Affaires étrangères, Paris, *Affaires diverses politiques,* Angleterre, vol. 52. 65 Service Historique de la Défense, Vincennes, Guerre, 17YD8, OGP MacAdaras, 'Mémoire de proposition en faveur de Mr MacAdaras (James Dyer), lieutenant colonel à titre auxiliaire pour le grade de chevalier dans l'ordre de la Légion d'honneur, 7 mars 1871'.

would seem to have been part of the Fenian rebellion of 1867 and would have found refuge in France since[66]. At the beginning of the war, MacAdaras had presented himself to the Duc de Palikao, the war minister, and represented himself as the spokesman of the organizers of the Irish ambulance in order to obtain an interview with the Empress Eugenie and thus seek authorization to create an Irish brigade.[67] In August 1870, he received permission from the French government to raise a brigade of 6,000 men or more in Ireland. MacAdaras also agreed to advance the funds necessary to recruit these volunteers.[68] Later he would maintain that he was owed expenses totalling 115,000 francs.[69] At the beginning of September 1870, the Second Foreign Regiment was established at Cherbourg in the hope that five Irish battalions would be recruited for it.

MacAdaras then embarked to Dublin in order to start recruiting. Having no news from him and still being sceptical of his plans, the French government sent no training or administration officers to Cherbourg to form a cadre for this regiment, but it did send 600 chassepot rifles and 6,000 uniforms. After the surrender of Napoleon III, the government of course changed, and, when MacAdaras was informed of the change of government by a telegram, he hurried back to France, where he obtained from the new government an agreement that the previous arrangements would be maintained under the same conditions. On 10 October, MacAdaras presented himself to General Lefort with a note from Minister Gambetta, which stipulated that 'M. Dyer is authorized to create a regiment of Irish volunteers under his own command'.[70]

MacAdaras pretended he had managed to recruit over 9,000 men and that these were ready to embark for the Continent. But only 216 of them actually landed at Caen (in October).[71] The French authorities had organized nothing to welcome them. They were sent to Cherbourg, where there was also no facilities ready to take care of them or feed them. Their conditions were so bad that the English consulate in Caen had to intervene to provide them with food and essentials. In the meantime, MacAdaras was away in Tours, close to the government. He was made lieutenant-colonel of the Second Foreign Regiment but when he came back, MacAdaras found his soldiers in a state of rebellion. According to a military report, 'the Irish that are in Caen behave rather badly and voice doubts about the intentions of M. Dyer and his honour'. On 30 October 1870, the Infantry Office reached the conclusion that

66 As note 63 above. **67** John Devoy, op. cit., p. 337. **68** Service Historique de la Défense, Vincennes, Guerre, 17YD8, OGP MacAdaras, Letter of MacAdaras to the war minister, s.d. [1871]. **69** Pierre Mitau, *Le Général MacAdaras* (Paris, 1894), p. 5. **70** Service Historique de la Défense, Vincennes, Guerre, 17YD8, OGP MacAdaras, report of the Garde Mobile, 30 October 1870. **71** The military file of MacAdaras gives several estimations: a manuscript letter with no date and no signature alludes to 300 Irishmen. A letter of 25 March 1871 mentions 200 but the rolls of the Second Foreign Regiment gives

it would be better to abandon the formation of this specifically Irish unit and to merge its members into another foreign regiment or simply send them back home.

There were also indications that the French authorities harboured a mistrust of MacAdaras. On 29 October 1870, the war minister, due to reports of the local authorities in Caen, warned the Ministre de l'Intérieur that 'the man named Dyer MacAdaras, claiming to be an Irishman, has been authorized to recruit men of this nationality for the Second Foreign Regiment. The civilian authorities seem to have information about that man that are of a nature to raise suspicion. According to them, Mr Dyer MacAdaras is not known in Ireland and would seem to have lived a long time in Germany.'[72] But the military file on MacAdaras contains none of these allegations, and he never was questioned by the authorities.

MacAdaras himself reported the aftermath of this story: 'When I came back, thus, the discontent was general. No clothes, no food, no wages and no indemnities. Most of them, in the face of this situation, wanted to go back to their country. On their way back, they met the other troops coming to France and told their countrymen what they had endured. Needless to say, they all turned round and went back.'[73] MacAdaras even put the blame on the English authorities and accused them of having hindered his plans by deterring new recruits and putting his agent in jail in London. He explained that, in spite of the publicity made of that episode, that he managed again to gather 9,000 men into seven companies in Liverpool where he sent an agent. But the very next day he learned that the French government has disbanded his brigade.

MacAdaras returned to Tours, where he was attached to the headquarters of the XXI Corps, under the command of General Jaurès. Until 14 January 1871, he took part in the Army of the Loire's campaign and was later mentioned in a positive light for his behaviour during this time.[74] This period saw his participation to the engagements that took place between 8 and 12 December 1870 at Vendôme, Poilly, Marchenoir, and at Le Mans from 8 to 14 January 1871. On 20 January 1871, General Loysel, who commanded the Army of Le Havre, asked MacAdaras to form a new 'Irish Brigade' at Le Havre, awarding him the rank of auxiliary general of brigade (brigadier-general). MacAdaras immediately went Le Havre with his orderly, Captain Vincent.[75] In February 1871, General Chanzy promoted MacAdaras' project to the war office, pointing that his previous Irish brigade had been disbanded 'for political reasons' and adding that 'the English government would not object to the creation and coming to France of this unit, because of his change of policies'.[76]

only 83 names. 72 Ibid., letter of the war minister to the minister of the interior, 29 October 1870. 73 Ibid., letter of MacAdaras to the war minister, 1871. 74 The positive comments of Generals Riu, Colin and Henrion-Bertier are recorded in Mitau, *MacAdaras* (Paris, 1894), pp 9 and 13–15. 75 Mitau, op. cit., p. 10. 76 Archives du Ministère des

The Armistice prevented this new initiative taking form. MacAdaras thus was never formally appointed as a general but he kept the title all his life.[77] Did MacAdaras succeed where the others failed? While he never raised enough recruits to form an Irish brigade, he did succeed in gaining the sanction of the French authorities. This is most likely due to the fact that he was the first potential Irish commander to contact the French authorities. This was before the capture of Napoleon and the fall of the Second Empire. The new French government had only to ratify a previous arrangement. Furthermore, in the autumn of 1871, the Republican government needed to recruit up to 600,000 men to face the needs of war.[78] Any potential recruiting scheme had to be favourably viewed in these circumstances. MacAdaras was also a combination of boldness and pugnacity, and refused to be brushed off by government officials. The rest of his career showed that he was an ambitious adventurer and that he never let any obstacle hold him back in his quest to hold a major political office.[79]

By the end of the war, the Irish had given considerable aid to the cause of France. In addition to large quantities of clothes, blankets, canvas, food and medical supplies, the Irish people had sent more than one million francs to help the victims of the conflict.[80] In the field, medical and military support made important contributions to the French war effort. This contribution was recognized in the fact that many Irish, not just J.P. Leonard, were later awarded French medals.[81] Due to the long-term tradition of the Franco-Irish connection, the Irish became involved in this conflict on the side of the French. The various attempts to re-form the Irish brigades were symptomatic of the will to retain the historical bonds between the two peoples. If this symbolic action was present from the beginning of the war, its political significance was referred to after the war.

THE EFFECT OF THE WAR ON THE FRANCO-IRISH RELATIONSHIP

After the war, most of Irish involved in the conflict went back to their native land. In Ireland, some nationalists used the concept of the historical Franco-Irish friendship to maintain pressure on England. From 16 to 28 August 1871,

Affaires, Paris, *Agents' Papers*, '079-Gambetta', vol. 24, official telegram from Chanzy to the war minister, Le Mans, II–1871. 77 OGP MacAdaras, 17Yd 8, note for the war ministry cabinet, 3 December 1889. 78 Jean Delmas, *Histoire militaire de la France*, ii (Paris, 1992). 79 See J. Julienne, 'General MacAdaras: an adventurer in the service of the Irish revolutionaries in France', *Irish Sword*, 23 (summer 2003). 80 Bibliothèque Nationale, Paris, n.a.f.24695, 'Correspondance de Dupanloup', vol. 24, letter of Leonard to Dupanloup, 23 March 1873, Paris. 81 NLI, MS 22538, P.J. Smyth Papers, letter of M. de Rémusat, 29 August 1871 to M. Smyth: by decision of 29 August 1871, P.J. Smyth was

a French delegation visited Ireland to formally thank the Irish people. It was composed of French and Irish personalities, such as the Comte de Flavigny, president of the society for the rescue of the wounded, and Ferdinand de Lesseps. Arriving in Dublin, the delegation then went to Cork, Bantry, Glengarriff and Killarney. At each stop on this trip, it was enthusiastically welcomed by large crowds but also by local officials. Receptions gave occasion for nationalistic speeches, and speakers used the idea of a 'Franco-Irish friendship' to praise exalt the Irish nation as a partner of a great power in its time of need.

Of course, the French delegates were well aware that their visit would be used for political purposes. The Comte de Flavigny noticed that the newly-elected MP, P.J. Smyth, had used his connection with Irish ambulance to be elected.[82] Before the delegation arrived, De Flavigny asked Smyth to use his influence 'to prevent unfortunate complications', as he knew that nationalist speeches were sometimes punctuated with calls for French military aid. This caution also led the delegation to meet high-ranking government officials, including the viceroy, Lord Spencer.[83]

On 17 August 1871 a scandal broke out, spread by press both in England and in France. During a banquet attended by 500 people in the exhibition hall in Dublin, Patrick Bulfin, the lord mayor, proposed a toast to the health of the queen. This stirred a concert of whistles and shouts. The French ambassador in London, the Duc de Broglie, condemned these 'meaningless' and 'deplorable demonstrations'.[84] As a result of press coverage of this event, the new French president, Adolphe Thiers, gave personal instructions that the Comte de Flavigny should award no medals other than to those Irish who were loyal to the English authorities.[85] The French consul in Dublin, M. Livio, was indignant at the way the French delegation had been hi-jacked by Irish nationalists, and reported as to the irritation of the Protestant community.[86] In England, the repercussions of the French visit to Ireland gave rise to a series of condemnations in the press. These included newspapers such as *The Times*, the *Pall-Mall Gazette* and the *Daily News*.[87] The various phases of this episode were followed in the French newspapers also.

awarded the legion de honneur at the recommendation of the French foreign affairs minister, M. de Rémusat. Dr Charles Ronayne, surgeon of the Irish ambulance, was awarded the 'médaille de 1870–71' by the French government as well as a Red Cross decoration. See Roynayne, 'The reminiscences of an ambulance surgeon', pp 38–9. 82 NLI, MS 22539, 'Attempt to form an Irish Regiment in the service of France during the Franco-Prussian war, 1870', letter of J. de Flavigny to P.J. Smyth, Paris, 14 August 1871. 83 A. Duquet, *Irlande et France* (Paris, 1872), p. 95. 84 NLI, MS 20678, Monsell of Tervoe Papers, letter of de Broglie to Monsell of Tervoe, 31 August 1871. 85 C. Bloch, *Les relations entre la France et la Grande-Bretagne, 1871–1878* (Paris, 1955), p. 24. 86 Archives des Affaires étrangères, Paris, *Correspondance politique des consuls*, Angleterre, n° 45, letter of 26 January 1871. 87 A. Duquet, op. cit., p. 170.

If the commitment of the Irish to the cause of France was mainly due to the tradition linking the two peoples, it was also a way for the Irish to show their hostility to England by refuting her right to legislate for Ireland. During a meeting organized in January 1871 by the nationalist MPs, the speaker that opened the session, P.J. Smyth, explained:

> We are coming here to honour the French republic and, as a people, to separate ourselves from any share of the foreign policies of the British government and to affirm the right of the Irish breed to be canvassed about their feelings and opinions when the existence of France is questioned. Ireland, indeed, may not be a 'big nation' but if England only lifted the restrictions created by the Foreign enlistment act and tomorrow 20,000 Irish men would win a new Fontenoy battle for France or will be killed for her protection … . We are a distinctive nation; the government we are submitted to can act the way it wants, we, as a nation, acknowledge the French republic.[88]

The speaker made a link between two distinctive elements, the Franco-Irish friendship and the Irish nationalists claims for self-government from England. Charity and military help was a way for the Irish nationalists to remind the French they still were their allies, as they have been in the past when they were united against their common enemy, England, who was seen as now hindering the 'Franco-Irish friendship' by its recruiting legislation. But it is also all legislation as applied by the English government in Ireland that is denounced. The Irish nationalists leant on the historical connections with France to claim their national sovereignty. P.J. Smyth later referred to the expansionist tendencies of Prussia and compared them to English imperialism in Ireland. France and Ireland were promoted as being linked in adversity against a common enemy, imperialism, as practiced by England as well as Prussia.

In Ireland, the Franco-Prussian conflict was also a symbol of the religious dimension of the Irish problem. By the 'alliance' to France, nationalist Ireland was supporting a Catholic nation against a Protestant country – Prussia. To oppose Prussia was also a way to oppose England, which may partly explain the violence of the demonstrations against the Prussian consulate in Ireland. While most of the Catholic newspapers were on French side, the daily newspapers of the northern part of the island openly support Prussia.[89] This

88 Archives des Affaires étrangères, Paris, *Correspondance politique des consuls*, Angleterre, n° 45, letter of the 26 January 1871. 89 The consulate in Belfast reported that 'the *Newsletter*, a conservative newspaper, is frankly against us: it's the Orangists' newspaper who see, or pretend to see, this conflict as a war of religion'. Archives des affaires étrangères, Paris, *Correspondance politique des consuls*, Angleterre, n° 45, letter of 23 July 1870.

religious sub-text of Ireland's understanding of the Franco-Prussian war seems to have been widespread in Ireland. In Co. Cork, Catholics demonstrated in sympathy with the French defeat while the Protestant community took it a providential sign in favour of Protestantism.[90] Through these demonstrations, Irish Catholic and Protestant communities found another arena for their own ideological conflict. The intensity of the reactions raised in Ireland due to the Irish contribution in the war reflected the depth of division in Ireland. In Ireland, the Franco-Prussian war crystallized different strands of political and religious tensions.

In this way, Napoleon III's policies resulted in war and the defeat of France but had also allowed elements in French and Irish society to become more closely linked. Once again and effort had been made to unite and fight a common enemy. The two peoples became closer when the Irish engaged in active support of France during the Franco-Prussian war. In France, this Irish assistance returned the Franco-Irish link to popular memory. In Ireland, agitated by Fenianism and also growing agrarian unrest, it reached a broader scale and became a symbol of the fight of the nationalists against England.

Immediately after the end of the war, the Irish enthusiasm was moderated by the activities of the Commune de Paris. The Irish perception of the Communards' insurrection as being revolutionary, violent and anti-clerical served as a counterbalance for wider sympathy towards Catholic France. Moreover, the balance of forces had changed in Europe. The increasing of power of the unified Germany, and the French defeat and the loss of Alsace and Lorraine, provoked a deep current of unrest and change that had an influence on the Franco-Irish relationship. France was no more a potential military ally, but remained for some time as a political ally. Irish nationalists acknowledged this change and evolved a new and elaborate a strategy from the 1880s. This was based on the seduction of French public opinion as well as the France's political and cultural elites.

90 Comerford, 'Anglo-French tension', pp 167–8.

The French invasion that never was: the Deuxième Bureau and the Irish republicans, 1900–4[1]

JÉRÔME AAN DE WIEL

In 1984, Christopher Andrew and David Dilks, specialists in the history of the British secret service, made the following observation:

> Secret intelligence has been described by one distinguished diplomat [Sir Alexander Cadogan] as the 'missing dimension of most of diplomatic history'. The same dimension is also absent from most political and much military history. Academic historians have frequently tended either to ignore intelligence altogether, or to treat it as of little importance. Historians have a tendency to pay too much attention to the evidence which survives, and to make too little allowance for what does not. Intelligence has become a 'missing dimension' first and foremost because its written records are so difficult to come by.[2]

This remark is particularly pertinent for the topic of this chapter: the relations between the Deuxième Bureau, France's military intelligence service, and Irish republicans between 1900 and 1904, a period that covers the 2nd Anglo-Boer War in South Africa. In 2001, a book dealing with the international impact of that war on foreign powers does not mention Ireland once, according to its index. This is rather surprising as Irish nationalists and republicans openly supported the Boers against the British. In the chapter on France, it is stated that Théophile Delcassé, the French foreign minister decided that the war must be stopped.[3] Was this really the case? Equally it could be asked if Delcassé represented the views of all the members of the government. The archives of the Service Historique de la Défense (the archives of the French army) in Vincennes reveal that the Deuxième Bureau had made contact with Irish republicans in order to prepare for an invasion of Ireland. This could have resulted in the encirclement of Great Britian while thousands of its soldiers were away in South Africa. This chapter will first focus on these contacts and invasion plans, and show how the British secret service reacted to the French threat. Then, it will be explained why Germany

1 This chapter is based on the author's book, Jérôme aan de Wiel, *The Irish factor, 1899–1919: Ireland's strategic and diplomatic importance for foreign powers* (Dublin, 2008).
2 Christopher Andrew & D. Dilks (eds), *The missing dimension: government and intelligence communities in the twentieth century* (London, 1984), pp 1–2.
3 Pascal Venier, 'French foreign policy and the Boer War', in Keith Wilson (ed.), *The international impact of the Boer*

later filled France's role for the Irish republican movement. Finally, the following question will be answered: Was the invasion of Ireland seriously contemplated by the French, or were all these landing plans in fact mere routine reconnaissance operations or theoretical war games?

A brief reminder of the international situation is necessary before detailing the contacts between the Deuxième Bureau and the Irish republicans. Relations between Britain and France were strained, to say the least. In 1898, a crisis broke out in the Nile Valley in eastern Africa when a British colonial expedition under the command of General Herbert Kitchener met a similar French mission led by Major Jean-Baptiste Marchand. Eventually, the French were obliged to withdraw. It is not clear if the incident could have led to a war between the two countries, but the French press spoke of a national humiliation.[4] In 1899, the 2nd Anglo-Boer War began in South Africa and, as could have been easily predicted, French public opinion was largely in favour of the Boers. Thereupon, the Dreyfus affair took place during which French officers unjustly accused a Jewish fellow-officer of having provided Germany with secret information. Across the English Channel, public opinion was largely anti-French, so much so that the French embassy in London had to be protected and that French citizens were openly insulted in the streets of the capital.[5]

In Ireland, however, the French were in a much better position. At that time, the Irish Nationalist Party was struggling for home rule and was supported by nationalists among the population, the overwhelming majority of which was Catholic. They were opposed by the unionists, most of whom were Protestants and did not support autonomy and favoured the maintenance of Ireland within the framework of the United Kingdom. The nationalists favoured the Boers, and there were support committees throughout the country. The Quai d'Orsay received a letter addressed to 'the great French Republic' from the Cork Transvaal Committee, informing it that although Ireland was 'cruelly persecuted', she was always 'on the side of Justice and Humanity'.[6] The French ambassador in London reported that Irish nationalist deputies at Westminster openly mocked British defeats in South Africa.[7] Under these circumstances, it was hardly surprising that the French military attaché in London, Colonel Dupontovice de Heussey, suggested to the ambassador, Alphonse de Courcel, that he should secretly finance the activities of Irish nationalists. The attaché was in touch with a

War (Chesham, 2001), pp 65–78. 4 P.M.H. Bell, *France and Britain, 1900–1940; entente and estrangement* (London, 1996), pp 9–10. 5 Laurent Villate, *La république des diplomates; Paul et Jules Cambon, 1843–1935* (Paris, 2003), p. 216 & p. 213. 6 Quai d'Orsay, Paris; correspondance politique et commerciale 1897–1918, nouvelle série, Transvaal-Orange, no. 11, French consulate in Dublin to Quai d'Orsay, 18 Nov. 1899. 7 Ibid., vol. 22, Paul Cambon to Théophile Delcassé, 11 Mar. 1902.

member of the Nationalist Party, codenamed 'XXX'. On 21 July 1898, he wrote to de Courcel:

> XXX would like to bring the [French] government to enter a kind of secret treaty with this party: the government would provide funds and the Irish Nationalist Party would create all sorts of problems for the British government's projects. I have often thought of these plans and I cannot help thinking that they are an option well worth considering. We could organize the scheme abroad and keep the man and the funds under our control. Everywhere, England is trying to cause us difficulties [colonial matters, Dreyfus affair]. I feel that with the Irish we would be well able to stop the [British] colonial movement.[8]

Dupontovice de Heussey's suggestion did not fall on deaf ears, and the Quai d'Orsay sent money to the Irish. Indeed, in 1899, Paul Cambon, the new French ambassador in London, received an 'important personality of the Nationalist Party' in his office. Cambon did not mention his visitor's name as he had the terrible habit for historians of never revealing sources and personalities in his correspondence. The Irishman had come to thank France for her support, and sums of money were mentioned. Then, he put forward a plan for an alliance between the United States, where millions of Americans of Irish descent lived, France and Ireland. This alliance was of course directed against Britain, which was at that moment in time trying to operate a rapprochement with Germany. Cambon sent a report on his unusual meeting to Théophile Delcassé, the French foreign minister.[9] It is not known how Cambon personally reacted to this alliance plan. The same goes for Delcassé, but in his case one can hazard an educated guess. It was rather unlikely that Delcassé ordered Cambon to follow this alliance offer up. The French minister feared the danger of a possible rapprochement between Britain and the countries of the Triple Alliance (Austria-Hungary, Germany and Italy).[10] If this happened, France would be strategically encircled. According to him, the time had come to negotiate with the British even if, for a brief period, they seemed to have in mind a rapprochement with Germany. Moreover, France had signed a military alliance with Russia in 1892 directed against Germany, but it was far from certain that these two powers would be able to pit themselves successfully against an Anglo-German alliance.[11] Delcassé was aware of this ominous possibility, and little by little he began to turn his eyes towards Britain and toy with the idea of making an ally out of her. The

8 Pierre Guillen, *L'expansion, 1881–1898* (Paris, 1985), p. 283. 9 Quai d'Orsay, Paris; Grande-Bretagne, politique intérieure, question d'Irlande, 1897–1914, vol. 4, p. 47, Paul Cambon to Théophile Delcassé, 10 June 1899. 10 Villate, *La république des diplomates*, p. 217. 11 Bell, *France and Britain 1900–1940*, pp 14–15.

painful souvenir of Fashoda did not enter in his calculations. Consequently, Irish separatists were, in a manner of speaking, put on the Quai d'Orsay's waiting list.

However, unlike diplomats, some French military officers were not yet willing to envisage an entente or alliance with the British. For them, nationalist Ireland still remained the best way to destabilize a dangerous foe. The Fashoda humiliation had not been forgotten, and public opinion was still bellicose. In other words, now seemed the ideal time to prepare for an attack on Britain, all the more so since her soldiers were bogged down in South Africa. British public opinion could not understand why an army of 450,000 men (over a period of three years) was not able to beat groups of Boers. As most of the soldiers were away, the public began to believe that the country was defenceless. Alarmist rumours about invasions began to spread. The British believed that the French and the Russians would be the invaders.[12] Although it is possible to speak about a collective psychosis, the public's fear was not without foundation, and invasion rumours were not only circulating in London. Somehow, the possibility of a French landing in Ireland had set tongues wagging. In Paris, the German ambassador, Count Georg Münster zu Dernburg, learnt about it and sent a report to the foreign minister in Berlin, Prince Bernhard von Bülow. The latter was sufficiently intrigued to ask for the opinion of his ambassador in London, Count Paul von Hatzfeldt. Hatzfeldt was not impressed and dismissed the rumours as baseless.[13] He was wrong.

France had a consul in Dublin but no military attaché there, yet the French military archives in Vincennes reveal that the French had sent a military intelligence team to Ireland in order to examine several possible landing places. The spies regularly sent extremely detailed reports to the headquarters of the Deuxième Bureau in Paris. Their reports contained sketches of the British coastal defences, regional maps, topographical studies, assessments of the quality of the roads, analyses of various nationalist and unionist groups, analyses of nationalist public opinion, and estimations of the quality and the quantity of British troops based in Ireland. One French mission lasted for an entire year.[14] The names of the French spies and their Irish contacts were never mentioned in the reports. Nevertheless, annotations and stamps show that they were sent to the president of the Republic and several ministers. It is now obvious that the Boer War generated many French missions in Ireland between 1900 and 1904. Their aim was to evaluate the possibility of invasion with the support of Irish republicans. The invasion of the country would lead

12 Christopher Andrew, *Secret service; the making of the British intelligence community* (London, 1985), p. 34. 13 Wolfgang Hünseler, *Das Deutsche Kaiserreich und die Irische Frage, 1900–1914* (Frankfurt am Main, 1978), p. 119 & footnote 2, p. 119. 14 Service Historique de la Défense (SHD), Vincennes; attachés militaires, 7N1230, folder 6, report dated 30 Sept. 1903.

to the encirclement of Britain. There was nothing really new here as such a plan had already been envisaged by revolutionary France in 1796 and 1798,[15] except that this time the French could depend on their own military intelligence instead of assessments done by Irish republicans.

The first two reports sent to Paris in 1901 were estimations of the various political organizations in Ireland. The Deuxième Bureau had kept a close watch on three nationalist groups. Firstly, there was Cumann na nGaedheal, a separatist movement that did not believe in home rule. According to the French agents, Cumann na nGaedheal had 15,000 members and incited people in the provinces side to seek French aid to oust the British. Secondly, there was the Gaelic League, the aim of which was essentially to revive the Irish language in the country. Although strictly speaking the league had no political aim, the French had correctly noticed that other nationalist groups tried to control it. The league numbered about 10,000 members. Thirdly, there was the United Irish League (UIL), which was mainly an agrarian organization which aimed at getting rid of Anglo-Irish landowners and achieving home rule. The league had 10,000 members and was in favour of the Boers in South Africa. The French agents were of the opinion that it was still too British but that it would still be possible to change it into a real separatist body.[16]

Of course, the Deuxième Bureau agents had realized that not only were the Irish divided between moderate and extreme nationalists but also between nationalists and unionists. In their reports, they pointed to the existence of the Orange Order which represented the 'English party in Ireland'. Despite the fact that the order had 50,000 men, the French were far from being impressed. According to them, the majority of its members were 'very vicious, brutal and ignorant workers from Belfast'. To conclude, there was an assessment of the troops that would oppose the French army. The Irish reserve battalions numbered 30,000 men, but the Deuxième Bureau deemed that '15,000 or 20,000 men could be taken away from the English in the three months following the beginning of the hostilities'.[17] The word 'hostilities' is striking here because it strongly suggests that the French spies were in fact reconnoitring the terrain for a future attack. This impression will be confirmed in the reports to come. According to the French agents, these 15,000 to 20,000 men were in fact nationalists and many of them supported the Boers.

15 John A. Murphy (ed.), *The French are in the Bay: the expedition to Bantry Bay, 1796* (Cork, 1997). 16 SHD, Vincennes; attachés militaires, 7N1230, folder 3, reports dated 15 Oct. 1901 & 20 Nov. 1901. 17 Ibid., report dated 20 Nov. 1901. Under the terms of the army reforms of 1881, the regiments of the old city and county militias were reorganized as reserve battalions of regular regiments. The South Cork Militia for example, became the 3rd Battalion, Royal Minister Fusiliers.

They also claimed that they had joined the British colours out of financial necessity or because they had been plied with drink by recruiting sergeants!

The Deuxième Bureau also noted the Irish regiments of the British army then based in Ireland: the Royal Inniskilling Fusiliers (3rd, 4th and 5th battalions), the Royal Irish Rifles (3rd, 4th, 5th and 6th battalions) and the Irish Fusiliers (3rd, 4th and 5th battalions).[18]

If the Irish were divided between themselves, so were the French. Théophile Delcassé, who controlled the Quai d'Orsay, was endeavouring to maintain the Franco-Russian alliance, to separate Italy from Germany and Austria-Hungary and to operate a rapprochement with Britain. These simultaneous aims were most difficult to attain during the various terms he served as foreign minister in five different governments between 1898 and 1905.[19] Nevertheless, he succeeded in reinforcing the military alliance with Russia and in assuaging the tensions with Britain. Despite all, there was actually a pro-British current of opinion in France, and in Britain some politicians, such as Joseph Chamberlain, began to think about a possible *entente cordiale* with the French. A few treaties were signed between the two countries, notably concerning Africa. It was the beginning of the so-called 'Delcassé system'.[20] Yet, some within the French government had not swallowed the humiliation of Fashoda and were not ready to support a rapprochement with Britain.[21] In fact, there was no unity of opinion in the French cabinet, which was a serious stumbling block in the elaboration of a coherent foreign policy. Moreover, it could be life-threatening to express one's opinion during cabinet meetings as on one stormy occasion the minister of war tried to strangle the minister of the navy![22] Therefore, it looked as if the British did not have much reason to worry about the French but rather about the Germans and their new fleet.

The year 1902 seemed to be auspicious for the Deuxième Bureau. The war in South Africa still dragged on but it was now clear that the Boers would not beat the British despite the humiliating defeats that they inflicted on their enemies earlier in the war. On 17 March 1902, the French military attaché in London sent an encouraging report to the government and the Deuxième Bureau in Paris. He began by writing that King Edward VII, the new British monarch, had decided to cancel his scheduled visit to Ireland, exasperated by the fact that three-quarters of the Irish farmers and almost all the Irish-Americans openly supported the Boers. The attaché indicated that the

18 Ibid., reports dated 15 Oct. 1901 & 20 Nov. 1901. **19** Bell, *France and Britain, 1900–1914*, p. 24. **20** Jean-Claude Allain, 'L'affirmation internationale à l'épreuve des crises (1898–1914)' in Jean-Claude Allain et al., *Histoire de la Diplomatie française* (Paris, 2005), pp 686, 688, 696–7 & 701. **21** Pierre Milza, *Les relations internationales de 1871 à 1914* (Paris, 2003), p. 118. **22** Eugenia C. Kiesling, 'France' in Richard F. Hamilton & Holger H. Herwig (eds), *The origins of World War I* (Cambridge, 2003), pp 260–2 & footnote 150, p. 262.

situation in the country was extremely tense, and he wondered whether the Irish question would sooner or later have international repercussions on Europe. The answer to his own question was '*oui et non*' On one hand, he believed the answer was 'yes', for he explained that the Irish separatists' strategy consisted of encouraging the Boers to persevere in their struggle, hoping that the British would become exhausted in the long run. If troubles broke out in Central Asia (Afghanistan and Persia) where Britain and Russia were rivals, then the British army would be in a desperate position, according to him. On the other hand, his answer was 'no'. He pointed out that a few factors stood in the way of a full-scale rebellion in Ireland, rebellion which in the right circumstances would inevitably lead to Britain's defeat on the international scene. He quoted two important factors: the lack of financial resources and the lack of armament among the Irish republicans.[23]

On the whole, the military attaché's report was a realistic evaluation of the situation, one which incited to caution. A month later, in April 1902, Paris received another report on Ireland and the news was good. Indeed, the 3rd Battalion of the Munster Fusiliers regiment, more widely known as the South Cork Militia, had come back from South Africa. When this battalion arrived in the small harbour town of Kinsale, it was disbanded because the Munster Fusiliers had not been exemplary in South Africa. As asserted by the French agent who got wind of this information, the Irish soldiers had never been sent to the line of battle and had been kept in their barracks under British surveillance.[24]

A short while later, in September 1902, the Deuxième Bureau agents in Dublin were given a very precise landing plan elaborated by Irish republicans. As usual, no names were mentioned in the subsequent French report. It would take to long to go into detail, but it was clear that the authors of this ten-page plan had meticulously studied the social, political and military conditions in the country. According to them, the best area for a French landing would be the southern coast, more precisely near Cork and Kinsale. A first landing of a small force would immediately head toward the west in order to create a diversionary movement. The Irish strategists expected the British army to follow the French in that direction. Then, a second landing would take place, but this time of a main force of 60,000 men. They would land in and around Kinsale. This force would head for Dublin. The plan also indicated the main objectives in the country and gave an estimation of the British forces.[25] All this seemed feasible in theory, but two remarks spring to mind. Firstly, it was far from certain that the small harbour town of Kinsale

23 SHD, Vincennes; attachés militaires, 7N1230, folder 5, report dated 17 Mar. 1902. 24 Ibid., report dated 18 April 1902. 25 SHD, Vincennes; attachés militaires, 7N1230, folder 5, report dated 20 Sept. 1902.

would be able to receive an army of 60,000 men and all its equipment, and let us not forget the important number of ships required for such an operation. Secondly, the republicans had not mentioned the Royal Navy's presence at all. Perhaps they believed that it was up to the French to look after this detail, a rather worrying detail having said this! It is of course true that the French navy had managed to elude the Royal Navy on two occasions in 1796 and 1798.

Was the Royal Navy at the very beginning of the twentieth century as strong as is generally believed? The answer seems to be no. Indeed, in 1889, after a series of manoeuvres, the Royal Navy had concluded that she would not be in a position to efficiently protect the merchant navy, essential for Britain's survival, against the French navy. This understandably greatly worried the heads of British naval intelligence. But, there was worse. Twelve years later, in 1901, nothing had really changed. First Sea Lord, Walter Kerr, did not think it was necessary to have more naval manoeuvres in order to prepare against a possible German attack. He complacently stated: 'It is no use speculating on what might happen in an unforeseen future.'[26] These were rather surprising words coming from a military mind, all the more since the future was indeed 'unforeseen': in 1901, it was not towards the North Sea that British eyes had to turn, but rather towards the English Channel! Royal Navy strategy remained remarkably vague and some officers understandably complained about it. When, in 1904, Sir John Fisher replaced Kerr, strategy was still in a cul-de-sac, at least to begin with. Fisher's global thinking was roughly summed up by his favourite aphorism: 'Hit first, hit hard, and hit anywhere.'[27] All this frankly beggared belief; all the more so since British politicians were fully aware of the threat posed by modern warships. Already in 1845, Lord Palmerston had said: 'The [English] Channel is no longer a barrier. Steam navigation has rendered that which was before impassable by a military force nothing more than a river passable by a steam bridge.'[28]

If the Deuxième Bureau had been aware of this and had reported it to Paris, it might well have encouraged some to envisage a bold strike on the United Kingdom to avenge Fashoda. What about Ireland's defence? The situation can be summarized accurately enough by one word: 'laughable'. Before 1890, military manoeuvres did not happen on a regular basis. On 18 August 1892, after some military exercises, General Garnet Wolseley wrote the following to the duke of Cambridge: 'The Artillery have been badly handled, generally losing many guns during the action for want of tactical handling. The Infantry straggle all over the country in an aimless fashion.

26 John Gooch, 'The weary titan: strategy and policy in Great Britain, 1890–1918', in Williamson Murray, MacGregor Knox, Alvin Bernstein (eds), *The making of strategy: rulers, states and war* (Cambridge, 1994), pp 285–6. **27** Ibid., p. 286. **28** G.R. Sloan, *The geopolitics of Anglo-Irish relations in the 20th century* (London, 1997), p. 115.

After two seasons here now, I am convinced that our senior officers require to be well grounded in tactics.'[29] In other words, in case of a foreign invasion, the British army in Ireland would not drive back the invaders into the sea. About ten years later, it was much the same story. In January 1901, the duke of Connaught told Lord Roberts: 'We are, as you know, very short of troops in Ireland just now and this is in itself a great encouragement to the enemies of England.' It was right for Connaught to complain at this time, but instead of pressing for reform he left Ireland for India for four months[30] It was only from 1905 onwards that Ireland's defence became 'part of a Command Defence Scheme which covered Home Defence for the whole of the United Kingdom' and that a new defence strategy was developed.[31] Also, there was the sensitive issue about Irish troops in Ireland. Could they be trusted by the British high command in case of war or invasion? This would preoccupy the British until the beginning of the First World War in 1914.[32] It had actually caught the Deuxième Bureau's attention.[33]

However, French military intelligence had not waited for the Irish republicans' landing plan as a report written in October 1902 shows. During the summer, the French had studied the best possible landing zones on the southern coast and had more or less reached the same conclusions as the republicans. They had spotted four areas, Ballycotton Bay, Courtmacherry Bay, Kinsale Harbour and Oyster Haven, and envisaged one massive landing or, on the contrary, a series of simultaneous landings probably to avoid a choking point in a single area. The Deuxième Bureau stressed that, if Dublin was the main objective, Ballycotton Bay would be the best option as nearby roads led to Fermoy, Lismore and eventually to the Irish capital. If Cork was the main objective, then not only Ballycotton Bay but also Courtmacsherry Bay and Kinsale would be the best options.[34] Furthermore, there was encouraging news. The number of Irish recruits for the British army had seriously decreased. In January 1900, the number was 37,316 men, but by April 1902, it had dropped to 16,837. The French explained that this was due, among other factors, to the activities of Inghinidhe na hEireann (Daughters of Ireland), an organization founded by the republicans Maud Gonne and Countess Constance Markievicz. It was in favour of separatism and boycotted every Irishman who joined the British army.[35] But, there was another factor which the French had not mentioned: fear of being drafted into the army.

29 Elizabeth A. Muenger, *The British military dilemma in Ireland; occupation politics, 1886–1914* (Lawrence, KA, 1991), p. 24. 30 Ibid., p. 112 & pp 73–4. 31 Sloan, *Geopolitics*, p. 136. 32 Muenger, *The British military dilemma in Ireland*; Muenger deals with this issue throughout her book. 33 SHD, Vincennes; attachés militaires, 7N1230, folder 3, reports dated 20 Nov. 1901 & folder 7, report dated 18 April 1902 ; 7N1231, folder 5, report dated 2 Mar. 1902. 34 Ibid., attachés militaires, 7N1231, folder 3, report dated 27 Oct. 1902. 35 Ibid.

During the Boer War, many young Irishmen had decided to emigrate as they were convinced that conscription would soon be implemented in order to replace the losses of the British army.[36] According to the French spies, British injustice in the country was a fact, and the Catholic clergy in Co. Cork would definitely support French soldiers as its members were true nationalists. Now was the time for landing. The Deuxième Bureau's report contained a surprising revelation:

> To come back to the real support that a landing corps would find in the country, I must write down the following answer that we got, passing off as other foreigners, every time we questioned nationalist notables on the chances of success of a Franco-Russian landing in Ireland: 'If invaders want to be followed, they must immediately give the Irish uniforms, armament, a military organization and officers, even foreign ones, and you will have excellent soldiers. If they only gather national guards, the invaders will have a useless mob and even a dangerous one if panic-stricken. History is there to show it.'[37]

The Russians in Ireland – was this a realistic scenario? It should not be forgotten that at that time there were tensions between the Russians and the British concerning Afghanistan, Persia and Tibet.[38] Russia was no friend of Britain. This means that a possible Franco-Russian joint-venture in Ireland was not as improbable as it appeared to be at first sight, especially all the more since in February 1901, Cambon, to his greatest surprise, had been informed by Delcassé that the 'chief of staff [of the French army] was in [St] Petersburg to elaborate defence plans, not only against Germany but also against England'.[39] Indeed, in July 1900, General Pendezec, the French chief of staff, had travelled to Russia to have talks with General Sakharov. Pendezec had then suggested the following plan: if France was attacked by Britain, then Russia would send 300,000 men into Afghanistan and threaten India; if Russia was attacked, then France would concentrate 150,000 men on her north-western coast to threaten southern England. The Russians had agreed with Pendezec's plan.[40] In the middle of the Boer War, one may wonder whether these plans had not been meant for an offensive, as it was most unlikely that Britain could attack France and Russia under those circumstances. Moreover, it was not the first time that imperial Russia showed interest in faraway Ireland. Indeed, in 1885, William O'Brien of the Nationalist Party had been

36 Patrick Maume, *The long gestation; Irish nationalist life, 1891–1918* (Dublin, 1999), p. 28. 37 SHD, Vincennes; attachés militaires, 7N1231, folder 5, report dated 27 Oct. 1902. 38 Milza, *Les relations internationales de 1871 à 1914*, p. 123. 39 Villate, *La république des diplomates*, pp 197–8. 40 Allain, 'L'affirmation internationale à l'épreuve des crises 1898–1914', p. 691.

approached by a Russian emissary in London. The Russian wanted to get Charles Stewart Parnell's approval for a 'Russian Volunteer Fleet' to ship over 5,000 Americans to Ireland and start a revolution there. But Parnell, leader of the party, had not agreed and had told O'Brien in no uncertain terms: 'The Russian may escape hanging – but you and I won't.'[41]

The British secret service knew that both the French and the Russians were in touch with Irish republicans. As seen above, French public opinion was largely in favour of the Boers. There were support committees such as the Comité français des Républiques sud-africaines (French Committee of South-African Republics) in which certain members of Action française, an ultra-nationalist organization, including the journalist and Member of Parliament Lucien Millevoye, were involved.[42] It was therefore no surprise that some Irish republicans lived in Paris and were active in pro-Boer circles. Maud Gonne was one of them. She edited a nationalist bulletin called *L'Irlande libre* (Free Ireland) and was Millevoye's mistress before she married the republican Major John MacBride, who had organized a unit of Irish volunteers to support the Boers against the British. The special branch of the Royal Irish Constabulary (RIC) knew that MacBride and Gonne were 'working in Paris with the object of obtaining permission of the French Government to the formation of an Irish brigade in that city'.[43] At first, Millevoye had not been particularly impressed by the Irish in Paris. In 1896, he had told Gonne: 'Your Irish revolutionists are only a set of *farceurs* [jokers].'[44] This was perhaps true at the time, but by 1900 there was a war being fought in South Africa. As to the British, they took no chances.[45] The special branch reported to London: 'Maud Gonne is one of the most dangerous conspirators we have at present to deal with.' It wanted the woman 'to be *shadowed* [*sic*]'. The Irish Chief Secretary's office ordered Major Gosselin of the special branch to find out 'who are the lady's associates and probable correspondents *in Paris* [*sic*] and *elsewhere* [*sic*] on the Continent [and] the sources from which her [financial] supplies are derived'. Gosselin went to see Sir John Ardagh, the director of military intelligence, who told him that he 'had a man in touch with Miss Gonne, and it was hoped and believed good results would accrue'. Gosselin opined that he did not believe that the French and Russian governments were financing the Irish republicans. But Ardagh was not so sure and answered:

41 Christy Campbell, *Fenian fire: the British government plot to assassinate Queen Victoria* (London, 2003), p. 159. 42 Bernard Lugan, *La Guerre des Boers, 1899–1902* (Paris, 1998), pp 224–52. 43 National Archives, Dublin; Chief Secretary's Office, Crime Branch Special 1899–1920, no 23489/S, reports dated 03 Dec. 1900 & 14 Sept. 1900. 44 Donal P. McCracken, *MacBride's Brigade: Irish commandos in the Anglo-Boer War* (Dublin, 1999), p. 78. 45 Sloan, *Geopolitics*, pp 119–21.

> Yes, that is perfectly true of the Russian Government itself – the Czar and his Chancellor – but you must remember there are several Secret Service funds in Russia and they act independently of each other – the Military Department, for instance, might be spending money and the Czar know nothing of it.[46]

Interestingly, Gosselin said that a man called Raffalovich of the Russian embassy in Paris had 'a sister [who] was married to a very prominent Irish agitator'. Ardagh knew this already and replied: 'Yes, I know the man well – a Polish Jew – and a very pronounced enemy of England.'[47] The Irish agitator in question was none other than William O'Brien, previously mentioned, whose wife, the wealthy Sophie Raffalovich, helped him financially in his home rule campaigns.

However, the plans drawn by the Irish republicans and the Deuxième Bureau had one common weakness: they did not include the obvious essential support of the French navy. The problem was that there was no real common strategy between the army and navy in France. The navy's ships were not the most modern and moreover, just like the Royal Navy, it was suffering from disorganization. The ministry for the navy did not even have a staff to study war plans. This would only be introduced in 1902. In fact, for the period under consideration, the French navy was being restructured.[48] The archives of the French navy in Vincennes do not contain reports regarding the Deuxième Bureau's plans. In other words, now was not the time to consider an invasion of Ireland.

Unfortunately for those politicians and military staff officers in Paris who were thinking of getting even with Britain after Fashoda, King Edward VII decided to hold out the hand of friendship to France. The first of May 1903 was the beginning of an earthquake in international relations in Europe which would badly affect nationalist Ireland. The British monarch, Edward VII, arrived in France for a state visit. Just before his arrival, the German ambassador in Paris wrote to von Bülow, who was now chancellor in Berlin: 'The closer King Edward's visit gets, the more French newspapers are opposed to the [Franco-British] rapprochement.'[49] At first sight, this looked to be the case as Edward was well and truly booed by the Parisian crowd who shouted 'Up Marchand!', 'Up the Boers!' But Edward had a remarkable sense of what would nowadays be called 'public relations'. To his great credit, he

46 Public Record Office (PRO) nowadays National Archives, London; CO904/202/166A, Maud Gonne's file, reports dated 30 Oct. 1900, 14 Nov. 1900, 20 Nov. 1900 & undated summary of Maud Gonne's activities, p. 131. All used CO 904 files are available on microfilm in Mary Immaculate College in Limerick and entitled 'The British in Ireland'. 47 Ibid. 48 Jean Doise & Maurice Vaïsse, *Diplomatie et Outil militaire* (Paris, 1987), pp 120–6. 49 Jean Guiffan, *Histoire de l'Anglophobie en France* (Rennes, 2004), pp 156–7.

managed to change an openly hostile public opinion into one of admiration and friendship. 'Long live the king!', shouted the Parisians when he returned to London. The Franco-British rapprochement was on the way, causing consternation across the Rhine. What were the reasons behind this drastic change in international relations? Briefly, the British were increasingly worried about Kaiser Wilhelm II and Admiral Alfred von Tirpitz's naval policy and that it would result in Germany being able to threaten Britain and her empire. Little by little British public opinion became more and more anti-German. This paved the way to the Entente Cordiale which was eventually signed by France and Britain on 8 April 1904. The Entente was not a military alliance, but it was clear that both countries would cooperate. For Irish republicans, this was a major blow as the French would no longer support them after decades of friendship. They had to find a new foreign power which could, one day, help them to get rid of British rule in Ireland. This power would of course be imperial Germany. The Irish newspaper *Kilkenny People* reported that John MacBride was hoping that, if the Germans invaded Britain, they would also send 100,000 rifles and also artillery to Ireland for the liberation of the country.[50] The republicans lost little time in contacting the Germans and their collaboration would be epitomized in the Easter Rising of April 1916 in Dublin, although the Germans were not exactly generous in their support.[51]

A minor incident but in fact very representative of France's change of attitude towards the Irish republicans took place in Fontenoy in Belgium soon afterwards. Fontenoy was an extremely important symbol of the ancient Franco-Irish military connection. It was there that in 1745, Marshal de Saxe at the head of the French army beat the English, Austrian and Dutch troops under the command of the duke of Cumberland. For a long time, the course of the battle was undecided, but towards the end of the fighting de Saxe ordered regiments of Irish exiles (the so-called Wild Geese) forward, which allowed the French to win. By the 1900s, Fontenoy had become an important annual commemoration for the Irish nationalist community, and it was on the battlefield that contacts were made and maintained as Irish and French regularly met during ceremonies. But, since the signing of the Entente Cordiale, Fontenoy had suddenly become a diplomatic embarrassment for the Quai d'Orsay. On 3 August 1907, the French consul in Tournay sent a report on the annual commemoration of the battle to Count d'Ormesson, the French ambassador in Brussels. He stressed that Irish nationalists had the intention of inaugurating a monument for the fallen Wild Geese (a Celtic high cross still visible today in the centre of Fontenoy). The consul reminded the ambassador that André Géraud, a specialist in Irish questions in the Quai d'Orsay, had advised that French citizens should no longer participate in the ceremonies.

50 Hünseler, *Das Deutsche Kaiserreich und die Irische Frage, 1900–1914*, pp 124–5. 51 aan de Wiel, *The Irish factor, 1899–1919*, see chapters five and six.

He wrote: 'It was during the king of England's visit to France, leading towards the Entente Cordiale. Monsieur Géraud's reservations seem to have been definitely well-founded as the whole occasion was essentially an Anglophobe display.' A few days later, d'Ormesson answered, saying that Géraud's instructions were still valid.[52] Irish nationalists, of all shades of opinion, had become aware of this change in the French attitude towards them. Anatole le Braz, a French Breton writer, was in Ireland in April-May 1905 and was able to ascertain the importance of Fontenoy for the nationalist psyche. He reported the following anecdote. Apparently, the lord mayor of Dublin had invited the French consul to attend a ceremony to commemorate the battle. But the consul was embarrassed and tried to explain that he had to decline lest Britain should be offended. The lord mayor got fiercely upset and said: 'Yes, if England had invited you to the commemoration of Waterloo, you would have gone!'[53] *Touché*, indeed!

Finally, time has come to answer the all-important question: at the end of the day, were these invasion plans of Ireland serious or were they routine reconnaissance missions regularly carried out by the Deuxième Bureau or even elaborate theoretical war games? There are hints. Concerning the French general staff, its officers lost all interest in Ireland almost overnight after the signing of the Entente Cordiale in April 1904. This is very clearly shown by the sudden lack of documents about Ireland after that date in the French army archives in Vincennes. Worthwhile documents can be found again from 1912 onwards when it seemed that Ireland was heading towards a civil war between the unionist Ulster Volunteer Force (UVF) and the nationalist Irish Volunteers. This in itself could be evidence that the landing plans were not simple theoretical exercises or mere reconnaissance operations, and that some French military officerss and politicians had indeed toyed with the idea of attacking Britain by way of Ireland during the Boer War. It was a unique opportunity to deliver a deathblow to a formidable colonial rival and to get even after Fashoda. This is certainly not far-fetched. Pride is a powerful motive for men and nations alike. However, in France, documents seem to be lacking that can back up this theory beyond any reasonable doubt. Perhaps they never existed. As Baron Hermann Speck von Sternburg, the German ambassador in Washington in the 1900s who was in touch with Irish-American republican groups, once said: '[Some intelligence is] better talked over than written.'[54] But what about archives in Britain? The National

52 Quai d'Orsay, Paris; Grande-Bretagne, politique intérieure, question d'Irlande, 1897–1914, vol. 4, Bossuet to d'Ormesson, 03 Aug. 1907 & Quai d'Orsay to Bossuet, 10 Aug. 1907. 53 Anatole Le Braz, *Voyage en Irlande, au Pays de Galles et en Angleterre* (Rennes , 1999), pp 147–8. The author is grateful to Professor Jean Guiffan (University of Rennes, France) for this reference. 54 Edmund Morris, '"A matter of extreme urgency": Theodore Roosevelt, Wilhelm II, and the Venezuela crisis of 1902: United States-Germany

Archives in London (the former Public Record Office) in fact contains a most revealing document. Major Gosselin of the Special Branch had succeeded in infiltrating Clan na Gael, an Irish republican organization in the United States. In December 1900, one of his agents took part in a secret meeting of the Clan in which John MacBride participated. MacBride had arrived in New York from South Africa by way of Paris. This is what the agent reported back to Gosselin:

> In his address before the various Camps assembled together in secret meetings, McBride [*sic*] alluded to the feelings of France, stating that recently he was brought to see the leading Minister of the Government in Paris by a well known Deputy, and the question asked him by the Minister was, what would the Irish be prepared to do in say six months or when called upon in conjunction with his country, and that, continued McBride [*sic*], is the question I have come out here to have answered, so that I can on my return show the Minister what we can do.[55]

The 'well known Deputy' was probably Lucien Millevoye. The 'leading Minister of the Government', a rather ambiguous term, might be René Waldeck-Rousseau as he was the Président du Conseil (prime minister) at the time. It would be pertinent to point out that it seems most unlikely that Waldeck-Rousseau, a man from the left, had clandestine relations with an ultra right-wing nationalist such as Millevoye, if it had been Millevoye of course. But, the fact is that the special branch's report fits perfectly well those of the Deuxième Bureau. It was the 'missing dimension', as very justly defined by Sir Alexander Cadogan and most judiciously elaborated upon by Christopher Andrew and David Dilks.

conflict over alleged German expansionistic efforts in Latin America', *The Naval War College Review* (spring 2002), www.findarticles.com, consulted on 4 October 2004.
55 PRO, London; CO 904/208/258, MacBride file, Major Gosselin's report dated 2 Jan. 1901, concerning Clan na Gael's meeting of 16 Dec. 1900.

Irish impressions of the French during World War One

SIOBHAN PIERCE

> That morning I had to go to the British zone and at last met with British troops. The first I saw were a small detachment of Irish Guards, enormous, stolid, in perfect step. What made them look even bigger and more dignified than usual was that they were being led to their unknown destination by a poor old, stumbling, shuffling, untidy little French Territorial who had to break into a trot every few minutes to keep ahead. The effect was extremely comic.[1]

The above quote is from from *Liaison 1914,* Edward Spears' famous and lyrical account of his time as a British Army liaison officer to the French. These particular lines primarily hint at good collaboration and frequent contact between the men of the two allied armies during World War I. While the tone of the above quote may appear slightly disparaging towards the French, Spears had spent a lot of his life in France, was a Francophile and more likely than most to depict the French in a sympathetic light. Still a young officer in 1914, he was 28 when war broke out, but he nonetheless would have been extremely knowledgeable about the British military system, as he had first joined the British forces, when he enlisted with the Kildare Militia at the age of 17. Spears was of the opinion that in 1914, 'to our men, the French people were represented by the hard bitten and often grasping peasants of Northern France. Most of them never saw a real French soldier, and thought the posts of old and shabby Territorials guarding bridges and railways were typical of the French Army.'[2]

Conversely, Spears had a very different war to the majority of men in the front line, in that he was most frequently among high-ranking officers as he was attached to the French Grand Quartier Général (General Headquarters). Spears' book is a useful reference for discovering an officer's view of the men of the French Army and as a staff officer's view of the relations between the men and their opinions. On the other hand, Spears' opinion is that of just one officer and many hundreds of thousands of men had very different experiences of the war. Therefore, other surviving accounts, letters and diaries of the war, could provide insights into what the soldiers thought of their French allies.

1 Edward Spears, *Liaison 1914* (London, 2nd ed., 2000) pp 119–20. 2 Ibid., p. xxxii.

It is estimated around 140,000 Irishmen served in the British Army in World War One, and a significant proportion of these would have served on the Western Front, in Northern France and Flanders and even those who served in the Eastern Mediterranean battle grounds would have served alongside the French.[3] Did the Irishmen fighting in World War I do so out of the conviction they should protect France? What did the Irish who served in France alongside the French say about their allies and the country they were fighting in? Did the soldiers record their impressions of the French and do these impressions and records contradict Spears' assumptions?

There are many reasons soldiers enlist in armies and, in particular, why Irish soldiers fought in World War One. Even before the war began a large number of Irishmen were in the British Army. In 1913, it is estimated that around 9% of that army were Irish.[4] Some of the factors which had led to these men joining up, and other men from urban centres to enlist during the war, were economic reasons, while other motives represent the normal driving factors for soldiers enlist, to sign up for a life of adventure, to fight alongside friends or to simply leave the life they are living. Other motivations can be due to patriotic convictions and a belief in the political or ideological reasons for which a war is being fought. During every war, one of the key tasks of governments is to ensure that the population understand that this is a just war, they are on the side of 'right', and the government's recruiting posters 'must see to it that everything is circulated which establishes the (war as the) sole responsibility of the enemy'.[5] The British government in particular encouraged soldiers to empathize with their fellow Europeans and to fight to protect small nations from the Germans. Some privately written Irish sources contain enthusiastic references to going to fight the Germans. On 12 August 1914, in a letter written on the day he got the news they are heading for France, Edward Meehan of the 2nd Battalion Royal Irish Rifles wrote to his parents about his fellow soldiers' enthusiasm for the war: 'There is great excitement here with the movements of thousands of troops. Everyone seems to be anxious to get at the Germans. I felt that my heart is as big as a mountain. I felt that it is my duty to go to help to put down such beasts as the Germans are.'[6]

Edward Meehan's letter reflects a militarily jingoistic view when he enthusiastically talks about tackling the Germans. Meehan was in the army at

3 Alongside the Irish contingent in the British army, Irishmen also served in the Royal Navy and Royal Flying Corps (later the RAF). It is estimated that over 200,000 thousand Irish men served in various services in total. 4 Keith Jeffrey, 'Irish military tradition and the British empire in an Irish empire?', in Keith Jeffrey (ed.), *Aspects of Ireland and the British empire*, p. 95. 5 Quoted from H.D. Lassell, *Propaganda techniques in the World War* (London, 1927) in Mark Tierney, Paul Bowen, and David Fitzpatrick, 'Recruiting posters' in David Fitzpatrick (ed.), *Ireland and the First World War* (Dublin, 1988). 6 Siobhan Pierce, 'For her sole and separate use': World War I wills, *History Ireland*, 14:5

the outbreak of the war; thus he and the soldiers he refers to were serving prior to the war and were to become what were referred to as the 'Old Contemptibles'. These soldiers also were responding to news of the war and reports in English newspapers, and this energy and jingoism is communicated in the lines of Meehan's letter.[7] Meehan at least seems to be more anxious to fight the Germans than to protect the French and he appears to be not alone in this. A letter from Sligo soldier W. Ward explains his enthusiasm for attacking the Germans, but interestingly his impulse arises specifically from a desire to protect or gain revenge for the French people: 'If you saw this country with houses burned and men, women and children, lying dead on the streets you would kill the Germans on the spot.'[8] However, this letter was printed in a local paper between September 1914 and March 1916 and thus a caveat arises: it could have been printed for certain reasons, such as to encourage enlistment.

Established in 1915, the Central Council for Organising Recruiting in Ireland[9] was created to ensure Irishmen enlisted. Even after the initial months of the war, the official propaganda orchestrated by the Council concentrates on France's neighbour, Belgium, which admittedly did suffer more. Indeed, one of the reasons the soldiers were encouraged to enlist was to assist poor little Belgium and one surviving recruiting poster encouraged men to enlist stating 'Irishmen! Remember Belgium.'[10] This recruitment poster depicts a brutish German soldier who has murdered a Belgian infant and mother. Other recruitment posters such as one stating, 'Be one of the 300,000, enlist today and have it to say you helped beat the Germans', specifically called on Irishmen to enlist in Irish regiments so they could have the satisfaction of saying that they helped beat the Germans.[11] Another Irish printed poster in the collection of the National Museum of Ireland, 'An Appeal to Gallant Irishmen', urges Irishmen to assist 'heroic Belgium'.[12] It asks Irishmen to enlist as the Germans have violated cathedrals and churches in Belgium – thus utilizing, either willing or unwillingly, the historic religious link between the Irish and Belgians.

From research on the largest body of Irish printed recruiting posters in existence in Trinity College Dublin, only 31 posters out of over 200 pieces contain wording that appeals to soldiers to enlist by referring to German outrages and atrocities being committed. Interestingly, however, the motive to

(September/October 2006), 44–9 at 49. 7 Meehan went missing the autumn during a German counter-offensive on the front at Le Pilly. At that time the 2nd Royal Irish Rifles experienced fierce trench fighting lasting three days, but managed to rebut the Germans in their sector and counter attack. Edward Meehan was not seen after this ferocious fighting and was presumed dead from 26 October 1914: National Archives of Ireland, IA–16–61 Soldiers Missing and Nuncupative Wills. 8 James McGuinn, *Sligo men in the Great War 1914–1918* (Cavan, 1994), p. 61. 9 Only established in 1915. See Fitzpatrick et al., 'Recruiting Posters', p. 47. 10 Jeffrey, *Ireland and the Great War*, p. 11. 11 Imperial War Museum PST 13631. 12 National Museum of Ireland, NMIHH:1995.93.

enlist in order to halt atrocities is equally present on posters created in 1915 and in 1918. The religious rationale to enlist is one that was proposed by another poster in the Museum's collection, 'Cardinal Logue and the War',[13] infers the belief that the cause they would fight for on the Continent would be a religious and just one. One poster in the collection of the Imperial War Museum goes further to create a religious link between Ireland and France and addresses the effect the war was having on northern France. Printed in May 1915, 'The Isle of Saints and Soldiers' depicts a man tilling the soil, while hovering in the background is a ghostly St Patrick pointing at the ruins of Rheim's cathedral.[14] This poster, though, appeals to the notion of defending a Catholic country or (strangely), its buildings rather than its people. This image is in fact used on another poster with a different caption, 'Can you any longer resist the call?'[15] As Nuala Christina Johnson remarks, 'Saintliness and soldiering could cohabit especially in a world where martyrdom for one's religious faith had strong cultural resonances.'[16] Surprisingly then, it appears that the majority of recruiting information did not use religious symbols on recruiting posters. Out of the 203 posters and other leaflets in the Trinity collection, only two have St Patrick as a motif and only eighteen are appeals from Catholic clergy to enlist.[17] On the other hand, references to Ireland and France sharing a national religion crop up later in the war in newspapers, in articles such as those published anonymously in the *Daily Chronicle* by Major William Redmond, brother of John Redmond MP, both of whom would have wanted Irish people to back the war. The *Daily Chronicle* was a left-wing liberal newspaper, which was very much in favour of the war. In an article titled, 'Religion and the War', published on 26 February 1917, he opines that the war has made the French more religious, 'Faith has taken refuge more and more in the hearts of the people ...'[18]

Major Redmond wrote about the Catholicism of France, as a way perhaps of reiterating that the soldiers were fighting to defend a good religious country. He specifically mentions in this article the sacrifice of the French and notes at that stage that over 2,000 French chaplains had been killed in the war, illustrating and reminding the readers of the enormous loss of life among the French.

This mention of blood sacrifice is one, which occurs in many contemporary writings about the war, but it is interesting in that Redmond mentions it in relation to loss of life among the French. Other references to the French are more symbolic and incidental. There were depictions of the French tricolour as part of a backdrop of Britain's allies' flags on another Irish recruiting

13 NMI, NMIHH:1995.40. 14 IWM PST 13637. 15 Illustrated Figure 8, Nuala Christina Johnson, *Ireland, the Great War and the geography of remembrance* (Cambridge, 2003). 16 Ibid., p. 42. 17 Fitzpatrick et al., 'Recruiting posters', pp 51 & 53. 18 Major William Redmond, MP, *Trench pictures from France* (London, 1917) pp 128 and 132.

poster, 'If you are an Irishman, your place is with your chums under the flags.'[19] This poster printed in Ireland shows the flags of Britain's various allies including Ireland's tricolour. Interestingly, the design of the Christmas present, Princess Mary's gift box, the small brass boxes containing chocolate, cigarettes and other items sent to all soldiers and nurses serving in December 1914, refers to the allies of England. Depicted along with the profile of Princess Mary are the names of all the allied nations, with Russia and France being in pride of place, in roundels on the front either side of the princess' profile. Overall, then, the sense from surviving recruiting posters, newspapers articles and letters is that the men were going to fight the Germans, and save a Catholic country, rather than to assist the French soldier or people.

Once the Irish arrived on the Continent, what are the records noting about their interaction and opinion of the French and vice versa? The official battalion diaries are somewhat reticent on recording any encounters with the local population: as Rudyard Kipling notes in his *History of the Irish Guards, First Battalion*, 'It was not worth while to record how the people of Ypres brought hot coffee to the Battalion as it passed through the day before (October 20) and how, when they halted there a few hours, the men amused their hosts by again dancing Irish jigs on the pavements while the refugees clattered past'.[20] There are mentions of the French populace in some other Irish soldiers' accounts, where they note that the initial overall reaction of the French people to the soldiers was very positive. Early in the war in September 1914, Private Duffy of the Connaught Rangers gave an interview printed in the *Galway Express* stating, 'The treatment of the British soldiers by the French and Belgians was splendid. 'You would insult them', he said, 'if you offered them money for anything.'[21] Another reference to these early initial encounters is in the writings of Private John Lucy from Cork. He joined the Royal Irish Rifles in the British Army in 1912, when he was 20 years old and published his memoirs of the war *There's a Devil in the Drum* in 1938. He remembers in 1914 soldiers of the Royal Dublin Fusiliers getting exasperated with the cheering locals who were looking for mementoes and asking the soldiers for buttons off their uniforms.[22] William Orpen (1878–1931), the Irish-born official war artist, in paintings like *Adam and Eve at Péronne* and *Changing Billets, Picardy* insinuates about the nature of the sexual relations between soldiers and some local women. In other works (such as in the painting *Bombing Night*), soldiers are depicted civilly talking to, and at times saving, the lives of the inhabitants of the villages and towns near the front.[23]

19 IWM PST 13657. 20 Rudyard Kipling, *The Irish Guards in the Great War, the First Battalion* (Staplehurst, 1997), p. 53. 21 William Henry, *Galway and the Great War* (Cork, 2006), p. 150. 22 Myles Dungan, *Irish voices from the Great War* (Dublin, 1995), p. 18. 23 Robert Upstone (ed.), Catalogue of *William Orpen, Politics, Sex and Death* (London,

Logically, as the war progressed in this that ravaged area most of the encounters took place between the troops and the locals, were when soldiers were billeted in a village or during business transactions. Overall, the Irish soldiers' opinions of the French people were probably as varied as Malcolm Brown mentions, when referring to the entire British armies: 'Some soldiers report favourably on their relations with the local people, others much less so. The evidence is, perhaps inevitably, somewhat contradictory.'[24] A number of of the letters and references to the French public certainly paint them in a very positive light; others admire the dogged determination of the French to stay in the area. Sapper T. Byrne from Sligo noted in a letter later printed in an Irish newspaper, that 'The people out here do not seem to trouble about the war so much as the people at home; I have seen women selling chocolates a mile from the trenches and farmers working in the fields with the sound of rifle fire and shrapnel shells bursting now and then in the same field.'[25] In a letter in the *Connacht Tribune* printed on 9 October 1915, Private Arthur Kersaw Peters, of 78th Field Company Royal Engineers too wrote about buying from the locals noting, 'we are about 1½ miles from a village where we can purchase a few extras. The army rations get a little monotonous. It is wonderful how a few families have stuck to their homes even when a stray shell at any time may demolish them.'[26]

Some soldiers' letters home grumble about getting paid in francs. Private Enoch Bowen of the 9th Inniskillings, who died on 1 July on the Somme in 1916, wrote to his sister in October 1915: 'All I want out here is fags and some matches and they will do me over a month I would send you the money for them only the money we get here is no good in Ireland you could not get it changed so I am gathering it up and I will get it changed into English Money and I will send you some.'[27] Enoch at least was not intent on spending all his francs in France. Other soldiers also did not spend all their money in France; soldiers donated 'war money' and French, Belgian Austrian and German currency to the collections of the National Museum of Ireland as early as 1918, and in 1919, 1920 and 1921.[28] The objects in the National Museum's collection give testimony to this strong financial link, which arose naturally between local people and the soldiers posted in the area. Many museum collections have French silk-embroidered postcards. These brightly coloured postcards with numerous different designs were created in France, to be sold to the soldiers to post home. Some of the postcards even show an awareness of

2005). **24** Malcolm Brown, *Tommy goes to War* (Stroud, 2005), p. 101. **25** McGuinn, *Sligo men in the great War 1914–1918*, p. 56. **26** Henry, *Galway and the Great War* (Cork, 2006) p. 228. **27** National Archives of Ireland Soldiers Missing and Nuncupative Wills IA-16-61. **28** In the registers of the National Museum of Ireland it is noted alongside civilian donations that the following soldiers donated 'war money' during these years; Private J.T.L. Saxe, 2nd Canadian Division Motor Transport Company France, Private

the songs of the British army at the time: for example, one postcard on display in the National Museum depicts a cheery soldier and the words 'It's a long way to Tipperary.'[29] Other encounters between soldiers and civilians naturally might not be so friendly, and some of the British Army court-martials were relating to crimes against inhabitants. One case study alone of the 1st Battalion Irish Rifles shows that five cases or just over 1% of all court-martial convictions were for crimes against inhabitants. Four of these cases also were cases where the soldiers were also absent without leave, escaping camp or confinement, or drunk.[30]

In January 1916, the 7th marquis of Londonderry complained in a letter home, that he was 'frustrated with his role as glorified secretary whose duties include hosting two and a half hour lunches with French Officers'.[31] This is a unique frustration for a soldier during the war, but it is also true that while the majority of men may not have been enjoying long lunches, the soldiers did not spend their entire time at the front and depending on where they were located and activity at the front, they did get to spend some varying amounts time at rest in the small towns and villages of northern France. During this free time, the soldiers would naturally encounter French people and soldiers at places like *estaminets*, which were café bars.

One of the most powerful visual records of the affect of the war and the life of the soldiers in France were created by the aforementioned William Orpen, who had lobbied hard to become an official war artist, and arrived at the front in the dark days of April 1917. Orpen's images of the British soldiers in France would range from images of a ravaged landscape, peopled by devastated and shell-shocked civilians, and soldiers, such as the powerful painting *The Mad Woman of Douai*, to convivial scenes like *Dieppe*. This latter painting shows a scene in a square in Dieppe, in front of the Café des Tribunaux. In this painting various armies and ranks are shown – French officers at the café itself, some French colonial soldiers, zouaves, crossing the square, and German prisoners being led away by a French regular soldier; chatting to civilians are some British officers from one of the Highland regiments, wearing their distinctive kilts. As one would imagine, cafes and estaminets do often feature in the soldiers' accounts of their leave time spent away from the trenches during the war. What is interesting is that it is officers who are seated at this impressive-looking café and in the image socializing around the square.

Major William Redmond's anonymously written, *The Square of the Empire*, published on 14 April 1917, mentions soldiers of all the nationalities and colonial countries of the British empire meeting and socializing in the Grand

Murtagh, Captain J. Goss and Major H.W.K. Wait. **29** NMIHA: 2006.33 Kindly on loan from Andy Barclay. **30** James W. Taylor, *The 1st Irish Rifles in the Great War* (Dublin, 2002), p. 197. **31** Peter Martin, 'Dulce et Decorum: Irish nobles and the Great War, 1914–1919', in Adrian Gregory & Senia Pašeta (eds), *Ireland and the Great War: 'a war to*

Square of a 'fair sized French town some distance from the firing line'.[32] In Redmond's article he continually demonstrates the dichotomy of army life continuing in these towns as he refers to the officers' tea rooms and officers' club, but writes of the men all encountering each other as they are passing by each other in the square. Orpen's paintings and Redmond's letter seems to imply that it was only the officers in the French and British army socialized together. This appears to be backed up by entries in Kipling's *History of the Irish Guards* when he notes in October 1916 that 'a team of chiefly officers, greatly daring, played a rugby football match against a neighbouring French Recruit Battalion'. He adds there was a return match: 'just before their departure from Hornoy they played a soccer match against the 26th French Infantry, and the next day the C.O. and all company officers rode over to that regiment to see how it practiced the latest form of attack in the open.'[33] Sadly, no match scores are given, so no assumptions can be made about the relative athletic skills of the soldiers of either nation.

Among the accounts of rows between soldiers none seem to refer to the men fighting to defend their army's reputation. It appears more likely that it was imagined regimental slights that caused rows, than friction between armies. 'John Lucy's Royal Irish Rifles, had an 'old and sworn enemy' in a nearby English regiment, and he noticed how: 'The Englishmen in our own regiment forgot nationality and beat up their own countrymen in the supposed defence of the honour of their chosen corps.'[34] Historian Richard Holmes suggests that 'inter-regimental brawls were common',[35] and that 'there was inevitably friction between dominions with Poms v Aussies best known.'[36] For the men of the various armies what the Australians described as 'the bonds of mateship' were probably what influenced the most the socializing. These bonds were what kept the men's spirit alive in the trenches and one assumes these ties resulted in the men socializing as a small unit or with men known to them. During the war, the men were encountering many nationalities and cultures for the first time. Spears thought cultural differences were one of the reasons that men of the two armies failed to mix well when they met. He thought that the culinary tastes of the two armies demonstrated this point extremely well:

> When, owing to the sudden German onslaught on Verdun, the Tenth French Army was hurriedly relieved by the British, and during the movement the French Commissariat fed some of our men whilst we supplied some French units, complaints were endless. French and

unite us all'? (Manchester, 2002), p. 41. 32 Redmond, *Trench pictures from France* pp 139–47. 33 Kipling, *The Irish Guards in the Great War,* p. 175. 34 Richard Holmes, *Tommy: the British Soldier on the Western Front 1914–1918* (London, 2005) p. 116. 35 Ibid. 36 Holmes, *Tommy: the British Soldier on the Western Front 1914–1918*, p. 14.

> British both declared that they were starved. Our people could do nothing with the vegetables, which they were expected to devise sauces. They hated the coffee and threw away in disgust the inordinate quantities of bread served out. On the other hand the gorge of the French rose at the slabs of beef provided by us. They declared they could not face all this meat and clamoured for more vegetables, bread and coffee. As for tea instead of wine – puah! Had the arrangement continued it might have led to mutiny. Not that our men disliked wine. Soldiers in blue and soldiers in khaki had at any rate that taste in common.[37]

Intriguingly, this difference of opinion over food crops up in the histories of the Irish Guards, which notes the men's thought that 'A well spoken quiet crowd the French, but their rations are nothing at all.'[38] Spears maintained that language barriers were the other main reason the two armies did not socialize together even when they did meet. He too thought the soldiers had little in common. 'French, and British kept apart, principally of course because they could not understand each others language, but they had few common interests.'[39] There were, however, some men who could speak French, and in particular, some of the regular soldiers who had enlisted before World War I had a smattering of knowledge. When the French population were cheering the British army early in the war, 'Naturally the cheers were for "Les Anglais", a misapprehension corrected by John Lucy of the 2nd Royal Irish Rifles, "Nous ne somme pas Anglais, nous sommes Irlandais." They liked that and laughed with pleasure and then shouted: "Vivent les Irlandais" and we shouted back "Vive les France".'[40]

When Field Captain John Williams of the 1st Royal Irish Rifles was accidentally wounded and died on 20 September 1915, he had listed among his effects a French Phrase Book. Williams had enlisted in Co. Wexford into the 5th RIR in 1892 when he was 17.[41] He had fought in the Boer War, and as an experienced career soldier, he probably thought it useful to carry on his person a phrase book. Nonetheless the presence of the phrase book demonstrates an on the ground use for knowledge of French and a willingness to learn and improve any linguistic skills. In the service books of the soldiers, one of which is on display in the National Museum it was noted if the soldier could speak French.[42] Sergeant John F. McGuinn, 2nd battalion Irish Guards, in a letter home from France on 4 February 1916 wrote honestly of how much and when the soldier were normally utilising their French:

37 Spears, *Liaison 1914*, pp 69 and 70. 38 Kipling, *The Irish Guards in the Great War, the Second Battalion* (Staplehurst, 1997), p. 131. 39 Spears, *Liaison* 1914, p. 70. 40 Dungan, *Irish voices from the Great War*, p. 18. 41 Taylor, *The 1st Irish Rifles in the Great War*, p. 244. 42 NMIHA:2005.75.

> I dabble in French all day, with some success, and much mutual amusement to the Natives and myself. Tommy's knowledge of French is for the most part limited to 'Bon', 'No bon', 'comprise', 'n's plus' and their attempts are most amusing for example one man with a strong Irish accent starts, 'Monsieur, me no comprise, what – pause – ah what the B.H. is it.'[43]

The new soldiers who enlisted in World War One came from a broad range of backgrounds with a wide variety of knowledge and skills, and once again this makes it difficult to summarize the soldiers overall language skills. Certainly one Irish regiment had excellent skills as 'six officers and 225 men of the Royal Guernsey Militia, many of them French speaking, volunteered for the 6/ Royal Irish, as they had been so impressed by the Royal Irish Battalion which had been in garrison on their island.'[44] It appears then that the some soldiers would have had some linguistic skills and certainly, if the soldiers had been determined to communicate with one another, they could have done so. At the beginning of the war, before the elaborate trench systems of later years, the two allied armies had problems even identifying each other. Spears noted an incident in the beginning of the war, he was mistaken as a German by French soldiers and noted: 'Things began to look ugly when I had to forcibly to prevent an enthusiast from starting to collect my buttons as souvenirs. Happily for me, but much to the disappointment of the crowd, Bénazet appeared and delivered a neat little speech, under the impression of which we quietly slipped away.'[45] Spears was very concerned about this and envisioned the many dangers arising from the two allied armies not recognizing each other not least on the field of battle. He proposed that postcards depicting the British uniforms should be distributed among the French. This recommendation was followed and the postcards were circulated, but he does not add if the reverse was carried out. He was definitely of the opinion the two armies seldom encountered each other in the first years of the war and even wrote, 'It was not unusual during '15 or '16 for French battalions to be totally ignorant of the fact that the British Army was fighting within a mile of them. The demarcation between the two armies was rigidly maintained and even at the actual points of contact the men seldom mixed.'[46]

While Spears was writing from the perspective of the official battle plans laid down and regulations, as the war progressed there were many times where units of the two nation's armies worked alongside each other and overlapped. Some photographic evidence in the Imperial War Museum and Australian War Museum's collections show instances when the different

43 McGuinn, *Sligo men in the Great War 1914–1918*, p. 76. 44 Holmes, *Tommy*, p. 153. 45 Spears, *Liaison 1914*, p. 69. 46 Ibid.

nationalities encountered each other in the front and reserve trenches. These photographs show the necessary organized co-operation near the actual front between the allies. One photo in the Imperial War Museum collection depicts French soldiers at an artillery forward observation post working alongside a British Army telephonist during the Third Battle of Ypres at Langemarck on 16 August 1917.[47] Like many of the images taken on the Western Front it is somewhat staged, the soldiers being very conscious the camera is pointed at them. Another photograph, this time from the Australian War Memorial's collection depicts, French Artillery and Australian Pioneers dugouts in May 1918.[48] In the centre of the photo, there are soldiers wearing the distinctive French Adrian helmets and Australian wide brimmed hats milling around a horse-drawn wagon. While the soldiers are aware of a camera (in particular two French soldiers in the forefront and some Australians in background), the photo communicates a distinct lack of formality as well as the hard work and harsh conditions at the front. This photograph is more than likely a genuine photograph rather than staged.

One British Army soldier who verifies that artillerymen were often formally working with their allies' artillery was an Irishman, R.T. Grange, who recounted his experiences on the Somme in 1916 for the Somme Association records. He remembered as a young 20-year-old being 'not greatly pleased'[49] when he heard he was going to be attached as a signaller for Forward Observation Officers of the French artillery. This may have been more a case of the 20-year-old's wanting to remain fighting alongside fellow volunteers from Ulster than having a dislike of the French. Bob Grange does go on to state that he thought that the words of a French war correspondent who 'was sending dispatches to his newspaper about the Ulster attack that day (and) he described it, I think, very well' the men of Ulster had 'entered the Battle of the Somme as enthusiastic young sportsmen and emerged from it as professional soldiers'.[50] Grange's fellow volunteers would be decimated by German machine gun fire on the Somme, and perhaps his apparent anti-French attitude can be more attributed to a sense he should somehow have been fighting alongside them. Therefore, the evidence appears to point at more collaboration between men when not in units than one would deduct from Spears' comments. Indeed, the quote at the opening of this chapter recording the Irish Guards being guided to the front by a Frenchman also signifies some on the ground cooperation in 1914, be it fleeting and not involving complicated and long-term mutual aid. Kipling's history of the Irish Guards demonstrates well, how these battalions, worked continually alongside the French. The Irish Guards were frequently neighbours of the French during the war, and the guardsmen were impressed with the French and their

47 IWM Q2723. 48 AWM E02203. 49 Dungan, *Irish voices from the Great War,* p. 106. 50 Ibid., pp 127 & 129.

75mm artillery. At the Third Battle of Ypres in 1917, the men of the Second Battalion thought, 'Ye can hear the French long before ye can see them. They dish out their field gun fire the way you'd say it was machine guns.'[51] Not alone was the French artillery supporting the battles on the Western Front but they too were present in areas of operation such as Gallipoli. While overwhelmingly the Irish who were in Gallipoli were part of the regular British infantry, there too were Irishmen in the Australian and New Zealand Auxiliary Corps. Sligoman Ernest Egan of the Australian Mounted Troops was wounded in May 1915 and while convalescing on Malta wrote to his father of their landing at Cape Helles and their subsequent actions in May 1915. On 26 June 1915, the *Connacht Tribune* printed his letter and including an extract where he noted, 'The French Artillery supported us; their artillery is magnificent.'[52] In the collection of the National Museum items referring to the French artillery crop up in the register. A large donation from a lady who was most likely the daughter of an artillery officer, an assumption made by the nature of the large collection she donated, which included 18 pdr artillery shell heads, includes several artillery badges depicting gun carriages and the famous French 75' guns.[53] Given the working relationship between the British infantry and the French artillery, maybe it is not surprising then that one of the only strong impressions of the French conveyed in the written and material records of the Irish soldiers seems to have been an admiration for the power and professionalism of the French artillery.

In Salonika too the French were present fighting alongside the 10th Irish Division. During the retreat in December 1916 the 10th Irish Division covering the retreat found that 'the French transport moved with painful slowness and many halts on the steeper parts of the hill, so that when day broke a large proportion of the column had not yet reached the pass'.[54] According to Lt Col Jourdain's history of the Connaught Rangers near the end of the same retreat, when the Irish had reached Salonika and as they were 'waiting for an engine, they sheltered from the pouring rain in boxcars, resting and watching the French troops stream past. The Irishmen were deeply impressed by the cheerfulness of the French in these conditions.'[55]

Most of the references to the French in soldiers' private letters and diaries are cursory, merely noting the presence or overall actions of the French Army, sometimes noting a unit of the French army but they do not normally provide insights into if the soldiers thought the French army were fighting effectively. For example on the 23rd October 1918, Cyril Falls refers in his history of the Royal Irish Rifles refers to a joint action between two companies

51 Kipling, *The Irish Guards in the Great War, the Second Battalion*, p. 131. 52 Henry, *Galway and the Great War*, p. 201. 53 NMIHH:1971.36–74. 54 Tom Johnstone, *Orange, Green and Khaki: the story of the Irish regiments in the Great War* (Dublin, 1992) p. 182. 55 Johnstone, *Orange, Green and Khaki*, p. 185.

of the 1st Royal Irish and a French squadron of dragoons, the dragoon unit being the 28th French Dragoons.[56] One other tantalizing reference to another unit is when in July 1918 the Irish Royal Rifles 'relieved the 42nd Regiment of the French Army on the night of 7 July in the St Jans Cappel Sector ... The Commandant of the 42nd Regiment presented the battalion with a cow.'[57] This simple comradeship is also demonstrated in a series of photographs, taken in the last year in the war. In January 1918, the 5th Army of the British Forces was taking over responsibility for a 28-mile sector of the front, which had previously been the responsibility of the French.[58] The photographs were taken at this time at a place called Ham, on the Somme, near where the changeover in responsibility was to happen. It shows Irish soldiers from the Royal Inniskillings, proceeding to the front and meeting and exchanging cigarettes with soldiers of the French 6th Division.[59] What probably makes this photograph most poignant was that an entire company of the 121st field company of the 36th Division, which the Inniskillings were in, died defending a bridge near Ham later that year during the German Spring Offensive, the Kaiserschlacht, of 1918.[60] These photographs could then be depicting the experience of the majority of the men – just passing each other as they trudge back from the front or march to the trenches and in a situation that they encounter each other as a unit rather than as individuals.

Other records show more cooperation in the front lines between the two armies than perhaps the Brigade staff were aware; the Irish Guards had an arrangement whereby they were lending machine guns to their neighbours. On 1 February 1915, after the fierce fighting that led to Wexfordman Michael O'Leary later being awarded his Victoria Cross, 'it is distinctly noted in the Diary that two complete machine guns were added to the defence of the post after it had been re-captured. Machine guns were then highly valuable pieces of military equipment and when the French who were their neighbours wished to borrow one such article "for moral and material support" a Brigadier General's permission had to be obtained.'[61] Later in the war on 3 August 1917, the 2nd Battalion Guards found themselves not only manning their own front line trenches but also providing left flank support for the neighbouring French front line, which were more forward than the Guards. Some of the Irish Guards took the initiative and while the rest of the battalion endured heavy German artillery fire and a steady downpour of rain, the brigade diary notes that 'the only people who kept warm seem to have been Lieutenant Rae and a couple of platoons who got into touch with the French and spent the

56 Taylor, *The 1st Irish Rifles in the Great War*, p. 146. **57** Ibid., p. 138. **58** Tom Johnstone, *Orange, Green & Khaki*, p. 338. **59** Some of these photographs are reproduced in David Murphy's *Irish regiments in the World Wars* (Oxford, 2007), pp 19–20. **60** Ibid., p. 377. **61** Kipling, *The Irish Guards in the Great War, the First Battalion*, p. 77.

night making a strong standing patrol of two sections and a Lewis Gun at Sentier Farm, which is where the French wanted it.'[62]

Other references to the French occur when the armies are overwhelmed and the war's bloodiest death tolls are at their height. Mass casualties resulting from such actions often swamped medical resources and the soldiers do occasionally note when they see the unfortunate dead French soldiers. Earlier in the war at Ypres in 1914, Lieutenant Gibb of the 4th Royal Irish Dragoons records encountering other French victims of gas attacks: 'The Brigade were ordered up to support the French and the Canadians who were getting badly mauled around St Julien where they had come up against he main German attack and found the air poisoned with gas fumes. French and Canadians were to be found lying insensible along the road and during the first effects of this method of warfare there was no provision for treatment.'[63] Later in the war on 17 August 1916, Major William Redmond's article in the *Daily Chronicle* refers to the graves of the French which he encountered when fighting at the front:

> The writer came the other day upon the roughly made cross of wood which marked the grave of a French infantryman. His name and regiment were rudely carved on the cross, with the date of his death and beneath were the words, in French, 'Dead on the field of honour'. Those who were dear to this soldier may never have the opportunity of standing by his grave or tending it. They may not grieve, however, for Nature has done all that is required or wished for. A quilt of wildflowers covers this humble resting place, and red poppies and blue cornflowers nestle around the little cross and with every breath of wind nod and point to the words, 'Dead on the field of honour'.[64]

One of the few Irish oral recollections of the war, mentioning the French, was by an Irishman who was badly injured during the war. In an interview decades later Peter McBride of the 7th/8th Royal Irish Inniskilling Fusiliers describes remaining with the remnants of his division after being badly injured in his arm while retreating from the face of the ferocious Kaiserschlacht. On the fourth night after he sustained his injuries, while still with his platoon he goes on to state he only got basic medical attention and to safety when:

> A young French Soldier brought me down to this cellar. They got an ambulance around the next day and we were driven a long distance away. I remember there was a wounded German cavalryman in the same

62 Kipling, *The Irish Guards in the Great War, the Second Battalion*, p. 131. **63** Revd Harold Gibb, *Record of the 4th Royal Irish Dragoon Guards in the Great War 1914–1918* (Canterbury, 1925) p. 42. **64** Redmond, *Trench pictures from France*, pp 41–2.

> ambulance. The driver had someone beside him and he was telling him how he'd been captured the previous night by the Germans and recaptured by the French.[65]

The ambulance brought McBride to a field hospital where such were the numbers of injured he remained for 36 hours before he was checked. Unfortunately, by that time McBride's arm was in a condition where it had to be amputated. Excerpts from Major-General Arthur Solly-Flood's account of the chaotic retreat from Mons in 1914 mention meeting a French interpreter; he recalled: 'St Quentin is chiefly memorable to the writer because he was upon arrival he was able, with the aid of a French Interpreter, to take his boots off for the first time since leaving Mons.'[66] Unfortunately the boots would not fit back on and they had to be split; however later they were recycled as they were given to a grateful cyclist orderly. He also mentions taking cover with French Hussars during the retreat and the mingling of French and British soldiers who 'including field and heavy artillery, were marching in all directions'.[67]

It appears that the records created by Irishmen or from Irish battalions, mentioning the French civilians and army are quite scarce. The official recruiting posters do not refer to the war being necessary to defend the French; it is rather to assist the Belgians, who realistically were in a worse situation. The effect the war is having on the French is mentioned only twice in the approximately 225 surviving recruiting items in the main archives and collections, and then only to point out that it is a Catholic country under attack and suffering. French people when referred to by the Irish soldiers are mentioned in a non-judgmental, even an admiring, way. References among private letters that are printed in newspapers early in the war in 1914 view the French people positively. These letters, though, would have been in the mainstream press and at an early stage of the war when enthusiasm was high and empathy from the French and Belgian civilians strong. References in the accounts of soldiers like John Lucy are slightly more mixed but still positive, and overall, it appears they realized what a burden the war was on the local population. The economic links to the population and countries where the soldiers were stationed survive in the postcards and material bought and sent home by the soldiers.

What is surprising perhaps is the fact that references to French soldiers are so rare. It appears the soldiers certainly had the opportunity to meet when on leave and also that the linguistic skills required for a basic conversation in an *estaminet* would have been more likely among the soldiers than one would assume. We can conclude if and when they wished the soldiers had the basic linguistic skills to share a convivial drink or cigarette. Interesting too is the

65 Dungan, *Irish voices from the Great War*, p. 188. 66 Gibb, *Record of the 4th Royal Irish Dragoon Guards*, pp 28–9. 67 Ibid., p. 26.

fact that the references from all kinds of official and unofficial sources specifically mention that it is the officers of the two armies who socialize. The official histories rarely mention the fate of those French units who served alongside the British and who were cooperating with them in the various offensive and defensive battles in the trenches. It may be the officers creating these official records may have felt it was not their duty to record such encounters. However, this omission can be startling like on one occasion which Kipling admonishingly notes when:

> On 11 February 1915, for example, it is noted that the men had baked meat and suet pudding for the first time since the war began; on the 13th not one man was even wounded through the whole day and night; while on the 15th more than half the battalion had hot baths for the first time since January. The diary records these facts as of equal importance with a small advance by the French on their right, who captured a trench, but fell into a nest of angry machine guns and had to retire.[68]

It is often during the chaos of retreat, be it from Mons in 1914 or to Salonika in 1917, that the few impressions and references to the French are recorded. There were no doubt times when the two armies had derogatory remarks to say about each other, such as the entry in the Irish Guards 2nd Battalion diary on 10 October 1917, the 'day passed quietly until the afternoon when the same French 'seventy five' which had been firing short the day before, took it into it's misdirectional head to shell No.1 Company so savagely that they had to be shifted to the left in haste.'[69] These derogatory opinions are not often recorded, and this reproachful reference to the French appears to be a rare one. It does seem that, for the most part, for the ordinary soldier the French army is part of the background to the soldier's war, the French an army which happens to be a ally and that when the French soldiers are making an impact in the soldiers' psyche it is not as individuals but as units. This omission should be considered given the horrible experiences, short life-expectancy and harsh conditions at the front. Battalion and unit histories above all are official records of fact and not emotion, and maybe this somewhat explains the lack of remarks about the French army. Perhaps too we should not be too harsh when judging the apparent lack of empathy in the records for the French soldiers, as during war the soldiers were naturally more worried about their own personal survival than any other concerns.

68 Kipling, *The Irish Guards in the Great War, the First Battalion*, p. 83. 69 Kipling, *The Irish Guards in the Great War, the Second Battalion*, p. 152.

'I was terribly frightened at times': Irish men and women in the French Resistance and F Section of SOE, 1940–5

DAVID MURPHY

In his first BBC radio broadcast of 18 June 1940, General Charles de Gaulle, leader of the exiled French military forces in London, stated that 'whatever happens, the flame of French Resistance must not and will not be extinguished'.[1] In the weeks that immediately followed the French surrender in June 1940, individuals and small groups of people across France began to engage in acts of resistance against the occupying German forces. Their activities ranged from helping allied servicemen to evade capture, to printing clandestine papers and pamphlets. Some of the first Resistance groups engaged in intelligence gathering, despite the fact that they usually had no means of sending information to London.[2] Regardless of organisational problems and a lack of equipment, thousands of men and women across France were motivated by a desire to act against the occupying German forces.

Between 1940 and 1944, the various forces of the Resistance developed into an elaborate and complicated system. It is impossible in the space of this short article to fully outline the organization and activities of the Resistance and the different phases of its organisation. There are several excellent modern studies that can be consulted.[3] Within the wider Resistance community there were two distinct types of groups. The first was the system of networks or *réseaux*. During the course of the war, 266 *réseaux* were established in France and over 150,000 agents served in them. Each was founded with a specific purpose in mind. Some *réseaux* specialized in sabotage while others became involved in the organization of evasion routes or intelligence gathering. They were linked to the various Allied intelligence agencies including SOE, OSS, MI6 and the Free French BCRAM.[4] In general, these organizations were run on a strictly

1 Terry Cowdry, *French Resistance fighter: France's secret army* (Oxford, 2007), p. 4. 2 Information from M. André Heintz, veteran of the French Resistance and author of various books and articles including, with Gérard Fournier, *'If I must die': from 'Postmaster' to 'Aquatint'* (Cully, 2008). 3 See M.R.D. Foot, *Resistance: an analysis of European Resistance to Nazism, 1940–45* (London, 1976) and Pierre Lorain, *Secret warfare: the arms and techniques of the Resistance* (London, 1976). See also M. Degliame-Fouché, *Histoire de la Résistance en France de 1940 à 1945* (5 vols, 1967–81). 4 The Special Operations Executive was founded in July 1940 and among its various clandestine activities in Europe it organized groups that operated in France. The Office of Strategic Services or OSS was

compartmentalized basis. Security was paramount and, as a result of this, *réseaux* tended to be small groups. Yet some *réseaux* were large, such as Zéro-France which numbered over 1,000 agents, the majority of whom served on an occasional basis only. Also, the *réseaux* developed out of a military and intelligence background and some specialized in recruiting within a specific institution, such as Ajax, which recruited policemen.

The second type of Resistance group were referred to as *mouvements*. They tended to be larger and the forces that created them came from within French politics and society. The Libération-Nord movement for instance was founded by trade unionists and socialists. While they may have later received military aid from the allies, they were all founded and organized by men and women in France. Many published their own newspapers and political ideology was often crucial in the organization of the *mouvements*. Some groups were pro-Gaullist while others supported General Giraud or were aligned with the Communist Party of France.

In January 1943 the various groups of the wider Resistance community were brought under some form of central control within France through the organisation of a central resistance movement, the Mouvement Uni de Résistance (or MUR). The MUR confirmed de Gaulle as the leader of the overall Resistance movement and also unified the various paramilitary units into the Armée Secrète. These military forces included the Francs-Tireurs et Partisans Francais or FTPF of the French Communist Party and also General Frère's Organisation de Résistance de l'Armée or ORA. Others included the *Combat* and *Libération* movements, among others. In February 1944 the Armée Secrète was reorganized as the Forces Françaises de l'Intérieur or FFI. The organization of internal resistance forces in France was facilitated by German attempts to conscript large numbers of young men into forced labour schemes in Germany. By 1943, thousands of French men were hiding out in the countryside to avoid such service. They lived in the camps of the Maquis. These Maquisards would later serve as part of the forces of the FFI.

In the years that followed World War Two, an official organization was established in order to confer the status of resistor on those whose wartime Resistance activities were confirmed. The Service Historique de la Défense at Vincennes holds the official archives of this administration. This archive, which happens to be one of the largest archives on the French Resistance, is administered by the Service's Bureau de la Résistance. The records show that over 700,000 men and women applied for official recognition as members of the Resistance after the war and over 60% of these applications were successful.

the forerunner of the modern CIA and served as the American organizing body for resistance operations in France and also further afield. Britain's MI6 also supported clandestine activity in France while, in October 1941, the exiled French forces in London formed the Bureau Central de Renseignements et d'Action or BCRA, which also acted with

The majority of people who served in the Resistance were French. However, there was also a significant number of non-French who served in Resistance groups. It is possible to find records of Hungarians, Italians. Romanians, British, Americans and even Germans. A large contingent was drawn from France's large Polish community and just under 20,000 of these served in two large Polish movements – the PKWN (Communist) and the POWN.[5] The Spanish Communist Party organized a guerrilla group of over 3,000 fighters who served in the south of France. This was known as the 'XIV Corps'.[6] Members of the exiled Armenian community in France also served and even formed their own specific group, the 'Manouchian group', under the command of Missak Manouchian. The pre-war labour organization Main d'Oeuvre Immigrée or MOI also formed an armed wing of non-French resistance fighters.[7] It must be emphasized that, unlike the members of the International Brigades or the fascist legions who served during the Spanish Civil War, the majority of these non-French resisters were living in France in 1940 and decided to serve their adopted country. Also, many allied servicemen later served with the Resistance while evading capture.[8]

In 1941, the Irish prime minister, or taoiseach, Eamon de Valera, sent a memorandum to the Irish Department of External Affairs requesting an estimate of the number of Irish people who still remained in occupied France. External Affairs estimated that, at that time, there was between 700 and 800 living in France.[9] At this time, it is possible to positively identify over twenty Irish-born men and women who served during the war with the French Resistance, the Free French Forces (Forces France Libre or FFL) and the SOE during the war.[10] Those who served with Resistance groups in occupied France during the war were all living and working there in 1940. It is also possible to identify several other resisters who were born in France or in French territory and were of Irish ancestry.

the Resistance. 5 Julian Jackson, *France: the dark years, 1940–44* (Oxford, 2001), pp 494–5. See also *Revue du Nord*, 226 (1975). 6 Ibid., p. 495. See G. Dreyfus-Armand, 'Les Espagnols dans la Resistance: Incertitudes et specificities', in P. Laborie and J.-M. Guillon (eds), *Histoire et mémoire: La Résistance* (Toulouse, 1995), pp 217–26. See also D. Peschanski, 'Les Espagnols dans la Résistance' in P. Milza and D. Peschanski (eds), *Exils et migration: Italiens et Espagnols en France, 1938–1946* (Paris, 1994). 7 The MOI had been founded in 1923 by the French Communist Party with the purpose of organizing and representing immigrant labour. Its ranks were swelled in the 1930s by immigrants who were fleeing the forces of fascism in their native countries. In April 1942 the organization formed FPT-MOI as its armed wing. 8 David Schoenbrun, *Soldiers of the night: the story of the French Resistance* (1980). 9 NLI, De Valera Papers, 28 November 1941: Walsh to De Valera. 10 SHD, Bureau de la Résistance collection, Vincennes. Also War Office files on SOE, National Archives, London.

THE IRISH IN THE RESISTANCE

Across the range of Resistance activities, it can be shown that the Irish were most prominent in intelligence gathering and evasion networks. Perhaps the best-known Irish resistor was Samuel Beckett, who was living in Paris when the war broke out. He was recruited into the Resistance by his friend Alfred Péron and joined an intelligence *réseau* named 'Gloria SMH'.[11] Beckett was originally given the rank of *sergeant-chef* within the *réseau*, but would later be credited as having been a *sous-lieutenant*. Beckett used the codename 'Samson'. and had joined Gloria SMH in September 1941, remaining active in this group for a year.

During its history, over 200 people were associated with Gloria SMH.[12] There was an embryonic movement operating as early as November 1940 but it was formally established in January 1941 by Jeanine Gabrièle Picabia, who used the codename 'Gloria'. It is now certain that Beckett was known to her before the war. Picabia's co-organizer was Jacques Legrand, whose operating code was 'SMH'. Originally Gloria SMH was part-funded by both SOE and the Free Polish organization in London, but later it was taken over entirely by SOE.[13]

As a network, Gloria SMH's main task was intelligence gathering and it had a special focus on naval and maritime intelligence. While it was centred on Paris, the network had operatives in Normandy and Brittany and also agents who worked on the canals and for the French rail company (SNCF). Gloria SMH's main operatives worked in Lorient, St Nazaire, Bordeaux, Dieppe, Marseille and in other towns and cities across France, including Vichy. Its intelligence gathering activities, therefore, spanned both the occupied and unoccupied zones. Perhaps the *réseau*'s greatest coup occurred in March 1941 when two of its agents reported the presence of the German warships *Scharnhorst* and *Gneisenau* at Brest. These warships were joined in June 1941 by the *Prinz Eugen*. The intelligence provided by Gloria SMH resulted in a series of RAF bombing raids on these warships over the course of the months that followed, the *Gneisenau* being the worst damaged. The network also occasionally passed evaders over to evasion networks and, by 1941, was receiving 100,000 francs a month from London to support its work. Apart from sending back information by courier, it also operated a radio in France and, from 1942, had a radio operator based in Belgium.[14]

Within Gloria SMH, Beckett was responsible for typing up intelligence reports and then passing them on to the network's photographer so that they could be microfilmed and then sent to London. Throughout his service, he

11 In the autumn of 1926 Alfred Péron had taken up a post as a *lecteur* at Trinity College, Dublin. It was there that he first met Beckett. 12 SHD, Bureau de la Résistance, Vincennes, Samuel Beckett file, 16P 42711. 13 SHD, Bureau de la Résistance, Gloria SMH files, 17P 135. 14 Ibid.

carried out this work at his Paris apartment in the rue des Favorites. While Beckett later downplayed his role in the network, this was extremely dangerous work as intelligence material was constantly being carried to his flat so that he could translate it and type up the initial reports. To limit the amount of suspicious traffic to his flat, Alfred Péron was the usual courier. If they were questioned by police or Gestapo, Beckett and Péron planned to claim that they were working on a translation of his latest novel, which was entitled *Murphy*. It would seem that Beckett hid the piles of incriminating paperwork among his own papers.

For Beckett, the most dangerous phase of the operation entailed carrying the typewritten reports to Gloria SMH's photographer. If stopped at a police check and searched, the military value of the paperwork that he was carrying would have been obvious to even the dullest policeman. Beckett carried out regular runs to a man known to him as 'Jimmy the Greek' who lived and worked in an address in the Avenue du Parc de Montsouris (now the Avenue René Coty). This was André (Hadji) Lazaro.[15] Beckett continued this work until September 1942, his post-war citation for the Croix de Guerre stating that he had carried out this work to 'l'extrême limite de l'audace'.[16]

In August 1942, Gloria SMH was effectively blown. It had been infiltrated by Robert Alesch, a Roman Catholic priest was also serving as an agent of German intelligence, the Abwehr.[17] Due to Alesch's activities Gloria SMH's agents began to be arrested in late August 1942. Those arrested included Alfred Péron and Jacques Legrand.[18] Realizing that he had been compromised, Beckett tried to warn other members of the network, including the photographer who did not take the warning seriously and was arrested shortly afterwards. Beckett's wife, Suzanne, was briefly detained by the Gestapo while trying to warn another member of Gloria SMH. The couple fled to Roussillon where they lived a clandestine existence for the rest of the war. Beckett later became involved in a local group in the final months of the war.

The risks that Beckett took during his Resistance career were considerable. In total twelve members of his *réseau* were shot during the war while a further ninety were deported. Those deported were incarcerated in the infamous Fresnes and Romainville prisons, while others were sent to the concentration camps at Ravensbrück, Mauthausen and Buchenwald.[19]

The role of the courier was perhaps one of the most dangerous within the Resistance community. Every courier carried information that could be used as the evidence that would spell his or her doom if they were stopped and

15 Lazaro was known within Gloria SMH as 'Tante Léo'. 16 SHD, Bureau de la Résistance, Samuel Beckett dossier, 16P 42711. 17 Alesch worked for Abwehr III, which was under the command of Oskar Reile. 18 Péron died in May 1945, shortly after being liberated by the Swiss Red Cross. 19 James Knowlson, *Damned to fame: the life of Samuel Beckett* (London, 1996). This authorized biography of Beckett provides a comprehensive

searched by the German authorities. In 1942 an Irishwoman named Mary Giorgi (*née* Dewan) began working as a courier in North Africa in the run-up to the Allied landings of November 1942.[20] She was married to Louis Joseph Giorgi, a commandant in the gendarmerie, and the couple had been based in Oran in Algeria from around 1940. Both were active in a *réseau* designated as PSW-AFR, which was had been established by the Free Polish Forces.[21] This was a very cosmopolitan network, including not only French and Poles but also Algerians and some Italians who had been born in Algeria. Its members spoke a range of languages including French, German, Italian, Arab and English. In the months before the Torch Landings of 1942, the members of PSW-AFR gathered much useful intelligence on German, Vichy and Italian troop dispositions in Algeria, Tunisia and Morocco. Mary Giorgi acted as a courier of information that was to be copied, photographed or transmitted by radio. By 1942, she was the mother of three small children and while carrying out this work she was actually pregnant with her fourth child.[22] After the Allied landings, she continued to work for the réseau for a short time, while her husband later served with BCRA, the French military intelligence organization.

Many Resistance groups dedicated themselves to helping Allied servicemen evade capture and also to getting them safely back to friendly territory so that they could continue the war. The first evasion networks sprang up in 1940, and by 1944 a series of elaborate networks had been established across France. The men and women of these networks carried out the risky business of ferrying Allied servicemen across the occupied and Vichy zones, occasionally stowing them in safe houses until it was safe to move them along again. Some of the evaders were sent out of France through Brittany, being evacuated from isolated beaches. Others were sent overland to Spain or across the Mediterranean to North Africa. Sending evaders to Switzerland was also an option but was not favoured as any intercepted evaders would inevitably be interned.

One of the largest evasion groups was that known as 'Musée de l'Homme'. It was named after the museum of the same name in Paris where its HQ was based and it was unusual among Resistance groups in that it started as a *réseau* but developed into a *mouvement*, even printing its own newspaper. Its agents maintained evasion routes that initially sent evaders to the unoccupied zone of France but later expanded to create contacts with Brittany, Toulouse, Marseille and also with Spain and Portugal. One of the movement's earliest members was an Irish nursing sister named Katherine Anne McCarthy, known in religion as Sr Marie Laurence.[23]

account of Beckett's activities in the Resistance. **20** Mary Dewan was born in Newbridge, Co. Kildare, in 1898. **21** The full title of PSW-AFR was 'Polska Sluba Wywiadowcza Africa'. From 1940 there was a special sub-section of SOE designated as E/UP that administered the Polish-run networks in German-occupied French territory. **22** This fourth child was Charles Giorgi, born 20 March 1942 in Oran. **23** SHD, Bureau de la

Katherine Anne McCarthy was a Franciscan nursing sister and she had previously served as a nurse during the First World War and had been awarded the British Red Cross Medal. In 1940 she was serving as a nurse in the civilian hospital in Bethune. During the fighting that raged across France in the summer of 1940, she found herself with several wounded British and French soldiers in her care. As these men recovered, she smuggled them out of the hospital, and some of her earliest evaders made it through the lines to the beachhead at Dunkirk and from there were evacuated. She later passed recovered patients onto one of the fledgling local Resistance groups and it was through this activity that she became involved with the wider work of the Musée de l'Homme movement.

As the movement's activities expanded, Katherine Anne McCarthy became involved in intelligence gathering but her activities brought her to the notice of the Gestapo and she was arrested at the Bethune hospital in June 1941. At her trial she was condemned to death on the testimony of an informer. This sentence was commuted to deportation yet her case files show that she had been registered as a special enemy of the Nazi state.[24] During the next four years, she was sent to various camps including Anrath, Lubeck and Cottbus. In December 1944, she arrived at the death camp at Ravensbruck. Miraculously she survived and was evacuated by the Red Cross in April 1944. In the 1960s, she was living in retirement in Co. Cork. In a signed deposition to the Resistance Bureau in 1963, she left the injuries and wounds column blank.

Coincidentally, another woman named McCarthy worked in evasion networks during the war. This was the Kerry-born Janie McCarthy who was living in Paris and working as a language teacher when war broke out. A British passport holder, McCarthy destroyed her passport and became involved with the Resistance as early as November 1940 when she joined an evasion *réseau* named 'St Jacques'.[25] She would continue to serve in evasion networks until the liberation of Paris in August 1944, by which time she had served in four *réseaux*.[26] Throughout the war she used her apartment at 66 rue St Anne as a safe house and often carried out the dangerous task of travelling with evaders on the metro. On one occasion she was stopped at a police control while travelling with an American airman who could not speak French but convinced the police that he was in fact a deaf mute. Janie McCarthy managed to survive the war undetected by the authorities and died in Paris in December 1964.[27]

Résistance, Vincennes, Musée de l'Homme files, SHD 17P 173 **24** Ibid. See SHD Katherine Anne McCarthy files in Bureau de la Résistance, Vincennes, and also Archive of the Deported files in SHD, Caen. **25** All of the Irish resistors in France held either British passports of passports of the Irish Free State. Despite its neutrality, the Irish Free State was still technically a British dominion. Possession of either type of passport could have placed the holder in a dubious legal position with the German authorities. **26** Janie McCarthy served in St Jacques, Comete, Shelburn and Samson. The Comete network had been founded in Belgium. **27** SHD 17P 210 (St Jacques), 17P 105 (Comete), 17P 214

There were also Irish resistors who became involved in more open military activity. They included William O'Connor who was active in the movement known as 'Voix du Nord'. O'Connor was based in Douai where he worked as a gardener in the local British military cemetery. Like Katherine Anne McCarthy, he had also served during WWI and had married a French woman and settled in France after the war. He was active in the Resistance from July 1940 and carried out intelligence gathering; but he also acted as a courier and occasionally ferried weapons that were then cached for later use. In September 1943 he was arrested and was imprisoned in a number of camps including Aachen and Rheinbach. O'Connor's last place of imprisonment was the hard labour camp at Siegburg, which was run by the Todt organization. He was released in April 1945 and later noted in a report that during interrogation he had been tortured and, alongside other brutalities, his teeth had been broken.[28]

Other Irish resistors were more lucky and completed their wartime career without ever coming to the attention of the authorities. Sam Murphy, who was from Belfast, became involved with a Maquis group known as 'Veny' in October 1942. Veny operated over a wide area in the departments of the Lot, Tarn, Ariege and both the Hautes- and Basses-Pyrenées. Murphy's speciality was sabotaging German vehicles and also stealing them for use by the Resistance and he continued to operate with the Veny group until the area was liberated in 1944. Thereafter he served with a battalion of the Free French Forces.[29]

When Paris was liberated in August 1944, Resistance leaders noted the arrival of new recruits to their groups who arrived to swell their ranks just as the German army began to show signs of collapse. This pattern continued as Allied forces moved across France. Perhaps unsurprisingly, long-serving resistors often looked with ill-concealed contempt at these 'Johnny-come-latelies', who were often referred to as members of the 'RMA' or 'resistors of the month of August'. Others were christened as the FFS: 'Forces Françaises de Septembre'. It is possible to identify one Irish-born resistor who was active in the liberation of Paris. This was John Pilkington, who had been born in Dun Laoghaire in 1905. He was described in his file as a 'homme de lettres' and was living in Paris in 1940.When fighting broke out on 19 August 1944 he was involved in several actions before Allied troops reached the city. His records show that he served in an ad-hoc unit known as the 'Groupe Mobile Armée Voluntaire' and he took part in the fighting between 19 and 25 August. Despite Pilkington's late involvement with the Resistance, it would be unfair

(Shelburn), 17P 212 (Samson). **28** SHD, Vincennes, Bureau de la Résistance, William O'Connor file. O'Connor was born in Dublin in 1893. See also SHD 18P 44, Voix du Nord files and O'Connor file, Archive of the Deported, SHD, Caen. **29** SHD, Vincennes, Bureau de la Résistance, Sam Murphy file. See also 19P 46/2, Maquis Veny files.

to look upon him as someone who came late to the cause of France. His file also notes that in 1940 he had been a soldier in a reserve infantry regiment of the French army. It also notes that he had been arrested in July 1940 and was not released until November. During this term of imprisonment he had been interrogated and his teeth had been broken. Finally, his file suggests that he had carried out occasional Resistance work between 1940 and 1944.[30]

There are also Irishmen and women in the Service Historique's files who were imprisoned by the Germans but who would not appear to have been involved with an organized Resistance *réseau* or *mouvement* or even involved in Resistance activity. A second Irish nursing sister, Agnes Flanagan, was sent to Ravensbruck but her file indicates no reason for her arrest.[31] The same can be said of Obal Atkinson, who was deported to Buchenwald in 1944 but no reason for his arrest or deportation is given in his file.[32] In 1943 an Irish miner named Thomas Hayward was sent to the labour camp at Vught because he had refused to work for the German authorities.[33] So, alongside those who were recorded as officially being involved in Resistance groups, other Irish people living in France ran foul of the German authorities for different reasons or for reasons that remain unrecorded.[34]

THE 'WILD GEESE' IN THE RESISTANCE

Since the late seventeenth century, Irish men and women had left Ireland and made France their new home. Some of the earlier articles in this book chart the fortunes of this Irish community, often referred to collectively as the 'Wild Geese'. In 1940, two regiments of the French army maintained the traditions of the former Irish regiments in the French service. These were the 87e and 92e regiments, which maintained the traditions of the regiments of Dillon and Walsh respectively. While the 87e was disbanded after the fall of France in 1940, the 92e regiment still exists and includes among its battle honours that of 'Résistance Auvergne 1944'.[35]

It is perhaps not surprising that the names of a number of French-born citizens of Irish ancestry appear on the registers of resistors and also on those of the deported. To make reference to all these cases would require an exhaustive list but some examples are worth citing. There were for instance

30 SHD, Vincennes, John Pilkington file. 31 Archive of the Deported, SHD, Caen. Flanagan had been born in Co. Offaly in 1909. She was working as a nurse in Paris in 1940 before her arrest. She was repatriated by the Swedish Red Cross in 1945. 32 Ibid. Atkinson was born in Dublin 1907 and was a musician. It is not known at this time if he survived the war. 33 Archive of the Deported, SHD, Caen. 34 Patrick Sweeney, who was born in Mayo in 1916, was sent to Buchenwald in 1944. The reason for this remains unknown and it is not known at this time if he survived. Archive of the Deported, SHD, Caen. 35 See the entries for these regiments in David Murphy, *The Irish brigades,*

some resistors named Murphy, who were obviously descended from Irish stock.[36] There were also O'Briens, O'Connors, O'Callaghans and O'Connells. Two descendants of Maréchal MacMahon de Magenta served in the Resistance. It could be argued, therefore, that these men and women were part of a Franco-Irish military tradition that went back to the seventeenth century.

There are also examples of resistors who bear Irish names but were born in French territory and this would seem to indicate wider patterns of migration within the Franco-Irish community. There are several examples of resistors who had Irish names but were born in French Pacific territories and served in the Resistance there. These include Abraham O'Connor who was born in 1913 in French New Guinea and also Raymond O'Donoghue (b. 1907) and Albert O'Callaghan (b. 1920) who were both born in New Caledonia. All three men served in local *réseaux* during the war and fought against the Japanese.

The most impressive example of a Franco-Irish family who members became involved with the Resistance was perhaps the case of the O'Connells, of Indochina (now Vietnam). French territory in Indochina fell under Japanese control in December 1941 and was technically the territory of Vichy France, which was not at war with Japan. Despite this, members of the local communities began organizing Resistance networks. Initially they engaged in intelligence gathering but were also preparing for military action. From 1944 they engaged in open resistance and operated in conjunction with Allied forces in the region. The majority of these Pacfic *réseaux* were supported by the American OSS, which used US naval and air units to maintain contact and land supplies.

One of the Indochinese *réseaux* was named 'Plasson' and was established in September 1941.[37] It began as an intelligence *réseau* and one of its briefs was to identify the location of POWs who were being held by the Japanese. It never numbered more than 95 people and by late 1944 was engaged in acts of sabotage against the Japanese. Within Plasson it is possible to identify three people of Irish descent – members of the O'Connell family who were living at the Plantation O'Connell in Tay Ninh province when the war broke out. The patriarch of the family was named Daniel O'Connell and he had been born in Saigon, Indochina, in 1898 and was serving as chief inspector of forests on the outbreak of the war. During the war, he would emerge as one of the major figures within Plasson. He served alongside his younger sister, Marie Madeline O'Connell (b. 1900) and his son, Patrick O'Connell (b. 1921) who also served in Plasson. In August 1945 Patrick O'Connell was arrested and held by the Japanese but was liberated during the following month. All three were active from 1942 and Daniel O'Connell was later awarded the Croix de Guerre with

1685–2006 (Dublin, 2007). 36 The names of Alponse Murphy, André Murphy and Jacques Murphy appear on Resistance lists. 37 It was named after its founder Lucien Plasson, pseudo 'Vernet'.

two citations.[38] A final member of the family, Daniel's younger brother, Guy O'Connell (b. 1909), was actually in Paris when war broke out as he was employed by the Banque de Indochine. From 1941, he was active with a *réseau* named Hector.[39]

The case of the O'Connell family in Indochina was quite a striking example of people of Irish descent who became involved in Resistance activity in France and also in French territory overseas. There were also others who, at a further remove, were of Irish descent but had been born in England. One such person was Mary O'Shaughnessy who was born in Leigh in Lancashire of Irish parents and was living in France in 1940. She was arrested and sent to Ravensbruck for helping allied airmen to evade but managed to survive and later gave evidence at the Nurembrug war trials.[40] Another example was Brian Rafferty who was born in England and worked for SOE until being arrested and sent to a concentration camp. He did not survive.[41]

FORCES FRANÇAISES LIBRES, FFL

In the months that followed the fall of France in the summer of 1940, the elements of the French army that had been evacuated to England were re-organized into new units. The surviving units of the French army were later reinforced by troops drawn from France's colonies overseas. Perhaps the most famous of these were the men of General Leclerc's 2me Division Blindée, who had fought in campaigns in North Africa before arriving in England in preparation for the invasion of France in 1944. These forces formed the exiled army of anti-Vichy France. They are commonly referred to as Free French Forces or Forces Françaises Libres (FFL).

In London, the military organization was controlled by an exiled Free French government, headed by General de Gaulle. During the years of exile, an administrative system was created to serve both the government and army of Free France. A recruiting drive was organized in England in the hope of attracting administrative staff, interpreters and medical personnel. Four Irish women appear in the records of the FFL and they held various positions within this administration.

The most senior was Blanche Dassonville (*née* Clarke-Bathurst) who had married a French officer, Lieutenant Achille F.H. Dassonville, in 1923.[42] On

38 SHD, Vincennes, Bureau de la Résistance: files on Daniel, Marie Madeline and Patrick O'Connell. See also Reseau Plasson files: 17P 193. 39 SHD, Vincennes, Bureau de la Résistance, Guy O'Connell file. See also Reseau Hector files: 17P 139. 40 Account of Mary O'Shaughnessy on the BBC People's War website: http://www.bbc.co.uk/ww2peopleswar. 41 M.R.D. Foot, *The SOE in France* (London, revised ed., 2004), pp 194 and 253–4. Rafferty was executed in Flossenburg in March 1945. 42 Blanche Clarke-Bathurst was born in Portadown in 1897. During WWII, her husband would serve

the outbreak of the war, she was living in Paris but managed to escape to England where she joined the FFL administration in 1941 and was attached to the 'l'assistance sociale' section of the FFL army. She held a degree from a Swiss university and her file notes that she could speak English, French, Spanish and Arabic and had worked as a tour guide from 1926. Perhaps it is not surprising that she was attached to the general staff of the FFL. In 1942 she requested to be sent to the Levant, where she served for the remainder of the war.[43]

The new FFL forces were also desperately short of medical staff, and the records show that at least one Irish nurse joined their medical staff in 1941. This was Mary Whelan who had been born in Gort, Co. Galway, in 1912. A qualified nurse, she initially enlisted for two years but renewed her enlistment in 1943. During the war she served in the Levant and North Africa before being appointed as an adjutant of the Marseille military hospital in 1944. In 1945 she was demobilized with the rank of lieutenant in the FFL medical services.[44]

Two further Irish woman served in the FFL organization in London. Joanne Patricia Cloherty, who was born in Galway in 1917 and served as a 'conductrice civile'. Dublin-born Pauline Marie Cottin served as a translator and her file notes that she held both British and French nationality. Both joined the FFL organization in 1942. This small contingent of Irishwomen represented an interesting sub-section of the overall Irish resistance contingent. In volunteering to join FFL they helped provide some of the skills and services that the organisation so badly needed in 1941–2.

THE SPECIAL OPERATIONS EXECUTIVE, SOE[45]

In July 1940, Winston Churchill established a new special forces organization that he later stated he wanted to 'set Europe ablaze'. This was the Special Operations Executive or SOE. It was created in the aftermath of the fall of France and, alongside the commando units that were founded around the same time, its purpose was to bring the war back to Europe by sending raiding parties and sabotage groups into occupied territory. By the end of the war, the organization had expanded considerably and the SOE ran over 60 training camps in the British Isles. Between 1940 and 1945 it sent over 7,500 agents

as a colonel on de Gaulle's staff and would later reach the rank of general. 43 SHD, Vincennes, Bureau de la Résistance, Blanche Dassonville file. 44 SHD, Vincennes, Bureau de la Résistance, Mary Whelan/Blachais file. In 1942 Mary Whelan married a French officer, Lt Philippe Blachais, whom she had met at the military hospital at Camberley where she was then serving. 45 The author is grateful to Professor Eunan O'Halpin of TCD for his advice regarding SOE files in the National Archives, Kew.

into occupied Western Europe. It is impossible in this space of this article to give even a brief summary of SOE's activities in Europe.[46]

It must also be pointed out that all research on the SOE is hampered by the fact that many of the files relating to its work and personnel were destroyed in a fire in 1946. Interestingly, two of SOE's founding officers, Lt-Col. J.C.F. Holland and Col. Sir Colin Gubbins, had both served in Ireland during the War of Independence, Holland being badly wounded. Both had studied the methods employed by the IRA during the war, and they had also studied the methods of unconventional warfare that were employed in Russia, Spain and China during the inter-war period. In 1940 they decided to use the lessons learned from their own personal experience and their later studies against the Germans.

Among its various activities, SOE took on the responsibility for contacting Resistance groups in France. It would later provide weapons, money and logistical support. It sent both British and foreign agents into France where they acted as radio operators, intelligence agents and training officers. Within SOE there were a number of sections that dealt with activities in France. A section designated as DF organized escape routes while EU/P worked specifically with members of the Polish community in France, as has already been mentioned. An independent French section, known simply as F, worked with the non-Gaullist Resistance, while RF was established to work with those Resistance networks that were established by de Gaulle. RF generally used French operatives, while F section used both French and non-French for its operations in France.

The SOE has been likened to some form of select club as its members were recruited on an 'invitation only' basis. The recruiting officers identified suitable candidates who were sometimes civilians and on other occasions were men and women already in the armed services. Having been vetted at a series of interviews, prospective SOE operators were invited to join and they then embarked on a rigorous training schedule. Indeed, the whole process was so secret that on some occasions the trainees underwent gruelling training without initially being told what the intention was behind it. On one occasion, a female trainee is reported to have questioned the whole process because, as far as she was concerned, she had volunteered to serve as a bilingual secretary.

At this time it is possible to identify two Irish women and one Irish man who worked for SOE in France. They all worked for F section of SOE.[47] It is perhaps fitting here to first deal with William Cunningham, who was one of

46 For a detailed account of SOE see M.R.D. Foot, *SOE: the Special Operations Executive, 1940–1946* (London, 1999). For an account of SOE's activities in France, see M.R.D. Foot, *SOE in France* (rev. ed., London, 2004). 47 SOE operated a further French section that was designated as AMF. This was based in Algeria and operated for 20 months between 1943 and 1944. For full details of SOE organization, see M.R.D. Foot, *SOE in*

the most mysterious of the Irish members of either the Resistance or SOE's F section during WWII. In the Resistance files at Vincennes, there are actually files on two separate individuals named William Cunningham. It is almost certain that they relate to the same man.[48] The first file gives details of his birth in Dublin and the fact that he joined the French Foreign Legion in 1933, giving his civilian occupation as that of 'journaliste'.[49] Having re-enlisted in 1938, he was based at Sidi-bel-Abbés at the outbreak of the war and formed part of the Foreign Legion demi-brigade that was sent to France on the outbreak of the war. Cunningham was one of the French soldiers evacuated from Dunkirk and was in England by June 1940. His first file then ends on the basis that he was discharged on age grounds. The second William Cunningham file at Vincennes gives, what I believe, to be a fictitious date of birth of 1 January 1914. This new date of birth took three years off Cunningham's age and qualified him for active service; sometime thereafter he was inducted into SOE's F section. He had been a machine gun specialist and was now trained to be a saboteur, using the pseudo 'Paul de Bono'.[50] By 1943, he had been given the rank of second lieutenant within SOE.[51]

While serving with SOE, Cunningham took part in a *coup de main* operation designated as 'Dressmaker'. In the earliest phase of its existence, SOE organized several *coup de main* or sabotage operations, several of which failed due to poor targeting intelligence or the poor performance of operatives. A combination of these factors led to the failure of Dressmaker in August 1943. Two teams of two men each were dropped into France on the night of 17–18 August 1943. Cunningham was in the team led by G.L. Larcher, who was on his second mission for SOE. Larcher later described his 1a.m. parachute drop into occupied France:

> Coming down I could see 2/Lt Cunningham about 100 yards from me and as soon as I landed I took off my harness and striptease, folded my chute and went to meet him. I found him straight away. He had landed about 50 yards from a farm and the package was very near him. We thus lost no time at all in getting together. After the three 'chutes had been rolled up, we proceeded to find a place to bury them.[52]

The two teams that made up the Dressmaker operation were tasked with sabotaging tanneries in the towns of Graulhet and Mazamet – Larcher and

France, op. cit. **48** SHD, Vincennes, Bureau de la Résistance: William Cunningham files. **49** William Cunningham was born in Dublin, 25 May 1911, and was the son of Thaddeus and Mary Cunningham of 107 Stannaway Road. **50** Ibid. While part of the SOE organization, Cunningham's files also credit him with being part of Force France Combattant or FFC, the internal Resistance organisation in France. **51** National Archives, Kew, Operation Dressmaker file, HS 6/908/1. **52** Ibid. The 'package' mentioned in Larcher's report was a container that held their equipment, which included

Cunningham's target being Mazamet. On reconnoitring the targets, both teams found that they had been given fools' errands as neither factory was in production. Due to poor targeting intelligence, the whole enterprise was wasted. Larcher later reported, however, that Cunningham had proved himself invaluable due to his fluent French and his skill as an operative. Having stowed their explosives and weapons, the Dressmaker operatives headed for Spain; Larcher and Cunningham crossing the border somewhere near Perpignan. They then were taken to Madrid before crossing the border to Gibraltar. Having travelled by plane to Bristol, both men reported back to the SOE office in Orchard Court in London on 11 September 1943.

The failure of Dressmaker resulted in considerable fall-out in London. In his report, Larcher concluded: 'In my opinion, this kind of operation is being badly planned at the moment.'[53] His superiors felt that the blame lay with the operatives and scribbled notes on the file state that 'no effort was made to do the job' and 'we must get rid of Larcher at once'.[54] It would seem that Larcher was not used by SOE again and it is unclear if Cunningham was also affected by this negative fall-out. He was not mentioned in a negative sense in the final reports and the censure of senior SOE officers seems to have been reserved for Larcher. Nevertheless, Cunningham does not appear in any further of the surviving SOE files. I have not been able to find any record of his later wartime career.

SOE was unique among Allied intelligence organizations in that it trained women to work as operators in the occupied territories. Its women operators were inducted either into the First Aid Nursing Yeomanry (FANY) or the Women's Auxiliary Air Force (WAAF). Indeed, about 50% of the entire FANY contingent during the war was in reality made up of SOE operatives. During the war, the French sections of SOE sent 50 women to France of which 14 were initially members of the WAAF.[55] Both of the Irish women who worked for SOE in France, Mary Katherine Herbert and Patricia Anne O'Sullivan, had originally been members of the WAAF.

Mary Katherine Herbert was one of the earliest female SOE agents who was sent to France.[56] She was the seventh of the fifty women agents that worked in France for SOE during WWII and was the first to come from the WAAF. At the beginning of the war she had been working at the British embassy in Warsaw and then worked as a translator at the Air Ministry in London. She joined the WAAF in September 1941 and was inducted into SOE in May 1942. She was later granted a WAAF commission that was back-dated to 1941. When she began her training, official sanction for the use of women agents in the field had only just been passed. Having undergone

pistols and explosives. The 'striptease' referred to was a type of parachute harness used by SOE operatives. 53 NA, Kew, HS 6/908/1. 54 Ibid. 55 See M.R.D. Foot, *SOE in France*, op. cit. 56 Mary Katherine Herbert was born in Ireland in 1903.

various courses including the wireless operators course, she was scheduled to go to France in October 1942.

During the course of the war, SOE agents would be transported to France using Lysander aircraft and also various types of boat. Others would also parachute: Mary Herbert entered France by an unusual means. Having been transported to Gibraltar, she was put aboard a North African felucca and landed by sea on the night of 30 October 1942, on the coast between Marseilles and Toulon.[57] She was tasked with operating as a courier and wireless operator to a circuit designated as 'Scientist'.

Scientist operated over a large area with agents operating in the Vendée, the Gironde and Landes. It also maintained cells in towns such as Bordeaux and Poitiers. As a result of the extended territory of Scientist, which included areas in both the Occupied and Vichy zones, Mary Herbert often had to travel over large distances carrying messages and on other occasions money or wireless sets. On one occasion, a German naval officer gallantly helped her carry a heavy suitcase onto a tram. The suitcase contained a wireless set. Apart from intelligence gathering, from 1943 the members of Scientist became more active in sabotage operations and had a number of successes, including the destruction of a German naval radio station at Quatre Pavillions.

In June 1943 the Prosper network in Paris was blown and its members began to be rounded up by the Gestapo. This had security implications for Scientist, whose members had to go to ground. By this time Mary Herbert had embarked on an affair with the leader of Scientist, Claude de Baissac, and was expecting his child: In December 1943 she gave birth to a baby girl and now had the dual concern of providing a fake cover story and documents for herself and her baby. In February 1944 she was detained during a search of her apartment block in Poitiers and was detained for some weeks in the Poitiers prison while her daughter was sent to an orphanage. In the weeks that followed she not only did not break under interrogation but managed to convince the Gestapo of her false identity. On her release, she reclaimed her daughter and went into hiding until Bordeaux was liberated in September 1944. Later that year she was married to Claude de Baissac.[58]

Colourful characters would appear to have abounded in the ranks for the SOE. One of the most colourful was Dublin-born Maureen Patricia O'Sullivan, usually referred to as Patricia O'Sullivan in her files. Perhaps inevitably, she was known within SOE as 'Paddy'. She had lived in France and Belgium before the war and was able to speak both French and Flemish. In June 1941 she joined the WAAF and was originally used as a translator. As was the case with so many of SOE's operatives, her language skills came to the attention of recruiters and she was inducted into the SOE in July 1943.

57 Beryl E. Escott, *Mission improbable* (London, 1991), pp 49–59. One of her fellow-travellers during this trip by felucca was Odette Samson. 58 Ibid. See also M.R.D. Foot,

Training courses followed in parachuting; the use of small arms and also in the use of mortars. By the end of 1943 it was obvious that she was destined to work as a field agent and the comments of her instructors around this time were interesting. One instructor reported that she was:

> A pleasant, intelligent Irish girl with a mind and will of her own. Purposeful and determined once she is convinced that what she is doing is right. Likes change but could be relied upon to see any job through on which she has set her mind.[59]

This opinion was not shared by all her instructors, and some later comments were profoundly negative. One instructor commented that she was 'not particularly intelligent and does not seem to take her work very seriously'. Another rather ungallantly, claimed that she was 'a tough type of woman, at the moment growing quite a successful moustache'.[60] In short, reports on her progress varied widely and by the end of 1943, her aptitude for SOE work was being questioned. This was not helped by the fact that she went AWOL over a weekend period in February 1944. Perhaps the instructor who commented that O'Sullivan was 'anxious to do the actual work' was the most perceptive. Exhausted and frustrated by months of training, she wanted to go to France and do her job.

On the night of 24–25 March 1944 she parachuted into France, where she was assigned to a circuit designated as 'Fireman' and operated in the area of Limoges. Her training had not been completed but Fireman desperately needed a wireless operator and O'Sullivan desperately wanted to get into action. In the months that followed she served as the circuit's wireless operator while also training three new operators. To ensure that her location was not discovered, she was continually on the move, sometimes even travelling in daylight with her wireless set. On one occasion she found herself confronted by a German checkpoint but brazened her way through by flirting unashamedly with both the guards. Having made a date with each of them, she continued on her way, neither of the flustered Germans bothering to check her suitcase.

O'Sullivan remained in France in the run-up to the D-Day landings, continuing her work with Fireman. When the German garrison at Limoges surrendered on 20 August 1944, her mission was considered to be over and she was whisked back to London by plane. She was promoted to the rank of flight officer within the WAAF and further operations in Germany seemed likely but the war ended before she could be sent into the field again. In the immediate post-war years she remained in the WAAF, working as an intelligence officer in India and Ceylon (modern-day Sri Lanka).[61]

SOE in France, op. cit. **59** National Archives, Kew; HS 9/1427/1; Patricia O'Sullivan files. **60** Ibid. **61** Ibid. See also M.R.D. Foot, *SOE in France* and Escott, *Mission improbable*.

SOE continues to fascinate historians of WWII both due to the nature of its clandestine operations and also due to the eccentric nature of many of its operatives. The three operatives with Irish connections who worked with F Section in France all had brief but potentially deadly careers. As with the case of the Irish who served with FFL and are noted above, they represent an interesting sub-section of the wider Irish community that had connections with the French Resistance.[62]

THE IMPRISONED AND THE EXECUTED

The work of the Resistance and the SOE in occupied France was fraught with dangers. Some of the French Resistance networks lost large numbers of operatives for various reasons. Resistance groups that were established in 1940–1 saw particularly high rates of attrition in the years that followed. Some groups were infiltrated by German agents and their members betrayed. Other resistors were betrayed by collaborators, while others came to grief at checkpoints due to the fact that they were carrying incriminating material.

Many of the SOE operatives were not French natives and ran the risk of compromising themselves due to an unconvincing accent or perhaps blurting out something in English when nervous. One of Patrica O'Sullivan's instructors for example was concerned due to the fact the she spoke perfect French – but with a Belgian accent. Katherine Herbert was anxious that she might say something in hospital during the pain of childbirth. Clothes, luggage or makeup could also be give-aways to the sharp-eyed sentry or policeman. The operators of Dressmaker were warned by local Resistance that their clothes were too new and that they stood out as a result.

Every resistor and operative was waging a constant battle of nerves, many suffering nervous breakdowns due to the strain of being involved in this clandestine war. There was a whole range of forces ready to catch out the unwary. It is usual just to refer to them as 'the Gestapo' but they included the German intelligence service or Abwehr, the SD or Gestapo and also German military police and customs officials. Also, while some members of the French Police and Gendarmerie might be willing to turn a blind eye when confronted with something suspicious, the paramilitary force of Vichy, the *Milice,* would not.

It is estimated that over 100,000 French men and women died at the hands of the German and Vichy authorities. Thousands more were imprisoned, tortured, sent to concentration and labour camps but managed to survive.

62 Another Irish woman, Erica O'Donnell, worked with the SOE but was primarily based in London. She worked on the Czech desk and was involved in the training of Czech agents, including those who assassinated Obergruppenfuher Reinhard Heydrich in 1942. In March 1944 she transferred to F Section. NA, Kew, HS 9/442/4.

While the Irish contingent within the Resistance was extremely small, it can be shown that a large proportion of them ran foul of the authorities. Excluding Irish men and women who served in FFL or SOE, sixteen Irish men and women are known to have served in the Resistance in France or Belgium.[63] Of these, nine were arrested and later sent to either a prison, labour camp or concentration camp. The nine included 6 men and 3 women (see Appendix).

The three Irish women represent an interesting case. They were Mary Cummins (Belgian Resistance), Sr Katherine Anne McCarthy and Sr Agnes Flanagan. Two of the three were nuns and all three were sent to Ravensbruck. There is documentary evidence to show that both Cummins and McCarthy were involved in the Resistance. Also, Sr McCarthy was classed as a 'nacht-und-nebel' prisoner in the Ravensbruck files.[64] In the case of Agnes Flanagan, her file in France gives no indication as to why she was arrested. Also, she does not appear in the files of any Resistance *réseaux*.[65]

Of the six men, we cannot be certain at this time for the reason for the arrests of Obal Atkinson or Patrick Sweeney. In Sweeney's case, it must have been deemed to be quite serious as he was sent to Buchenwald. In neither case can it be ascertained if these men were later liberated. In the case of Robert Armstrong and Robert Vernon, we know that both of these men were executed by the Germans in 1944 and 1945 respectively. Robert Armstrong had served in WWI and was the gardener at the British cemetery at Valenciennes in 1940. He was active in the Resistance from October 1940 and served in a *reséau* designated as 'St Jacques', which specialized in helping downed allied airmen.[66] He was arrested November 1943 and, having been interned in different camps, was sent to Waldheim in December 1944 where he arrived on 18 December. He was executed on the same day.[67] Robert Vernon worked with the Alliance *réseau*, which was an intelligence *réseau* but also ran evasion networks. He was arrested in January 1943 and was later sent to Sonnenburg where he was executed on 30 January 1945, alongside other Resistance agents.[68] By the time of Vernon's execution, the Irish government had made representations through its legation in Berlin but succeeded in only gaining a stay of execution.[69]

The Irish men and women who survived their imprisonment were either repatriated by the Red Cross or liberated by Allied troops in 1945. Their

63 Included in this number is Mary Cummins, who served with the Belgian Resistance. 64 Sr McCarthy was prisoner number 85470 at Ravensbruck. Information kindly supplied by Dr Insa Eschebach of the Mahn- und Gedenkstatte, Ravensbruck. 65 Sr Agnes Flanagan file, SHD, Archive of the Deported, Caen. 66 Janie McCarthy also worked for St Jacques. 67 SHD, Vincennes, Bureau de la Résistance: Robert Armstrong file, 16P 17321. Further material held in the Deported Archives, SHD, Caen. 68 SHD, Vincennes, Bureau de la Résistance: Robert Vernon file, 16P 10908. Also Reséau Alliance files, 17P 721–2. 69 Eunan O'Halpin, *Spying on Ireland: British intelligence and Irish neutrality during the Second World War* (Oxford, 2008).

surviving files often refer to them being sick and suffering from the effects of malnutrition, ill-treatment and sometimes torture.

CONCLUSIONS

As far as I am aware, this article is the first attempt to address at any length the subject of the Irish in the French Resistance. The subject is one that must be examined using archival material as there is very little secondary literature. This perhaps accounts for the lack of attention that these men and women have received in the past as the archival sources on all of these cases are held outside of Ireland – in archives in France, England and Germany. At the same time, something very similar could be said for the subjects of virtually all of the articles in this book.

Can any conclusions be reached on the Irish who served with the Resistance? In many ways, it could be argued that the Irish contingent within the Resistance was simply too small to support any major conclusions. Yet despite the small number of Irish men and women involved, some interesting points can be made.

The number of Irish women involved in the Resistance, FFL and SOE is interesting. Over 50% of the total Irish contingent was made up of woman. This is vastly at odds with the numbers of women in the Resistance as a whole. Historians estimate that women made up around 8% of the total Resistance population. In the Bureau de la Résistance archives at Vincennes, women make up around 5–6% of the total number of documented cases.[70] It should also be pointed out that, in the majority of cases, these Irish women were not involved simply because their husbands were active in the Resistance. It has often been assumed in the past that women resistors simply followed the lead of their husbands and joined the Resistance. In some cases, such as that of Blanche Dassonville or Mary Giorgi, we can see that both husband and wife were involved in the Resistance. The majority of the Irish women resistors, however, were not married to Frenchmen at the time and indeed some would never marry at all.[71]

There is also an interesting connection with WWI as at least three of the Irish resistors had served in the earlier world war – Robert Armstrong, William O'Connor and Sr Katherine Anne McCarthy. While Sr McCarthy had remained in France after WWI and worked as a nurse, both Armstrong and O'Connor had married Frenchwomen and had been naturalized as French citizens. They were both working in British military cemeteries in 1940.

70 J.F. Dominé and Nathalie Genet-Rouffiac, *L'Arme féminine de la France* (Vincennes, 2008). 71 Ibid.

Within the wider Resistance community, it can also be shown that the Irish were active across a wide range in terms of both time and activity. Several were involved with some of the first Resistance groups and were active from 1940. Others joined between 1941–3 while at least one (John Pilkington) was involved in the final stages of the Resistance in 1944. In terms of activities, one can find Irish people involved in all of the main Resistance activities – sabotage, intelligence gathering, propaganda and the organisation of evasion networks. Outside of mainland France, it is also possible to identify Irish people, or people of Irish descent, working with the Resistance in North Africa and as far away as France's Pacific possessions.

There was no single *réseau* that contained all the Irish resistors. The only *réseau* or movement that had more than one Irish resistor was St Jacques, in which both Janie McCarthy and Robert Armstrong served. It is unlikely that they knew of each other's existence within the *réseau*, due to the fact that they operated in different areas.[72] There is one common feature, however, shared by all of the Irish who served in the Resistance, FFL or SOE – they all spoke French. The capacity to speak French and their Irish birth were the only common factors that linked all of those in this small Irish contingent.

If the Irish Department of External Affairs estimated that a maximum of 800 Irish people were living in France in 1940, then those who became involved in the Resistance represented around 3–4% of the total Irish population in France. During the course of the war it is estimated that 8–10% of the total French population was involved in the Resistance.

The Irish in the Resistance also represented the end of a military tradition that stretched back to the seventeenth century. The articles of this volume have addressed various phases of this Franco-Irish military connection, and within the Irish Resistance contingent there were echoes of these earlier phases. As has already been mentioned, there were descendants of the Wild Geese of the late seventeenth and early eighteenth centuries who served in the Resistance. There were also descendents of prominent ninteenth century Franco-Irish figures, the most obvious being the descendants of Maréchal MacMahon. Irish veterans of WWI – a war that saw a large number of Irish troops and nurses serving in France and Belgium – were also involved. So, despite the small number of Irish that were involved, there was a level of continuity with links to earlier phases of Franco-Irish military co-operation.

The issue of motivation is a difficult one to gauge. The surviving files on Irish resistors and SOE operatives do not specify directly why Irish men and

72 There was one *réseau* that was designated as 'Pat O'Leary' as this was the *nom de guerre* of its leader – a Belgian named Albert Guerisse. A doctor and former Belgian army medical officer, Guerisse had been evacuated to England in 1940 and used his Irish pseudo during his later wartime career. The *réseau* Pat O'Leary was one of the largest and most successful in WWII.

women rallied to the French cause. For many, it would seem to have been a case of being loyal to their adopted country and the families and friends that they had there. In terms of politics, the majority of Irish resistors were not involved with *réseaux* or *mouvements* that were linked to the French communist party. But it could be argued that some, including Samuel Beckett, were anti-fascist and had viewed the rise of the extreme right during the interwar years with considerable alarm. Their involvement in the Resistance could be seen as a natural progression from this and as a reaction to the overthrow of France by the forces of Nazi Germany.

The fate of some members of this Irish contingent remains a mystery at this time but it can be seen that many returned to their former lives at the end of the war. Those who were living in France in 1940, and survived, remained in France after the war. Sr McCarthy, who was in extremely bad physical shape on being liberated from Ravensbruck, returned to France following a period of convalescence in Ireland. In general, they did not write memoirs or leave accounts of their wartime activities, apart from some newspaper interviews. Some were later involved in veterans' associations or *amicales*, but we have no real indication as to the motivations of the majority of these people. As they have now all passed away, this must remain the case.

The subject of foreigners in the Resistance raises a number of interesting questions on the subject of national identity in wartime. Perhaps this is the ultimate question in terms of the Irish resistors and it is also one that must now remain unanswered. How did these men and women view themselves? As Irish people fighting for France? As naturalized French citizens doing their duty? As Irish citizens reacting in a personal way in opposition to Ireland's policy of neutrality? It could be argued that it was their French associations and loyalties that proved to be the most powerful driving forces. From a strictly institutional and legal point of view, the Service Historique de la Défense catagorizes them as having served France. Those who were executed are classed as being 'mort pour la France'. This is, perhaps, fitting.

In the post-war years, their services were recognized, and again this was predominantly by France. The majority of Irish resistors were awarded French decorations including the Croix de Guerre and the Médaille de la Résistance. Samuel Beckett for example was awarded both the Croix de Guerre and the Médaille de la Reconnaissance Française. Some, such as William O'Connor were also awarded certificates of valour. Those who served in the SOE were awarded both British and French medals. Patricia O'Sullivan, for example, received both the Croix de Guerre and the MBE. There is a whole series of memorials to the Resistance across France. These commemorate local groups and sometimes specific individuals. In the upper corridors of Les Invalides in Paris there are memorials to different communities of non-French men and women who served in the Resistance.

To date, those Irish who served remain un-commemorated either in Ireland or France.

Those who have engaged in a clandestine war, and survived, are usually men and women who know how to keep secret. The vast majority of people who served in intelligence and resistance work would seem to have returned as quietly as possible to their former lives after the war, making a minimum of fuss. Like many veterans, they were not too inclined to speak of their wartime activities in later years, let alone boast of them. This no doubt is yet another reason why the Irish resistors dropped off the historical radar in the post-war years. It is almost certain that, if asked, their comments on their wartime activities would have been brief. Patricia O'Sullivan of SOE later succinctly summarised her wartime career:

> I was terribly frightened at times, but there was a wonderful spirit of sharing danger with men of the highest order of courage, which made it a privilege to work for them.[73]

APPENDIX 1

Irish men and women in FFI or FFC, 1940–5

Name	DOB	POB	Resistance Réseau	Remarks
Robert Armstrong	07.10.1894	Edgesworthtown Co. Longford	St Jacques	Active 1940–3 Arrested 1943 Pseudo 'Bob' Sent to Waldheim Died in prison Executed (?) 18.12.1944
Obal Atkinson	16.07.1907	Dublin		Musician Deported to Buchenwald 17.01.1944 Reason for arrest unknown
Samuel Beckett	13.04.1906	Dublin	Gloria SMH	SOE-sponsored intelligence circuit. Active 1941–2. Pseudo 'Samson'. Rank of 2Lt. →

73 NA, Kew, HS 9/1427/1.P. O'Sullivan files.

Name	DOB	POB	Resistance Réseau	Remarks
William Cunningham	25.05.1911	Dublin	FFC/SOE	Took part in 'Operation Dressmaker' Aug. 1943. Pseudo 'Paul de Bono'
Mary Cummins	24.04.1905	Dublin	Belgian Resistance	Arrested and interned in different concentration camps including Ravensbruck.
Agnes Flanagan	16.05.1909	King's County		Nursing sister No reason recorded for arrest Sent to Ravensbruck Repatriated by Swedish Red Cross
Mary Giorgi	06.11.1898	Newbridge Co. Kildare	PSWAFR	Active 1942 In North Africa Served as a courier
Sheelagh Greene	14.05.1917	Ireland (?)	Jade Amicol	
Thomas Hayward	04.09.1894	Fermoy, Cork		A miner who refused to be sent for forced labour Interned in Vught Repatriated by Swedish Red Cross
Katherine Anne McCarthy	17.12.1895	Drimoleague Co. Cork	Musée de l'Homme	Irish Franciscan nursing sister Active from 1940 Arrested 1941 Interned in various camps and finally in Ravensbruck. Rank of 2Lt Liberated by Red Cross
Janet 'Janie' McCarthy	19.01.1885	Killarney Co. Kerry	St Jacques, Comete Shelburn, Samson	Active, Paris, 1941–4
Sam Murphy	11.10.1912	Belfast	Maquis Veny	Active 1942–4 →

Name	DOB	POB	Resistance Réseau	Remarks
William O'Connor	25.02.1893	Dublin	Voix du Nord	Served 1940–5 Arrested Sept. 1943 Deported and interned in Todt camp at Siegburg.
John Pilkington	10.05.1905	Dun Laoghaire	Groupe Mobile	Active, Paris, Armée Voluntaire, August 1944
Patrick Sweeney	17.03.1916	Mayo		No reason for arrest recorded Sent to Buchenwald 19.01.1944
Robert Vernon		Dublin (?)	Alliance	Active Jan. 1943 Arrested Jan. 1943 Executed 30.01.1945 at Sonnenberg

Irish men and women in Force France Libre (FFL), 1940–5

Name	DOB	POB	Unit	Remarks
Blanche Dassonville	02.07.1897	Portadown Co. Down	FFL forces	Served as admin. officer in England and the Levant, 1941–4
Joanne Patricia Cloherty	17.07.1917	Galway	FF government, London	Worked as a civilian administrator, 1942–3
Pauline Marie Cottin	19.09.1915	Dublin	FF government, London	Translator, 1942–
Rory O'Moore	26.08.1916	Killeshandra Co. Cavan	Legion Etrangère	Enlisted and deserted 1941
Mary Whelan	04.05.1912	Gort Co. Galway	FFL forces	Officer in FFL medical services Served North Africa and France 1941–5

Special Forces Executive (SOE), French sections, 1940–5

Name	DOB	POB	Aliases	Operation/Circuit
William Cunningham (see above)			Paul De Bono	Dressmaker, 1943 Rank of 2Lt
Erica O'Donnell	11.03.1940	Dublin		From 1942, Czech section March 1944, posted to French section Attached to etat major FFI
Maureen Patricia O'Sullivan	03.01.1918	Dublin	Josette Simonet Marie Claire	Barthelemy/ Fireman, 1944
Mary Katherine Herbert	01.10.1903	Ireland	Claudine	Scientist, 1942

Contributors

PATRICK CLARKE DE DROMANTIN has carried out extensive research on the integration of Irish Jacobite exiles in eighteenth-century France. He has published widely and his publications include *Les oies sauvages: mémoires d'une famille irlandaise réfugiée en France* (Bordeaux, 1995). He has also contributed to Thomas O'Connor and Mary Ann Lyons (eds), *Irish migrants in Europe after Kinsale, 1602–1820* (Dublin, 2003).

PIERRE-LOUIS COUDRAY is a doctoral student of the University of Angers, France, where he lectures on historical methodology in English studies and is a temporary junior lecturer (Attaché Temporaire en Enseignement et Recherche). His PhD subject is 'Les soldats irlandais au XVIIIe siècle: Jeux de miroirs entre France, Grande-Bretagne et Irlande' under the direction of Professor Jean Brihault of the University of Rennes II Haute-Bretagne.

NATHALIE GENET-ROUFFIAC is conservateur en chef du patrimoine at the Service Historique de la Défense, Vincennes. She is a graduate of the Ecole Nationale des Chartes and her PhD from the Ecole Pratique des Hautes Etudes was on the first generation of Jacobite exiles in Paris and Saint-Germain-en-Laye. She is the author of several articles on the Irish Jacobite exiles and on the Irish regiments in the French service. Her latest book is entitled *Le grand exil: les Jacobites en France, 1688–1715* (SHD, Vincennes, 2008).

HUGH GOUGH lectures in French history at University College, Dublin. His main area of research and publication is the political and cultural history of the French Revolution. He has also written on the growth of political journalism during the revolution, placing a special focus on the provincial press and the development of printing and distribution techniques. He has published on the historiography of the Terror of 1793–4 and has carried out research into the history of the guillotine, of executioners and of capital punishment during the revolution. Professor Gough has published widely and his publications include *Europe 1763–1960* (Dublin, 1974); *The newspaper press in the French Revolution* (London, 1988) and *The Terror in the French Revolution* (London, 1998).

LAVINIA GREACEN lives in Dublin. She is the author of *Chink: a biography* (London, 1989), which was a *Sunday Times* book of the year. Her second book was *J.G. Farrell, the making of a writer* (London, 1999), which was chosen as a

Spectator book of the year. She is currently writing a life of General Lally entitled *Voltaire's last case*, and she has recently edited the *Selected letters and diaries of J.G. Farrell* (Cork, 2009), to mark the 30th anniversary of Farrell's death. She is a member of the Biographers' Club.

PIERRE JOANNON is the former editor of *Études Irlandaises* and président d'honneur de l'Ireland Fund de France. He has studied the historical connection between France and Ireland over a period of decades and has been awarded honorary degrees from Irish universities and also honorary Irish citizenship for work in fostering a link between Ireland and France. He is an honorary consul for Ireland in France and has published widely. His publications include his monumental study *Histoire de L'Irlande et des Irlandais* (Paris, 2006), among many others.

JANICK JULIENNE obtained her PhD from Université de Paris VII in 1997. Her thesis was entitled 'La question irlandaise en France de 1860 à 1890: perceptions et réactions'. She has since published widely on this subject in France and in Ireland in journals such as the *Irish Sword*. She is currently teaching as a professor of history and geography in a lycée in Courbevoie.

SYLVIE KLEINMAN is a French translator and cultural historian who lives in Ireland. Her PhD thesis was entitled 'Translation, the French language and the United Irishmen: 1792–1804' (Dublin City University, 2005) and it examined the measures taken to overcome language barriers during the decade of partnership between revolutionary France and radical Ireland, and the strategic role of key Irishmen in France. She has worked for the Dictionary of Irish Biography (Royal Irish Academy) and is currently compiling the index for the complete *Writings of Theobald Wolfe Tone* (Oxford University Press). She is writing a cultural biography of Tone's life and travels in France and French-occupied Europe from 1796 to 1798. She was recently awarded an IRCHSS post-doctoral fellowship to continue her research.

NICHOLAS DUNNE-LYNCH is a writer and independent researcher with a special interest in the Irish involvement in the Anglo-French Wars of 1793–1815. Working mainly on the Irish in the British army, he has given papers at various conferences, including the Third Wellington Congress in 2006, and has published in the *Journal of the Society Army Historical Research* (JSAHR). His article in this collection was based on his in-depth research into the Irish Legion of Napoleon and he intends to develop this into a larger study.

GEORGES MARTINEZ was born in Oran, Algeria in 1941. He served in the French army in both armoured units and in special forces. He left the army in 1964 and began a career in banking and has since held various appointments in finance and industry. For many years he has devoted his spare time to history and he has published many journal articles. He is a member of the French Commission for Military History and also of La Sabretache and the Historical and Archaeological Society of Charente. He is the founder of a military studies group based in Poitou Charente.

DAVID MURPHY is a graduate of University College, Dublin and Trinity College, Dublin. He worked with the Royal Irish Academy's Dictionary of Irish Biography from 1997 to 2005, where he specialized in writing biographies of naval and military subjects. In 2004 he was awarded a fellowship to the West Point summer seminar in military history. He currently lectures in military history at NUI, Maynooth. His publications include *The Irish brigades, 1685–2006* (Dublin, 2007), *Ireland and the Crimean War* (Dublin, 2002), *Irish regiments in the world wars* (Oxford, 2007) and *The Arab revolt, 1916–18* (Oxford, 2008). From 2009 he will act as a visiting lecturer at St Cyr, the French cadet academy.

SEÁN Ó BRÓGÁIN lives and works in Donegal. He holds a BA (Hons) in Scientific and Natural History Illustration from Lancaster University. In the past few years, he has undertaken commissions for the Office of Public Works in Ireland and has completed paintings for the Battle of the Boyne Visitor Centre. He has also carried out work for An Post (the Irish postal service) and Osprey Publishing as well as completing private commissions.

ÉAMON Ó CIOSÁIN is a lecturer at NUI, Maynooth and Senior Research Fellow of the Irish Research Council for the Humanities and Social Sciences, 2008–9. He has published in various languages on Irish migration to France in the early modern period and its literary dimension, and his thesis 'Les Irlandais en France 1590–1685: Réalité et images' (Université de Rennes 2 – Haute-Bretagne) is due for publication. He is co-author with Alain Le Noac'h of a series of publications on Irish immigrants in parish archives from Brittany.

EOGHAN Ó hANNRACHÁIN is a native of Cork in Ireland and he divides his time between Ireland and Luxembourg. A senior official in the Irish Department of Finance, between 1986 and 2002 he was the financial controller of the European parliament. He has dedicated much of his spare time over many years to studying the history of the Wild Geese in France. To date, he has published over eighty articles on the Irish in France during the reigns of Louis XIV and Louis XV. These have appeared in journals such as the *Irish Sword*, *La Sabretache*, *Ríohct na Midhe* and *Hémecht*. His publications include articles on the Irish regiments, the Irish veterans in *Les Invalides* and also Irish prisoners in the Bastille.

SIOBHAN PIERCE holds a MA in modern history from University College, Dublin, where her thesis was entitled 'The Irish Army's participation in the U.N.'s Peacemaking Mission in the Congo, 1960–4'. She has been a member of staff of the National Museum of Ireland for the last six years and was one of the curatorial staff who developed the National Museum of Ireland's military exhibition 'Soldiers and chiefs: the Irish at war at home and abroad, 1550–2001'. She has previously written about the wills and last letters of the Irish soldiers who died in World War I and examined the economic and social profile of those soldiers from the city of Dublin who fought in that war. She is currently assistant education officer, developing educational programs on the history exhibitions at the National Museum.

JÉRÔME AAN DE WIEL is a lecturer in modern Irish history at the University of Rheims, France, and is currently visiting professor in the History Department, University College, Cork. He is the author of *The Catholic Church in Ireland, 1914-1918; war and politics* (Dublin, 2003) and *The Irish factor, 1899–1919; Ireland's strategic and diplomatic importance for foreign powers* (Dublin, 2008). His areas of research include ecclesiastical, diplomatic, military and espionage history concerning Ireland in the twentieth century.

Index